Historical Climate Variability a
in North America

Lesley-Ann Dupigny-Giroux · Cary J. Mock
Editors

Historical Climate Variability and Impacts in North America

Springer

Editors
Dr. Lesley-Ann Dupigny-Giroux
University of Vermont
Department of Geography
94 University Place
Burlington VT 05405-0114
USA
ldupigny@uvm.edu

Dr. Cary J. Mock
University of South Carolina
Department of Geography
Columbia SC 29208
USA
mockcj@sc.edu

ISBN 978-90-481-2827-3 e-ISBN 978-90-481-2828-0
DOI 10.1007/978-90-481-2828-0
Springer Dordrecht Heidelberg London New York

Library of Congress Control Number: 2009933459

Cover illustration: (left) Rooftop Instruments on the Alaska Building, Seattle, Washington, between 1905 and 1911 (courtesy of Pemco Webster & Stevens Collection, Museum of History & Industry); (center) The logbook from the HMS Racoon in 1813, photo from The National Archives, U.K.

Printed on acid-free paper

Springer is part of Springer Science+Business Media (www.springer.com)

Preface

When we decided to create *Historical Climate Variability and Impacts in North America*, we had several goals in mind. The first was to address a gap in the literature about the methods for and the value of using documentary data such as diaries and ships' logs in augmenting and enriching the instrumental record dating back to the 1600s in North America. Another key goal was to bring together communities of scholars working on central problems related to historical climate variability, but spanning the range of modern to paleo perspectives within climatology. As such, this book is unique in its contributions from synoptic, applied, dynamic and historical climatologists, as well as by historians and museum archivists. Finally, we wanted to produce a volume that would be instrumental in the formation of the next generation of historical climatologists. As such, this book is designed to be used in upper-level undergraduate or graduate level courses or by researchers in general.

This book complements existing monographs such as *Improved Understanding of Past Climatic Variability from Early Daily European Instrumental Sources*, edited by Dario Camuffo and Phil Jones and published by Kluwer Academic Publishers, which focuses on the use of early temperature records, measurement errors and trends in extreme temperatures. It extends such works by using both the instrumental record and documentary data from the last 200–400 years to paint a portrait of climate variability in North America through a variety of lenses. It differs from similar monographs in that there is less emphasis on individual modes of climate variability or teleconnections, and more on century-scale issues, extreme events, and processes. Its focus on North America, about which much less has been written as compared to Europe or Asia, is also unique. In addition, the qualitative and quantitative methods for examining climate variability described herein, will inform and enhance our understanding of current climate change science, with less emphasis on climate prediction and policy recommendations.

We would like to thank and acknowledge Gert-Jan Geraeds for his fostering of this book idea; all of our collaborators without whom this work would not have covered the four corners of North America; the countless historical societies, libraries, archives and other data repositories whose open doors led to the discoveries that await the reader in the subsequent pages. On some personal notes, Lesley-Ann Dupigny-Giroux would like to thank Steve Doty, Glen Conner and Tom Ross, who sparked her interest in historical records and who continue to augment the rich archives of the Climate Database Modernization Program of the National Climatic Data Center. Cary Mock gratefully thanks Merlin Lawson, Paul and Jeanne Kay, Richard White, and Mike Chenoweth for sparking his historical climate interests.

Burlington, VT, USA Lesley-Ann Dupigny-Giroux
Columbia, SC, USA Cary J. Mock

Contents

Contributors

Dennis B. Blanton Fernbank Museum of Natural History, Atlanta, GA 30307-1221, USA, dennis.blanton@fernbank.edu

Dorian J. Burnette Department of Geosciences, University of Arkansas, Fayetteville, AR 72701, USA, djburne@uark.edu

Michael Chenoweth Unaffiliated Independent Scholar, Elkridge, Maryland, USA, m.chenoweth@att.net

Glen Conner Department of Geography and Geology, Western Kentucky University, Bowling Green, KY 42101, USA, glen.conner@wku.edu

Stephanie F. Dodds Department of Geography, University of South Carolina, Columbia, SC 29208, USA, doddss@mailbox.sc.edu

Nolan J. Doesken Colorado Climate Center, Department of Atmospheric Science, Colorado State University, Fort Collins, CO 80523, USA, nolan@atmos.colostate.edu

Lesley-Ann L. Dupigny-Giroux Department of Geography, University of Vermont, Burlington, VT 05405-0114, USA, ldupigny@uvm.edu

Stuart Foster Department of Geography and Geology, Western Kentucky University, Bowling Green, KY 42101, USA, stuart.foster@wku.edu

David A. Glenn Department of Geography, University of South Carolina, Columbia, SC 29208, USA, David.Glenn@noaa.gov

Edward J. Hopkins Department of Atmospheric and Oceanic Sciences, University of Wisconsin-Madison, Madison, WI 53706, USA, ejhopkin@facstaff.wisc.edu

Rezaul Mahmood Department of Geography and Geology, Western Kentucky University, Bowling Green, KY 42101, USA, rezaul.mahmood@wku.edu

Douglas O. Mayes Department of Geography, University of South Carolina, Columbia, SC 29208, USA, dougandmichele@earthlink.net

Brent McRoberts Department of Atmospheric Sciences, Texas A&M University, College Station, TX 77843-3150, USA, mcrobert@tamu.edu

Cary J. Mock Department of Geography, University of South Carolina, Columbia, SC 29208, USA, mockcj@sc.edu

Jan Mojzisek Department of Geography, University of South Carolina, Columbia, SC 29208, USA, jmojzisek@ci.charlotte.nc.us

Joseph M. Moran American Meteorological Society, Suite 800, Washington, DC 20005-3826, USA, moranj@uwgb.edu

John W. Nielsen-Gammon Department of Atmospheric Sciences, Texas A&M University, College Station, TX 77843-3150, USA, n-g@tamu.edu

Karen L. Paar Liston B. Ramsey Center for Regional Studies and Archivist, Southern Appalachian Archives, Mars Hill College, Mars Hill, NC 28754, USA, kpaar@mhc.edu

David A. Robinson Global Snow Lab, Department of Geography, Rutgers University, Piscataway, NJ 08854, USA, drobins@rci.rutgers.edu

Heather Tompkins Geomatics Technician, Visual Heritage Division, Cartography, Architecture and Geomatics, Library and Archives Canada, Gatineau, Québec, Canada, Heather.Tompkins@lac-bac.gc.ca

James H. Tuten History Department, Juniata College, Huntingdon, PA 16652, USA, tutenj@juniata.edu

Introduction

Climate variability can be analysed via a number of methods including palaeoclimatology, synoptic analyses, statistical deviations, comparison with analogs and identifying the causal dynamics of observed patterns or events (Bradley, 1999). Documentary evidence represents another way of exploring the inherent shifts in storm tracks, changing frequencies of hydrometeorological variables as well as fluctuations in temperature and precipitation over space and time. Given that in general, society was quite vulnerable to climate variations in the historical past, and that some societies in particular were dramatically affected by extreme events through time, these data can also be used to reconstruct historical climate impacts (e.g., Diaz and Murnane, 2008; Endfield, 2008). Documentary sources include, but are not limited to, personal and professional diaries and journals; ship logs (both those during voyages or while docked in port); trade journals and ledgers; plantation records; newspaper articles, editorials and advertisements; and the station histories that complement the numerical records of either observers' lives, their geographic surroundings, challenges in taking observations or reasons for relocating a station. In this field known as "Historical Climatology", "Early Instrumental data", which are numerical meteorological data taken by adhering to standards quite unlike those used today, are also regarded as a special type of historical/documentary data (Chenoweth, 2007).

Piecing together the weather and climate of a place or an event of interest from such disparate and often incomplete records is painstakingly slow, and seldom complete. Missing pieces of a climate puzzle can come from very far afield, often in unlikely places. Each chapter in this book represents many years of assembling the climate picture of an event, natural hazard or extreme condition; an anomalous month or season; or the creation of a new methodology for working with these quasi-quantitative data. Most apparent, this book focuses on studies from North America, since the historical climate community here has only increased relatively recently (in the last decade and a half), and prior to this most such studies have focused on Europe and Asia.

The book is divided into four sections: 1) Case Studies, 2) Reconstruction of Extreme Events and Parameters, 3) The Role of Station History in Understanding the Instrumental Record, and 4) Methodologies and Other Analyses.

The regional case studies open with "The Great Flood of 1771: An Explanation of Natural Causes and Social Effects" by Blanton et al. that uses synoptic reanalysis and historical sources to explore the human and societal ramifications of a catastrophic flood along the eastern Atlantic seaboard in May 1771 on the eve of the American Revolution. The study serves to highlight the disproportionate toll of natural hazards on lower income populations as a result of mobility, housing locations and access to resources, a finding that is as true today as it was three centuries ago. This is followed by "Historical changes to the Tennessee Precipitation Regime" by Mojzisek and Mock who continue the thread of precipitation anomalies in the US Southeast by using a century's worth of data to show the decadal variability of the region with a maximum occurring in March and a minimum being observed in October. The study also highlights the teleconnective nature of preferred modes of variability such as the El Niño-Southern Oscillation and Pacific Decadal Oscillation and points to additional data needed to fully resolve regional and temporal patterns. The third chapter in this section is "Don't want to see no more . . . like that!: Climate Change as a Factor in the Collapse of Lowcountry Rice Culture, 1893–1920 by Tuten. It is an in-depth analysis of the intertwined socioeconomic, trade and hydroclimatic factors that led to the final collapse of rice farming in South Carolina between 1893 and 1920. Like the other chapters in this section, the role of human vulnerability to climate especially extremes, again comes to the fore. The section closes with "Climate in the Historical Record of Sixteenth-Century Spanish Florida: The Case of Santa Elena Re-examined" by Paar. Covering the earliest period in this book (1566–1587), a period with sparse documentary evidence, this chapter highlights the use of contextual analysis of the written record in the light of complementary findings from dendroclimatology and archaeology.

Section II is devoted to the reconstruction of extreme events and months/seasons. In "Historical Accounts of the Drought and Hurricane Season of 1860", Dodds et al. use synoptic mapping to explore the severe drought that gripped much of the central and southeastern US followed by a thorough updated reconstruction of the active 1860 hurricane season. The influence of La Niña conditions on this active pattern is also explored. The following chapter by Glenn and Mayes on "Reconstructing 19th Century Atlantic Basin Hurricanes at Differing Spatial Scales" continues the theme of synoptic scale reconstruction applied to the 1850 season. From this, steering factors and the intensity at landfall can be extracted, thereby expanding eyewitness accounts that tend to be concentrated in highly populated areas. In revisiting "The Sitka Hurricane of October 1880", Mock and Dodds combine land based records with pressure records from the USS Jamestown to reconstruct the hourly passage of an extratropical storm near Alaska, that was one of the strongest events in western North America in the last few hundred years. The final two chapters in this section deal with cold extremes. In "Daily Synoptic Weather Map Analysis of the New England Cold Wave and Snowstorms of 5–11 June 1816", Chenoweth revisits the Year without a Summer, focusing on the reconstruction of daily weather maps in early June. Ship logbooks were highly instrumental in the discovery of a hurricane in Florida that coincided with the snowstorms and freezing weather in the US during this time. Finally, in "March 1843: The Most Abnormal Month Ever?" Neilsen-Gammon and McRoberts use daily and monthly reconstructions for the central and eastern US to determine the anomalously southern storm track that was instrumental in bring severe, winter-like weather to the southern parts of the United States.

The third section on the role of station history in understanding the instrumental record encompasses crucial metadata issues to be considered when using historical records. The section opens with "Weather Station History and Introduced Variability in Climate Data" by Conner which outlines the importance of the history of a station, which if unaccounted for could lead to spurious conclusions from the data. Such station histories include changes in observers, observational procedures, time of observations, station moves, instrumentation and exposure changes. This is aptly followed by Hopkins and Moran's "Monitoring the Climate of the Old Northwest: 1820–1895". This chapter focuses on the observational networks and station histories of today's states of Ohio, Michigan, Indiana, Illinois, Wisconsin and Minnesota and highlights the linkages between observed climate data and human activity, a topic that was as important in the 19th century as it is today. This section closes with Foster and Mahmood's "Spatial Metadata for Weather Stations and the Interpretation of Climate Data" which presents the use of GeoProfiles (the documentation of the spatial characteristics of an observing station's topography, exposure etc.) within a geospatial environment. They argue that the data from poorly sited stations can still be used if the digital and descriptive record of their exposure and representativeness are understood and incorporated in the analyses.

The final section of the book is devoted to methodologies and other analyses. In "A seasonal warm/cold index for the southern Yukon Territory: 1842–1852", Tompkins applied content analysis to journals from the Hudson's Bay Company to create a hierarchical coding scheme by which normal and extreme weather conditions could be identified. By showing that winter warming in the Yukon may be related to the end of the Little Ice Age, this technique shows the applicability of using documentary evidence to fill in palaeoclimatic gaps in the record. The theme of index creation is continued in "Backward seasons, droughts and other bioindicators to explore climate variability" by Dupigny-Giroux where a new indicator-based drought index is developed for northern New England. This chapter also highlights the use of phenoclimatic fluctuations and analyses backward seasons in terms of their influence on frost, sugar maple production and drought. The volume ends with "The challenge of snow measurements" by Doesken and Robinson which delves into the historical and present-day challenges of measuring snow properties and the biases that are introduced when such measurements are poorly made.

The challenge of achieving a full understanding of climate and the atmosphere is a common theme in climate science, including those among the historical climate community. Similarly, given the limits of climate determinism, comparable tough challenges also exist in reconstructing the full nature of societal impacts from climate change. Therefore, data quality and rigorous historical climate reconstruction methods, some of which are borrowed from history, are critical. Many of these aspects are unique contributions from this volume. We hope that the volume provides key frameworks for any scholars interested in studying historical climate change, but also stimulates such further work on North America.

References

Bradley, R.S., 1999. Paleoclimatology: Reconstructing Climates of the Quaternary. Academic Press, San Diego.

Chenoweth, M., 2007. Historical climatology. In: Elias S (ed.) Encyclopedia of Quaternary Science. Elsevier, U.K.

Diaz, H.F. and Murnane, R.J. 2008. Climate Extremes and Society. Cambridge University Press, Cambridge, UK.

Endfield, G.H., 2008. Climate and society in colonial Mexico: a study in vulnerability. Blackwell, Malden, MA.

Part I
Regional Case Studies

The Great Flood of 1771: An Explanation of Natural Causes and Social Effects

Dennis B. Blanton, Michael Chenoweth, and Cary J. Mock

Abstract Floods are extreme events associated with severe weather that have plagued human populations almost everywhere they have lived. The causes and effects of a catastrophic flood occurring on the southern Atlantic seaboard in May 1771 were examined through a combination of historical climatological sources and synoptic meteorological analysis. Reconstruction of the event reveals the unique weather pattern responsible for an unusual early-season flood of this magnitude. Compilation of historical records allows for an assessment of the human toll and social response. Comparative analysis exposes aspects of an increasingly predictable outcome.

Keywords May 1771 · Virginia · Flood · Atlantic seaboard

1 Introduction

Annual river cycles can have the positive effect of setting a rhythm for streamside inhabitants, when a predictable pulse of rising and falling water recharges natural systems and facilitates commerce. Occasionally extreme events such as great floods punctuate human history with other kinds of effects. In the most catastrophic situations, entire communities can be lost or moved, economic impacts can have severe social and political implications, and official policies can be enacted to safeguard against future disasters. Alternatively, significant cultural change can be remarkably limited following severe floods in spite of destruction and hardship.

D.B. Blanton (✉)
Fernbank Museum of Natural History, Atlanta, GA 30307-1221, USA
e-mail: dennis.blanton@fernbankmuseum.org

M. Chenoweth (✉)
Unaffiliated Independent Scholar, Elkridge, Maryland, USA
e-mail: m.chenoweth@att.net

C.J. Mock (✉)
Department of Geography, University of South Carolina, Columbia, SC 29208, USA
e-mail: mockcj@sc.edu

L.-A. Dupigny-Giroux, C.J. Mock (eds.), *Historical Climate Variability and Impacts in North America*, DOI 10.1007/978-90-481-2828-0_1,

This chapter describes the causes and effects of a catastrophic flood on the south Atlantic seaboard early in US history, on the eve of the American Revolution. In Virginia it has been referred to as the Great Fresh of 1771, regarded as perhaps the most catastrophic flood in the region's history. Our purpose is to offer a more complete account and analytical explanation of the flood itself, and to explore its social, economic, and political implications. The historical sources we draw on are collected not only from Virginia where prior attention has been focused, but from wherever relevant observations were recorded, extending as far southward as Florida and as far northward as Massachusetts. We also bring to bear the perspective of synoptic climatology to reconstruct the weather pattern that accounts for this unusual late spring event. Finally, we attempt to digest existing historical perspectives from colonial Virginia to offer a fuller interpretation of the flood's meaning at this formative stage in colonial history. The larger goal is to use this case as a means of exploring the place of natural climatic events in human affairs.

2 Historical Accounts of the Event

A variety of sources allow us to establish the basic facts surrounding the May 1771 flood including personal correspondence and journal entries, official records and correspondence, and newspaper accounts. The great weight of the material is attributed to observers in eastern Virginia, either in the vicinity of towns like Richmond at the falls of rivers, or toward the coast in Tidewater (Fig. 1). The same is true of the information from South Carolina, where most reports come from the more populous Low Country districts.

Period newspaper sources attribute the flood to heavy and prolonged rain in the Appalachian Mountains (Bland, 1771; Virginia Gazette, 1771a). Knowledge of the weather in the highlands 50 or more miles (80.47 km) west of major fall line towns arrived in the centers of population only after the water receded, however, such

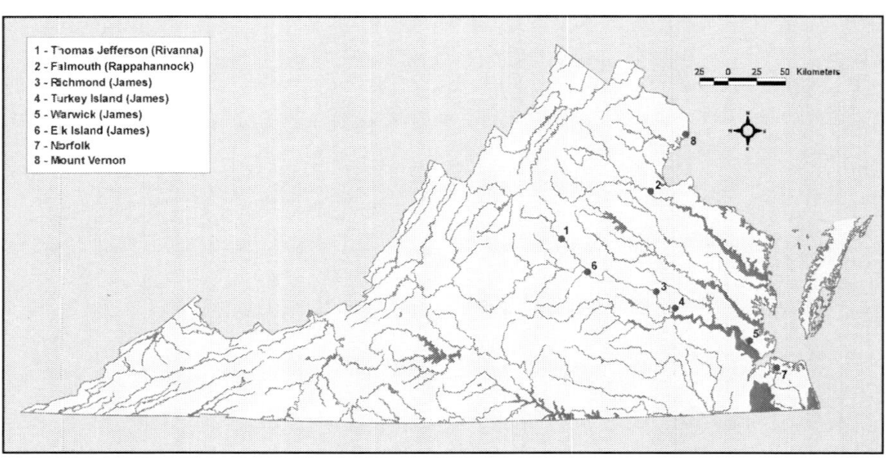

Fig. 1 Locations of Virginia place names

that residents of eastern Virginia were startled by the flood, appearing as it did under "serene" skies (Bland, 1771). ("Fall line" towns were those situated at the physiographic transition between the Piedmont uplands and the Coastal Plain, usually marked by shoals and falls in streams where they pass over resistant Piedmont bedrock.) Many locales, even east and south of the mountains, clearly experienced a wet spring that created conditions conducive to the eventual flood. As one example noted, some flooding was already being reported in South Carolina in the latter part of April and early May after an "uncommon quantity" of rain had fallen (Rudisill, 1993). Persistent wet weather was described as 10–12 days of constant rain by "waggoners" arriving in Richmond from Augusta, Georgia. Farther north at Mount Vernon, George Washington recorded generally threatening skies and showers from May 19 to 28 (Jackson and Twohig, 1976) (Table 1). Similar weather

Table 1 Flood and weather chronology

Date	Location	Weather	Flood Status
May 3	Pee Dee River, SC	–	"River still rising"
May 8	Savannah River, GA-SC	–	"High"
May 12–16	Cambridge, MA	Cloudy, rain, thunder and lightning	
May 13–17	Norfolk, VA	Clear	–
May 14–26	Pied.-Mtns. SC to VA	Heavy Rain	–
	South Carolina	"Uncommon" abundance of rain	
May 18–22	Norfolk, VA	"Squally", rain	–
May 19–21	Cambridge, MA	Rain, clouds, thunder and lightning	
May 22	Savannah River, GA-SC	–	"Still high"
May 23	Pee Dee River, SC	"fine rain"	–
May 23–24	Yadkin River, NC	–	Bottoms flooded
	Norfolk, VA	Clear	–
May 25	James River, Richmond, VA	"Serene"	First obs. rise in river
	Cambridge and Salem, MA	Rain	
May 25–26	Norfolk, VA	"Squally"	–
May 26	Rivanna River, W. Pied., VA	–	Peak flood stage
	James River, Richmond, VA	No rain	"Rapid rise to 16"/hr
	Pee Dee River, SC	"wet day"	"rising very fast"
May 27	James River, Richmond, VA; Rappahannock River, Falmouth, VA	No rain	Maximum flood stage
	James River, Richmond, VA	No rain	Begin recede at sunset
	Pee Dee River, SC	–	"rising very fast; Very deep water"
May 28	Pee Dee River, SC	Raining	"Higher than ever"
	Cambridge and Salem, MA	Clouds and rain, Showers	
May 27–31	Norfolk, VA	Clear	–
May 30	Pee Dee River, SC	–	"waters begin to fall"
May 31	Pee Dee River, SC	–	"water fell fast"
June 13	Rappahannock River, VA	–	Water still "red"

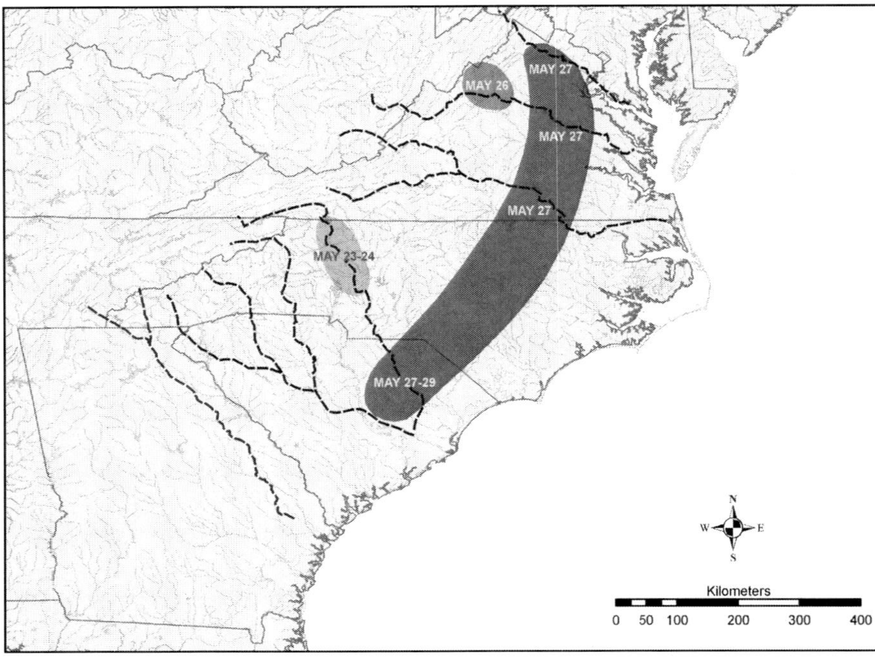

Fig. 2 Approximate timing of flood crests on the Atlantic slope

was registered by a vessel operating near Norfolk in the lower Chesapeake Bay. Rainfall was never great there but sporadic showers did occur with gale force winds on May 25th (Public Records Office n.d.) (see Table 1). Based on several reports that put the first visible rise of the water at Richmond on May 25, the deluge in the Virginia mountains would have occurred between about May 14–26. The peak flood stage at all of Virginia's fall line towns was on May 27 (Fig. 2) (Virginia Gazette, 1771a, 1771b; Van Horne, 1975; The Farmer's Register, 1836). Geographically the extent of flooding appears to have been vast, extending at least through the Carolinas and over Virginia (see Fig. 2). Rivers specifically noted at flood stage in late May were the Savannah, Pee Dee, Wateree, Congaree, Yadkin, Roanoke, Rivanna, James, Pamunkey, Mattaponi, Rappahannock, and Shenandoah, ordered from south to north respectively.

An expected west to east cresting trend is revealed by Thomas Jefferson's remark that waters on the Rivanna River, a tributary of the James in the western Piedmont some 50 air miles (80.47 km) upstream from Richmond, reached a peak level a day earlier on May 26th (Betts, 1945). Yet not all eastern Virginia streams were flooding. Conspicuously absent are comments about high water in the Chickahominy River basin near Williamsburg. One newspaper source specifically observes that the "Appamattox has been little or nothing effected, which proves that the rains must have fallen up in the high country" (The Virginia Gazette, 1771b). The common feature of these unflooded streams is that none drains large segments of the Piedmont or mountains. No mention of Potomac River flooding has come to light either, other than a brief note that its Shenandoah River tributary was out of its banks (The Maryland Gazette, 1771), a fact taken to indicate that the rain-producing weather system had weakened considerably in northern Virginia. (George Washington's diary, for example, includes no comments about Potomac flooding around Mount Vernon in May, but later he was summoned to Williamsburg to deal with official, flood-related proceedings in July (Jackson and Twohig, 1976:29–31, 39–41).)

Farther south rain and swollen streams are noted a few days earlier (The Farmer's Register, 1836). The Savannah River at the border between South Carolina and Georgia was described on May 22nd as "still high" (Hawes, 1988). By May 23–24th the Yadkin River in the western Piedmont of North Carolina was already out of its banks, leaving floodplains "like a sea" (Fries, 1922). Otherwise, details of the flood chronology from South Carolina are generally consistent with the Virginia sequence. While the Carolina accounts are less numerous and detailed, we can establish that waters on the lower reaches of the Pee Dee, Wateree, Congaree, and Savannah Rivers peaked by May 28th. For example, the lower Pee Dee rose very rapidly on May 26–27 and on May 28th was described as "higher than ever" (Rudisill, 1993).

The reported time lapse between the first perceptible rise and the flood crest at Richmond is 60 h, and the rate of flooding between May 25th and the time of the crest on May 27th was at times very rapid. Although one account from Richmond cites peak rates of five feet per hour, more believable reports say up to 16 in. (40.6 cm)/h (The Virginia Gazette, 1771b). More than likely the most rapid rise at the Virginia falls was occurring on the 26th. By the 27th, the river was rising no more than 2 in. (5.08 cm)/hour and by sunset that day the water was observed to be falling (The Virginia Gazette, 1771a). In low country South Carolina, the flood chronology is quite similar. On the lower Pee Dee, water levels rose "very fast" on both May 26 and May 27, even as much as 15 ft (4.57 m) in 24 h on the second day (Rudisill, 1993; The Farmer's Register, 1836). On May 30th the waters began to fall and by the 31st came down "very fast." A single diary entry by Col. Landon Carter for June 13, 1771 describes lingering effects in Virginia. He observed that the Rappahannock River "seems not to have lost its red Colour which it got in the vast fresh in May last" (Greene, 1987).

Reported maximum flood levels in 1771 are impressive and at least in Virginia are not matched in official records (Camp and Miller, 1970; Bailey et al., 1975). At Richmond the James River floodwater maximum was marked at 40 ft (12.19 m) above normal level (Bland, 1771; The Virginia Gazette, 1771b). At Osborne's about 10 miles (16.09 km) below Richmond the water level reached "five or six lengths of shingles upon the roofs" (The Maryland Gazette, 1771). Given in relative terms, the peak level at the Fall Line was most commonly noted as 20–25 ft (6.09–7.62 m) above the previously observed "old record", probably set in May 1766 (The Virginia Gazette, 1771a). A single account measures the relative height as 10 ft (3.048 m) above the flood levels recalled uncertainly from August "1720 or 1724" (The Pennsylvania Journal and Weekly Advertiser, 1771.) (More than likely this is a reference to high water associated with an August 1724 hurricane cited in various documents (Ludlum, 1963:20–21).) A "mighty fresh" is noted even earlier on the

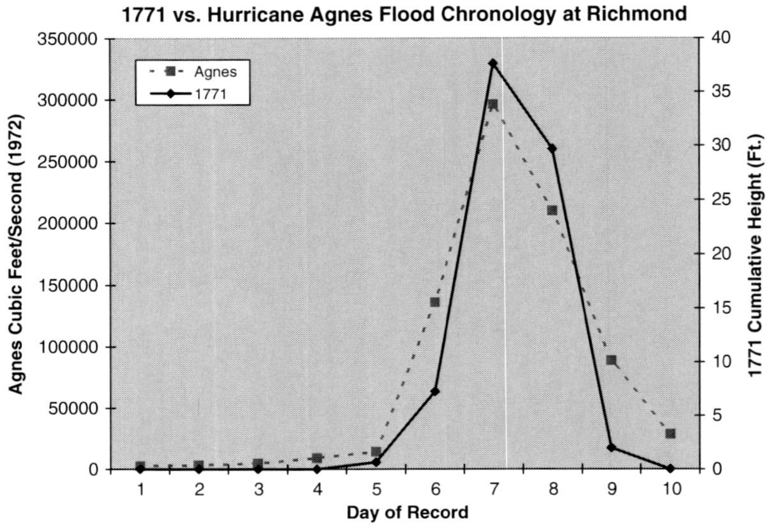

Fig. 3 Comparison of the 1771 and 1972 hurricane Agnes flood chronologies (1972 hurricane Agnes data from Bailey et al. 1975)

James River in April 1685, one that flooded the home of William Byrd I at the site of today's Richmond and exceeded what then was the highest observed river level by 3 ft (0.91 m) (Virginia Magazine of History and Biography, 1908:351). On the Roanoke River, the 1771 flood is cited as 9 ft (2.74 m) above the previous all-time high mark (The New London Gazette, 1771a).

Major late spring flooding of the James River of the magnitude observed during the 1771 event is almost unheard of. Official James River flood records kept at Richmond in the twentieth century list the highest mark of 36.5 ft (11 m) in June, 1972 associated with Hurricane Agnes (Fig. 3). The next two highest levels are also associated with tropical storm systems, 30.8 ft (11.7 m) in November, 1985 associated with Hurricane Juan and 28.6 ft (8.7 m) in August, 1969 associated with Hurricane Camille (Bailey et al., 1975; Camp and Miller, 1970). Two major spring floods in the modern era occurred in successive years and made the top ten list for the James River at Richmond. Both are earlier spring events and represent snowmelt effects, as probably did the April, 1685 flood. In March, 1936 the river level reached 26.5 ft (8.1 m) and in April of the next year it peaked at 25.2 ft (8 m). Comparatively then, the 1771 level eclipses the highest officially recorded tropical storm flood at Richmond by a small margin, but it tops the highest recorded spring flood level by about 14 ft (4.26 m).

The credibility of the Piedmont-Fall Line accounts of water levels seems to be good. This is not the case for reports from places like Warwick and Osborne's in the tidal reaches of the James River. The roof-high level given for Osborne's is remarkable knowing that the town sat on a bluff elevated 50 ft (15.24 m) above the mean river level and that such a mark on even the average colonial dwelling would

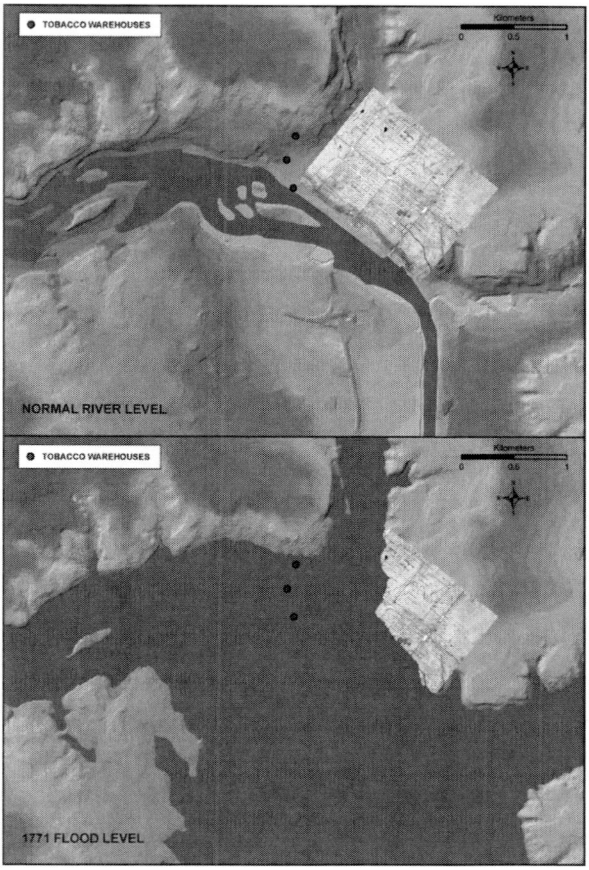

Fig. 4 Graphic simulation 1771 flooding at Richmond, Virginia compared to normal James River level (Historic map overlay from Mayo and Wood 1737)

be at least 10 ft (3.048 m) above the ground. Perhaps the account is referring to a river-side building below the main town. Similarly, a 40-foot crest above normal river levels at Warwick, more than 60 miles (96.56 km) below the Richmond and the Fall Line, is dubious. Excellent documentation for the 1972 Hurricane Agnes floods, which all but matched the 1771 extreme, establishes that tidal reaches experienced much lower crests than sections at and above the Fall Line. With Hurricane Agnes, for instance, the maximum river level above the projected normal was no more than 8 ft about 10 miles (16.09 km) below Richmond, Virginia, and there was no discernible rise at all as far down as the former site of Warwick (Davis, 1974:A112).

Using the Richmond area as an example, a historical plat of the town was draped over a topographic depiction and river levels were manipulated relative to modern mean values. Figure 4 shows the current normal river level and the estimated 1771 maximum level as estimated from the current 10 ft amsl level at the lower falls (Mayo and Wood, 1737). Within the roughly 12 km^2 area shown in the figures, an estimated 632 hectares would be inundated by the 1771 flood, exceeding the 574 hectares covered by the Agnes flood level. Note in the image the inundated locations of key colonial-era tobacco warehouses estimated from historical sources (Sheldon, 1975).

Physical damage from the flood was not only severe but lasting, to the point that it represents a true geological event. Indeed, an inscription on a monument memorializing the flood east of Richmond declares that the flooding rivers "changed the face of Nature, And left traces of their Violence that will remain for Ages" (Gibbons, 1976). Hundreds of acres were scoured of tilled soil, countless tons of which were re-deposited elsewhere. From 6–12 ft of sand were laid down in many locations along the James River, especially from the Richmond area upstream (Bland, 1771; Cabell n. d.; The Virginia Gazette, 1771b). Elsewhere, pavements of stones were left behind. Region-wide wholesale reconfiguration of floodplains occurred as entire islands were "torn to pieces" and as many channels were opened as were closed (The Farmer's Register, 1836). One correspondent astutely noted that such damage was proportional to stream size. Vast numbers of logs and other debris was swept downstream to accumulate in reeking piles as far away as estuarine embayments like Albemarle Sound (The New London Gazette, 1771b). There is also little doubt that the severity of this flood was exacerbated by runoff from an unprecedented extent of cleared land in the upland portions of Virginia.

3 Synoptic Reconstruction of Weather Conditions in May 1771

Understanding of the meteorological events that contributed to the Virginia floods in May 1771 requires a sufficient number of weather reports over a wide area of the eastern half of the United States. The plotting of these data using modern meteorological analytic techniques to determine the approximate locations of areas of high and low pressure and cold and warm front boundaries (see Table 1).

Daily weather records were obtained from seven ships of the English Royal Navy at six locations. These records were obtained from the UK National Archives, London from original ships' logbooks. These records include HMS *Halifax* moored near Halifax, Nova Scotia, HMS *Deal Castle* moored in the East River, New York City; HMS *Magdalen* moored in the Hampton Roads, Virginia area; HMS *Martin* moored at Cape Fear, North Carolina; HMS *Bonetta* moored at Charleston, South Carolina; HMS *St. Lawrence* moored at Nassau, Bahamas; HMS *Zephyr* moored at Pensacola, Florida and its relief ship, HMS *Lowestoft*, which arrived at Pensacola in mid-May 1771 from Jamaica. In addition, private weather records kept by Thomas Thistlewood at Savanna-la-Mar, Jamaica, George Washington at Mount Vernon, Virginia (Jackson and Twohig, 1976) instrumental weather records for Cambridge and Salem, Massachusetts (Winthrop n.d., Holyoke n.d.) were used as sources of wind and weather reports. Newspaper accounts and the Pugh Diary, from a publication of the St. David's Society, Florence, South Carolina were also consulted to provide sporadic daily weather accounts from other areas.

The data were mapped to produce a daily series of synoptic weather. The map series, informed by modern theories of synoptic weather analysis were used to locate major weather features such as areas of low and high pressure, the boundaries between air masses (cold and warm fronts), and the movement of these features through time. Although the weather network is very limited, the available data set limits the possible range of weather at a given site, which is further narrowed by principles of continuity in the day-to-day displacement of the mapped weather features. A similar process, with an even sparser network, has been successfully employed in reconstructing the weather events during the Spanish Armada invasion of England in 1588 (Douglas et al., 1978).

Our analysis indicates that warm weather at the beginning of the month gave way to cooling on May 4–5 as a low pressure area moved offshore from the mid-Atlantic Ocean northeastwards to the Canadian Maritimes. High pressure quickly moved over the Atlantic coast. Pressure gradually fell in New England until 10 May as a rather seasonable pattern set in. The station network is too restricted to clearly delineate most weather features at this time. It appears that by May 8, an area of low pressure formed off the mid-Atlantic coast along a frontal boundary and moved slowly to the northeast.

By May 11 pressure was low as an old low pressure center slowly meandered over the Gulf of St. Lawrence area. In the meantime, a cold easterly wind had set in at Mount Vernon and Norfolk, Virginia. The weather stayed unseasonably cold and cloudy from May 11 to 15 all along the east coast. The pressure rise at Cambridge, Massachusetts on May 12–13, appears to reflect a cold Canadian high centered to the north of New England. On May 12, a low pressure center was probably forming to the southeast of New England over the Atlantic Ocean. As this low deepened on the night of May 12–13, it brought a strong northerly flow into Virginia and colder weather on May 13–14. A reinforcing trough of low pressure brought more cold air into New England on May 15, corresponding with the pressure minimum. Cooler air gradually slipped into Virginia from May 17 to 19. On May 18, a cold front pushed across New England and was followed by a ridge of high pressure remaining to the north over Canada. George Washington observed cooler weather at Mount Vernon on May 18 and May 19 as the leading edge of the cooler air appears to have stalled in the mid-Atlantic region. Again, the station network is too sparse to definitely pinpoint the low and high pressure centers. Also, the frontal positions are approximate.

It would appear that during the period May 11–15, a slow-moving cold front and a northeasterly fetch to the winds helped to induce the beginning of the near two-week wet spell over the mountains of the southern Appalachians reported by traveler accounts. Mount Vernon had some rain on May 8, as a low pressure trough was centered over the area that day, and a thunder storm occurred on May 9 as warm air moved back briefly into the area. Otherwise, it was dry in eastern Virginia until May 19 when some thundershowers moved through the Norfolk area. These storms were along the north edge of much warmer air that was just beginning to make a major surge to the north. During the night of May 19–20, thunder and rain was reported at Mount Vernon as this warm air moved north. This warmer air would reach Cambridge and Salem, Massachusetts on the night of May 21–22, with rain and thunder preceding the warm air on the afternoon of May 21.

By May 22 high pressure was probably becoming stronger over the western Atlantic and an area of low pressure was moving very slowly over the southern Mississippi River valley. High pressure probably prevailed over central North America and lower pressure over the Canadian Maritimes and Québec. As warm tropical air moved into the eastern US, there was an enhanced southeasterly flow implied over the Southern Appalachians, which would lead to enhanced rainfall amounts due to orographic uplifting of the moist air over the mountains. With troughs and frontal boundaries in the region providing additional lifting to the air mass, the stage was set for frequent showers, at times heavy, over a large region. By May 23, it is believed that a warm frontal boundary was in place over the Southern Appalachians, with warm air overrunning slightly cooler air at the surface. An east to southeast wind flow at the surface was favorable to enhanced rainfall in the mountains of the Carolinas and Virginia. In the meantime, winds at Pensacola, Florida which had blown from a southerly direction in the afternoon and evening of May 22, had turned light northerly and at times variable with constant showers of rain that would continue through the night of May 25–26. Although winds were southerly in the Carolinas, the shift to light variable northerly winds at Pensacola indicates a trough of lower pressure on an axis in the eastern Gulf of Mexico.

In Jamaica, the weather was stormy at Savanna-la-Mar on May 23 as persistent south to south-southeast winds blew day and night with hard squalls of wind. This indicates that pressure was above average to the east and below average to the west, defining a trough from the northwest Caribbean to at least the northwestern Florida coast ahead of surface and upper level low pressures moving in from the region of Texas. Winds remained persistently in the south through May 25 and the weather grew less disturbed. Sometime on the nautical day of May 24 (noon May 23 to noon May 24 civil time) a ship was wrecked on the southwest coast of Cuba in a gale. This very likely is a tropical cyclone located in the northwest corner of the Caribbean Sea and probably moving northward ahead of the approaching trough of lower pressure to its northwest.

By May 24, the weather was fair and unremarkable with a southerly flow over eastern Virginia. Very warm air on a southwest wind was blowing at Cambridge and Salem, Massachusetts. Dry weather with southerly winds blew in the Carolinas and southeast Virginia while it was a rainy day with north winds at Pensacola. On May 25, the morning temperature at Cambridge was 22 degrees above the average for the month, and along with May 24, was the warmest day of the month. By evening, a cold front would move through and drop temperatures to more seasonable levels. However, this cold front did not reach Virginia. Strong winds from the southwest blew across Virginia on the 25th, with fair skies. At Point Comfort, Virginia, strong gales (equivalent to winds of about 30–35 miles per hour [13.4–15.6 m/s]) blew from the west-southwest. At Mount Vernon, "very fresh" southwest winds blew in the morning but in the afternoon the wind was from the east with rain. By the morning of May 25, the constant rains had ended at Pensacola, Florida and the weather was described as "moderate and fair", and the north winds of the afternoon became variable in direction overnight as high pressure over the southwest Atlantic began to build back into Florida and the Bahamas.

With no wind shift or rain at Point Comfort, Virginia the evidence points to an area of low pressure moving across central Virginia during the night of May 25–26. The low moved quickly to the northeast ahead of the cold front that was gradually moving toward the area. But this cold front stopped moving south as is evident in a brief warming at Cambridge and Salem, Massachusetts on May 27. The weather had cleared at Mount Vernon on the afternoon of May 26 and the weather was "warmer" than it had been that morning in the rain-cooled air from the east. Also on May 26, the day was described as a "very wet day" along the lower reaches of the Pee Dee River in South Carolina. At Pensacola, Florida the weather was moderate and fair with east-southeast winds prevailing.

On the morning of May 27, the wind was again from the south at Mount Vernon and rain was falling. At the same time, a southwest wind blew at Point Comfort, Virginia and no rainfall was reported. It was at this time that the James River reached its peak level as the floodwaters from upstream, first noticed on May 25, inundated the river valley under fair skies. Warm weather and rain continued at Mount Vernon until near mid-day on the 28th when skies finally began to clear under northerly winds following the passage of a cold front. A low pressure center had moved offshore from the west and was bringing much colder and wetter weather to Cambridge, Massachusetts and towards Halifax, Nova Scotia. Point Comfort in Virginia remained ahead of the cold front until the evening of May 28. Also on that day, rain continued at Pugh's residence in South Carolina, and the Pee Dee River was "higher than ever".

The cold front pushed through the entire region by the morning of May 29, abruptly ending the weeks of rainy weather in the Southern Appalachians and bringing unseasonably cold air in its wake. The mornings of May 29 and 30 were the coldest of the month at Cambridge, MA with frost reported each morning.

4 Possible Linkages with Tropical Systems

The phenomenal floodwaters of this event are evidence of an immense fall of rain on the higher reaches of the James River. As noted, the closest modern analog to the 1771 flood in Virginia, both in terms of timing and severity, is the flooding associated with Hurricane Agnes in June 1972. The James River crested at 36.5 ft (11.12 m) at Richmond that year, just shy of the 38–40 ft (11.58–12.19 m) maximum cited for 1771 at the same location. Both are considered early season floods, with Agnes occurring just one month later in the year than the 1771 event in late May. Hurricane Agnes and its effects in the Middle Atlantic and Northeastern states is one of the best-documented events of its kind in those regions. Scientists used the storm as an opportunity to collect quality data for a rare, extreme weather event. There are differences in the two cases, of course, but a comparative analysis allows for a better understanding of the weather that generated the great colonial-era "fresh."

One question concerning the 1771 flood is whether it was the result of a tropical storm. Hurricane Agnes, for example, was recognized as a relatively weak but very large, early-season tropical storm, but its duration and severity increased owing to the original system's absorption into an extratropical low pressure system. Exacerbating the flood threat was above-average Middle Atlantic rainfall, especially in Virginia, during May and mid-June, 1972. Agnes was first recognized over the Yucatan peninsula on June 14, 1972 as a tropical disturbance. The storm intensified over the Caribbean and Gulf of Mexico and briefly reached hurricane force on June 18th. Landfall occurred the next day on the Florida panhandle and for several more days the tropical depression followed a northeasterly, overland path. This tropical system was very large, covering some 10,000 square miles. It moved across Georgia for most of June 20th, with rain already beginning in southern Virginia. By the next day, the center was over the Carolinas and rain had extended northward to Canada. It was on the 21st that a secondary low was spawned, traveled on more or less a parallel path and absorbed Agnes on June 22nd with the lowest recorded pressures of the storm occurring over Maryland. A related effect was that the heaviest rainfall of the storm fell on June 21–22. The system stalled over the northeast the next 48 h where it dominated the weather for several more days.

Record rains from Virginia to New York accounted for the extreme, regional flooding. The 24-hour Virginia maximum just south of Washington, D. C. was at least 10 in. (25.4 cm) on June 21st, with totals over a large, contiguous area reaching at least 15 in. (38.1 cm). In Virginia the most severe flooding occurred along streams immediately south and west of Washington, and in the James River basin. Extreme water levels occurred in the upper Roanoke drainage as well, but the downstream effects were moderated by dams.

In 1771, we have evidence of a tropical storm or hurricane present in the northwest Caribbean on May 23 and 24, which appears to have moved north through the Yucatan Channel and towards northwest Florida. The persistent north winds and rainy weather at Pensacola, Florida suggests a trough of low pressure from higher latitudes linking up with the tropical cyclone to its south and upper level winds probably pulled the storm northward towards Florida, passing to the east of Pensacola and staying well to the west of Charleston, South Carolina and Cape Fear, North Carolina, as revealed in the persistent southerly winds at these locations. While the intensity of the tropical cyclone cannot be gauged from available data, it represents an area of concentrated moisture being pulled north into an already moist flow of air that had prevailed for several days before its likely landfall in northwest Florida during the early morning hours of May 25. From here, the center moved north and northeast towards central Virginia. The low apparently moved northeast from central Virginia to the Atlantic and probably dissipated or was absorbed into the cold front moving east from New England on May 27 and 28.

The 1771 scenario then has some similarity with 1972 in that a surge of moisture from the tropics was drawn into the southern Appalachian Mountains. Each storm was enhanced by orographic lifting by the mountains such that very heavy rains fell on the mountains and eastern slopes in a very short time. The center of the low pressure passed over Virginia on May 25–26 and the rise in the rivers was coincident with the passage of the low. Rainfall probably preceded the low pressure center's arrival by at least a day. Unlike the event in 1972, in 1771 much of eastern Virginia did not share in the heavy rainfall, nor did the latter storm involve the merging of a tropical storm by a developing secondary low pressure system. Instead, the center of tropical moisture moved out to sea. Therefore, Pennsylvania and Maryland did not suffer from major floods in 1771 as they did in 1972.

5 Historical Socioeconomic Implications

Of all states affected, Virginia sustained the greatest losses from the 1771 flood, and the physical damage compounded already-existing economic and political strains. By 1771 Richmond, Virginia was coming into its own as an economic hub as were other fall line towns, although tidewater ports such as Norfolk still enjoyed prominence (Sheldon, 1975, Selby 1976). Positioned at the edge of the Piedmont, these burgeoning interior river ports served as the markets for business linked to expanding western communities. Virginia's capital was still in Williamsburg at the time but would relocate to more pivotal Richmond in 1780. Although wheat production had begun to rival the emphasis on tobacco, the latter was still very much a critical commodity (Sheldon, 1975; Gray, 1933). In Richmond, tobacco inspection stations and warehouses were clustered around the mouth of a creek known as Shockoe Bottom and nearby at Byrd's (see Fig. 1). Just across the James River from Richmond other warehouses were at the new community of Rocky Ridge (Sheldon, 1975; Dabney, 1976). This location gave direct access to the tidal head of the James River and was, thus, a key transhipment point. Other warehouses were located about seven miles (11.2 km) upstream, at the head of the falls, in Westham. Analogous facilities were present at the falls of the Rappahanock in Falmouth and at Dixon's in King George County, and on the Potomac River at Quantico (Van Horne, 1975; The Virginia Gazette, 1771b). Secondary inspection stations and warehouses were situated at other river communities such as Warwick and Osborne's on the James River. Tobacco was packed into large barrels called hogsheads for shipment and large quantities of it accumulated at the principal warehouses in anticipation of export. So, at the time of the great flood a significant portion of the colony's wealth was collected at riverside warehouses, and each of the noted facilities was affected.

By the decade of the 1770s, Virginia was exporting about 70 million pounds of tobacco, a figure representing about 40% of all tobacco traded from the colonies (Selby, 1976). Virginia sought trade only with Britain in exchange for a ready market and defense. The trade was based mainly around the activity of prominent planters who essentially served as the middlemen for tobacco exports and imported goods. Debt levels ran higher in Virginia than in most colonies but it still enjoyed relative affluence. There was no formal system of banking and little currency to circulate so loans between individuals were not only the norm but actually necessary. Exchanges often took the form of barter with expectations of extended credit. In Virginia, a key element of this informal system was use of tobacco as a medium of exchange. Losses of any magnitude to this veritable currency could only erode the economic welfare of many prominent citizens if not the colony itself.

The third quarter of the eighteenth century in Virginia, and elsewhere in British North America, is recognized as a particular time of ferment. Myriad challenges to the maintenance of a comfortable order have been noted by historians. Already mentioned was a precariously mounting level of debt in the colony and overextended credit in Britain. Sharp ups and downs in the regional economy also characterized the period between 1760 and 1780, in part due to fluctuations in crop yields related to other weather conditions (Selby, 1976; Billings et al., 1986). Financial crisis in Britain translated into assertion of stronger demands on the colonies. Especially well known is Britain's notorious passage of Stamp Acts in 1764 to cover French and Indian War debts. In Virginia, a 1766 scandal involving misappropriation of public funds further undermined confidence in the government. Passage of the Townsend Acts in 1767 was another measure intended to generate new revenue from import tariffs. Virginians reacted to a persistent tea tax in 1770 with a non-importation strategy. During the 1760s, a swell of evangelical fervor began to challenge the old Anglican establishment. In fact, a notable Separate Baptist movement distracted the colony beginning in May 1771. On the whole, the decade or so leading up to 1771 was a time of relative discontent and weakening of the general economic and social order.

In this context a major flood became a true disaster. An estimated 150 persons drowned in Virginia and regionally economic losses were great. Along with prized floodplain soils, crops and facilities of every description were washed away or damaged. Richard Bland wrote in a letter that, "Promiscuous heaps of houses, trees, men, horses, cattle, sheep, hogs, merchandize, corn, tobacco and every other thing... were seen floating upon the water" (Bland, 1771). Prominently mentioned in the records are losses to the wheat crop in Virginia and rice and indigo in South Carolina (Van Horne, 1975; The Farmer's Register, 1836; The Virginia Gazette, 1771a, 1771b; The Maryland Gazette, 1771). Fencing, mill dams, tobacco houses, livestock, and dwellings were also widely lost. Shipping in the James River suffered extensively. Vessels moored in the river were battered by drifts of logs and debris, at times piling up to the level of the bowsprits (The Virginia Gazette, 1771a, 1771b; The New London Gazette, 1771b). The current was so swift that many large vessels lost anchors and were swept downstream, while others were grounded beyond recovery.

The two most profound impacts were on the prices of land, much of which was devalued by the flood's effects, and in losses of stockpiled tobacco in Virginia (Van Horne, 1975). As described in a newspaper description,

> All the tobacco at Schockoe inspection [warehouse] is damaged, and it is imagined that there were not less than 1600 hogsheads at it. Byrd's, near all the ground tier is damaged, supposed 600 hogsheads. Three fine large granaries, lately built... are carried away, with sundry valuable goods in them; two are totally lost... The merchants at Rocky Ridge likewise had their warehouses near the river carried away, and 300 and odd hogsheads of tobacco damaged (The Pennsylvania Journal and the Weekly Advertiser, 1771).

All told, approximately 5,000–6,000 hogsheads of tobacco, or more than two million pounds, were reported as lost to the flood (Manarin and Dowdey, 1984). These losses are attributed to the inundation of warehouses at river landings where tobacco was stored before shipment.

For these reasons, a major flood in 1771 – particularly from the Virginian perspective – could only have negative implications. As one would anticipate within the context that has been outlined, great loss of tobacco wealth made for dire circumstances both personally and publicly. The gravity of the flood's impacts is portrayed vividly in official records associated with the short term of William Nelson, acting governor of Virginia 1770–1771. The calamity required an unusual emergency session of the Virginia Assembly and precipitated an official petition to the British government for flood damage relief. Just two and a half weeks after the water receded Nelson attempted to convey some sense of the flood's impacts to the crown:

> The total loss of about four thousand hogsheads of Tobacco at the several Inspection houses on the James and Rappahannock Rivers for which the Publick stand engaged to pay, will greatly affect if not wholly ruin the Credit of many Merchants here as well as the Principals in Great Britain, if a speedy remedy is not applied for the relief of the Sufferers; and it would be very injurious also to the welfare and credit of the Colony in general, if some early provision is not made.
> The publick loss which I have mentioned though it amounts to 40 or L50,000. Is trifling when compared to the sufferings of Individuals; their Lands being destroyed, for above an hundred Miles up on both sides of James River; their houses, Tobacco, Corn, Stocks of Cattle, Horses, Hogs, Sheep etc. being swept away by the Torrent, besides the loss of some people in their houses. In short it is by far the most dredful Catastrophe that hath happened to Virginia since its first Settlement by the English.... (Van Horne 1975:144–145)

Other implications were spelled out in related correspondence:

> ...without a speedy Interposition in their Favour, the Credit of the Merchants would be entirely sunk; which could not fail to affect that of all others connected with them, & must have a very baneful Influence on the Trade of this Country: That in the Course of their Dealings, it has become usual for them to pay the Tobacco Dues imposed by raising the Salaries of the Clergy, and other Levies, for their Customers, who are taught to depend upon them with Security, and provide no other Means; but by this fatal Stroke they are disabled, not only from making that Use of the Tobacco they had provided, but also from purchasing other Tobacco, and consequently from complying with the Expectations of their numerous Collectors, who may demand what they will, for such Dues as are by Law payable in Tobacco (Ibid.:149).

The Virginia assembly managed to muster £30,000 in flood compensation during a special session held between July 11 and 20, but this figure fell short of allaying hardship in most quarters (Jackson and Twohig, 1976:39–41). The calamity did not discriminate. According to Thomas Jefferson, many leading planters had not recovered their losses from the flood by the time of the revolution. Not unexpectedly, the hardest hit were the less privileged class. A period account reveals, "Many people have suffered greatly... more especially as all the corn, which numbers of poor families entirely depended upon for subsistence, is carried down the stream" (The Pennsylvania Journal and the Weekly Advertiser, 1771).

While the Great Fresh of 1771 made a difficult period even less comfortable, its lasting socio-economic fallout was not always profound. Perhaps a few people of means were ruined but the general sense is that most of the affected elite recovered relatively quickly. It is debatable whether any significant official aid ever reached the truly needy underclass, social relations being what they were in the colonies. The language in official correspondence addresses the losses of the planter and merchant classes most explicitly, and the mere fact that prominent representatives could be rallied for a special government session probably speaks as much to their own self interests as to the larger public good.

At the community level, Richmond and even smaller towns in Virginia were not abandoned or significantly redesigned. In fact, over ensuing decades the riverfront was thoroughly built up with industry and businesses. One of the rare, documented planning responses was a petition to move most if not all tobacco warehouses to higher ground (Manarin and Dowdey, 1984:120). In a similar spirit, the plantation economy endured for many more decades in much the same cultural landscape. The risk of a river-side location was, in most cases, found to be worth taking given the prospects of short-term payoffs. Altogether, this outcome reinforces the view that fundamental cultural change is a rare outcome of disaster (Hoffman, 1999). Instead, there is a stronger tendency toward persistence and maintenance of the status quo. Just as this historic case reveals, some of the more rapid reaction was toward protection of prominent social and economic interests, revealing even more clearly the relations of power in the colonial South.

The 1993 flooding of the Mississippi River serves as a useful comparison for the historic 1771 flood. Exhaustive studies of this modern event establish that the strongest responses are political ones with relatively short-sighted goals (Wright, 1996:272). The most effective action is invested in near-term recovery, especially of an economic sort, rather than on long-term preparedness. Another parallel is the observation that socioeconomic costs of the 1993 flood were heaviest on the less privileged sector: "the riskiest thing to be was poor" (Wilkins, 1996:224, 230–232). A disproportionate portion of the flood-damaged housing was in lower-income areas, or, put another way, a greater proportion of lower-income housing was sited in floodplains. Yet despite massive media coverage and public reviews of flood hazards, the general public remained poorly informed about the causes of flooding and options to reduce hazards. Instead, a sense of resignation to the threat remained so strong that only a minority were prepared to make – or were financially capable of making – major lifestyle changes (Ibid.:230–232). Economically, the modern Mississippi flood had a range of significant effects, with implications for the agricultural economy being the strongest. Losses in productivity due to scouring and sedimentation on floodplain farms, for instance, were significant and even lasting in some places. Conversely, however, area farmers on unflooded lands enjoyed substantial benefit as grain prices rose several percent for the year or so afterward (Changnon, 1996:283–284). It was also determined that drought conditions in the region several years prior had a more significant negative effect on the agricultural economy than the more spatially confined 1993 flood.

All things considered, one has to wonder how the 1771 event might have contributed to the revolutionary ferment in the colonies. Embedded in eroding relations with England was a desire to maintain a degree of independence but the flood's effects worked to undermine that objective. Flood recovery costs were such that colonial appeals for official aid had to be made to the British government. Yet economic decline in Britain placed greater burdens on the colonies for revenue. In that context, neither side appears to have been comfortable with the thought of bearing the other's burden. The objective here is not to build the case that the flood played a pivotal role in mounting colonial strains, but rather to introduce it as one among many factors that deserve to be taken into account. It does stand as a perfect example of a random natural event that introduced additional, complicating circumstances into a human context.

6 Closing Thoughts on Extreme Weather Events and Human Affairs

The 1771 flood, in comparison with the modern Hurricane Agnes flood in 1972 and the 1993 Mississippi River flood, serves to reveal some consistency in human response. Major flood events, rated probabilistically as 100-year or 500-year occurrences, appear to seize the attention of most victims only temporarily. The strongest response is predictable – rapid restoration of essential services like water and food supplies, sanitation, shelter, and basic transportation. Much urgency is also placed on the resumption of commerce to minimize economic disruptions. Arguably, it is this interest that ultimately consumes the most time, effort, and money. Consequently, it also becomes the most politicized aspect of flood recovery.

Compared to other natural catastrophes, major floods rank lower in human costs than several other hazards such as hurricanes or even droughts. The principal reason that flood effects are less severe is the confinement of the most direct impact to floodplains. Upscale housing and central commercial districts have rarely been situated within floodplains, whether historically or in modern times. A corollary, however, is that lower-income housing and commercial areas are more routinely sited in flood zones. In the eighteenth century, slave and tenant housing, along with grittier commercial facilities, would have been commonly located in lower-lying areas. The universal result is a disproportionate impact of floods on the less privileged classes.

Wholesale cultural change then, is not the usual outcome of major flood events. This is so because widespread effects are often short-lived and spatially limited, the conventional wisdom says they are infrequent, and the harder-hit under-classes often do not have the luxury of alternative places to live. Sweeping and lasting cultural responses are most likely to occur only when the natural catastrophe coincides with times of significant societal stress, whether of an economic, political, or other sort.

Acknowledgements This material is based upon work supported by the National Science Foundation under Grants ATM-0000881 and ATM-9904383. Any opinions, findings, and conclusions or recommendations expressed in this material are those of the author(s) and do not necessarily reflect the views of the National Science Foundation.

References

Bailey, J.F., J.L. Patterson, and J.L.H. Paulhus (1975) Hurricane Agnes Rainfall and Floods, June–July 1972. Geological Survey Professional Paper 924. US Geological Survey and National Oceanic and Atmospheric Administration, Washington, DC

Betts, E.M. (Editor) (1945) Thomas Jefferson's Garden Book, 1766–1824, With Relevant Extracts from His Other Writings. The American Philosophical Society, Philadelphia

Billings, W.M, E.S. John, and W T. Thad (1986) Colonial Virginia: A History. KTO Press, White Plains, New York

Bland, R. (1771) Letter from Richard Bland to Thomas Adams, August 1, 1771. Adams Family Papers, Mss 1, Ad198a12-167, Sec 6 (B) Folder 2 of 7, The Virginia Historical Society, Richmond

Cabell, N. (Compiler) (n d) Letter from John Howard (Botetourt County) to Dr William Cabell (Amherst County), June 6, 1771. Collection of Papers Relating to Virginia's Agricultural History, Library of Virginia, Misc Reel 287. Richmond

Camp, J.D. and E.M. Miller (1970) Flood of August 1969 in Virginia. Water Resources Division, Geological Survey, U S Department of the Interior, Richmond, Virginia.

Changnon, S.A. (1996) Losers and Winners: A Summary of the Flood's Impacts. In The Great Flood of 1993: Causes, Impacts, and Responses, pp. 276–299, Edited by S.A. Changnon. Westview Press, Boulder, Colorado

Dabney, V. (1976) Richmond, The Story of a City. Doubleday and Company, Inc., Garden City, New York

Davis, J. (Editor) (1974) The Effects of Tropical Storm Agnes on the Chesapeake Bay Estuarine System. The Chesapeake Research Consortium, Inc, The Johns Hopkins University, Baltimore, Maryland

Douglas, K.S., H.H. Lamb and C. Loader (1978) A meteorological study of July–October 1588: The Spanish Armada storms. Climatic Research Unit, CRU RP6, University of East Anglia, Norwich, England

Fries, A.L. (Editor) (1922) Records of the Moravians in North Carolina, 1752–1771, Vol. 1. Edwards and Broughton Printing Company, Raleigh, North Carolina

Gibbons, J. (1976) Flood Marker of 1771. Henrico County Historical Society Magazine 1:21–22

Gray, L.C. (1933) History of Agriculture in the Southern United States to 1860, Vol. 2. Publication No 430, Carnegie Institute of Washington. Waverly Press, Inc, Baltimore, Maryland

Greene, J.P. (Editor) (1987) The Diary of Colonel Landon Carter of Sabine Hall, 1752–1778. The Virginia Historical Society, Richmond

Hawes, L.M. (Editor) (1988) The Journal of the Reverend John Joachim Zubly, A M, D D, March 5, 1770 through June 22, 1781. The Georgia Historical Society, Savannah

Hoffman, S.M. (1999) After Atlas Shrugs: Cultural Change or Persistence after a Disaster. In The Angry Earth: Disaster in Anthropological Perspective, pp. 302–326, Edited by A. Oliver-Smith and S.M. Hoffman. Routledge, New York

Holyoke, E. Unpublished weather journal. Houghton Library, Harvard University, Cambridge, Massachusetts.

Jackson, D. and D. Twohig (Editors) (1976) The Diaries of George Washington, 1732–1799, Vol. 6. University Press of Virginia, Charlottesville

Ludlum, D.M. (1963) Early American Hurricanes, 1492–1870. American Meteorological Society, Boston

Manarin, L.H. and D. Clifford (1984) The History of Henrico County. University Press of Virginia, Charlottesville

Mayo, W. and J. Wood. (1737) A Plan of Richmond. Colored manuscript map, 29 × 46 cm. Library of Virginia Map Collection, Richmond.

Rudisill, H.F. (Editor) (1993) The Diaries of Evan Pugh (1762–1801). St David's Society, Florence, SC

Selby, J.E. (1976) The Revolution in Virginia, 1775–1783. The Colonial Williamsburg Foundation, Williamsburg, Virginia

Sheldon, M.P.B. (1975) Richmond, Virginia: The Town and Henrico County to 1820. Ph.D. thesis, University of Michigan. Ann Arbor, MI

The Farmers' Register (1836) "The Great Flood of 1771, in Virginia and South Carolina", No. 8:506–507

The Maryland Gazette (1771) Extract of a Letter from a Gentleman in Virginia, dated June 5. June 20, 1771, page 2

The New London Gazette (1771a) Williamsborgh, May 30. Volume VIII, June 28, 1771, page 1

The New London Gazette (1771b) Williamsburgh, (Virginia). Volume VIII, July 26, 1771, page 4

The Pennsylvania Journal and the Weekly Advertiser (1771), No. 1488, June 13,1771; Dateline Richmond Town, May 13, 1771

The Virginia Gazette (1771a) Williamsburg, May 30. May 30, 1771, page 3

The Virginia Gazette (1771b) June 6, 1771, page 3

Thomas Thistlewood Papers, Monson Collection 31/58, Lincolnshire Archives, Lincoln, UK

Van Horne, J.C. (Editor) (1975) The Correspondence of William Nelson as Acting Governor of Virginia, 1770–1771. Virginia Historical Society. The University Press of Virginia, Charlottesville

Virginia Magazine of History and Biography (1908) No. 16: p. 351

Wilkins, L. (1996) Living with the Flood: Human and Governmental Responses to Real and Symbolic Risk. In The Great Flood of 1993: Causes, Impacts, and Responses, pp. 218–244, Edited by S.A. Changnon. Westview Press, Boulder, Colorado

Winthrop, J. Unpublished weather journal. Houghton Library, Harvard University, Cambridge, Massachusetts.

Wright, J.M. (1996) Effects of the Flood on National Policy: Some Achievements, Major Challenges Remain. In The Great Flood of 1993: Causes, Impacts, and Responses, pp. 245–275, Edited by S.A. Changnon. Westview Press, Boulder, Colorado

Historical Changes to the Tennessee Precipitation Regime

Jan Mojzisek and Cary J. Mock

Abstract The low-frequency variability of precipitation in the Southeast United States is described using historical records exceeding 100 years for 9 stations. Four out of the nine historical stations belong to the climate regime with a dominant winter precipitation. Seasonal precipitation indices were computed by dividing the rainfall total for each season by the annual amount. This step allowed a direct comparison between winter and the rest of the year's precipitation characteristics without the impact of the absolute annual amount. Abnormally wet winters are evident during the earliest part of the record prior to 1860 and at the end of the nineteenth century for the majority of stations inside the core high winter precipitation region. The 1950 decade suggests a period of dry winters. The strongest signal in temporal variability of the winter-to-spring precipitation ratio occurred around 1977 for the entire region west of Appalachians.

Keywords Southeastern United States · Teleconnections · Decadal variability · Tennessee precipitation regime

1 Introduction

Understanding the nature and causes of temporal and spatial variations in precipitation has tremendous benefits for society, including the southeastern United States which experiences periodic floods and severe drought (Hirschboeck, 1991). Precipitation trends and rainfall fluctuations at interannual and decadal time scales are fundamentally important because they control water supply and modulate events

J. Mojzisek (✉)
Department of Geography, University of South Carolina, Columbia, SC 29208, USA
e-mail: jmojzisek@ci.charlotte.nc.us

C.J. Mock (✉)
Department of Geography, University of South Carolina, Columbia, SC 29208, USA
e-mail: mockcj@sc.edu

L.-A. Dupigny-Giroux, C.J. Mock (eds.), *Historical Climate Variability and Impacts in North America*, DOI 10.1007/978-90-481-2828-0_2,

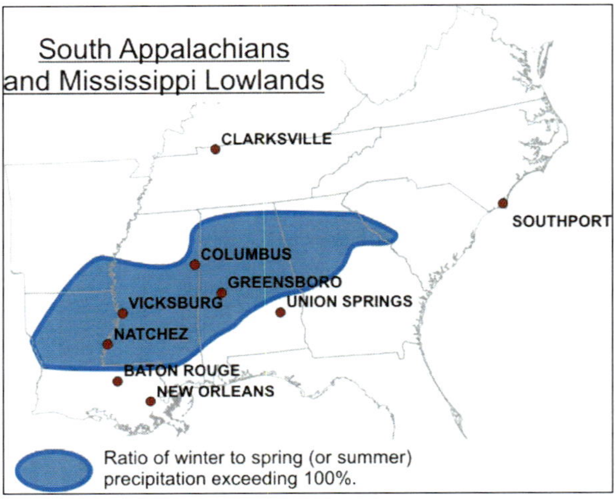

Fig. 1 "South Appalachians and Mississippi Lowlands" region with winter precipitation maximum as defined by Mojzisek (2002)

such as floods and droughts. Several studies have focused on unique types of winter precipitation of the southern United States, which are often referred to as the Tennessee precipitation regime (Soule, 1998). This region of high winter precipitation maximum in the Southeast United States was previously recognized in several studies (Henry, 1897; Trewartha, 1961; Lydolph, 1985; Henderson and Vega, 1996). For example, Trewartha (1961) delineates the region where the winter precipitation exceeds the spring precipitation, with the northern boundary along the Kentucky-Tennessee border, encompassing the entire state of Tennessee and stretching along the northeast/southwest axis as far as southwest Mississippi and the northeastern part of Louisiana. A more precise designation for this region of high winter precipitation (Fig. 1) is located further to the south of the original location, and is named the `South Appalachians and Mississippi Lowlands` (Mojzisek, 2002).

The importance of lengthy and continuous climate series, which has been long advocated by paleoclimate scientists, has been highlighted in recent years by the increased debate about anthropogenic climate change. The knowledge of natural climate variability requires climate series, which can be reconstructed from data found in historical documents and documentary sources. Historical data sources range from individual daily weather records, such as personal diaries and almanacs, to monthly and seasonal means that often vary in length and accuracy (Bradley, 1976). Stations with historical records extending at least 20 years are rare and many stations contain records shorter than 5 years (Mock, 2000).

2 Methodology & Study Area

A long-term precipitation record was constructed for nine stations (Fig. 2) to assess temporal trends and periodicities of precipitation in the Southeast United States. Instrumental data for the southeastern United States span a period from 150 to over 200 years. This study used the best data sources provided by abstracted data

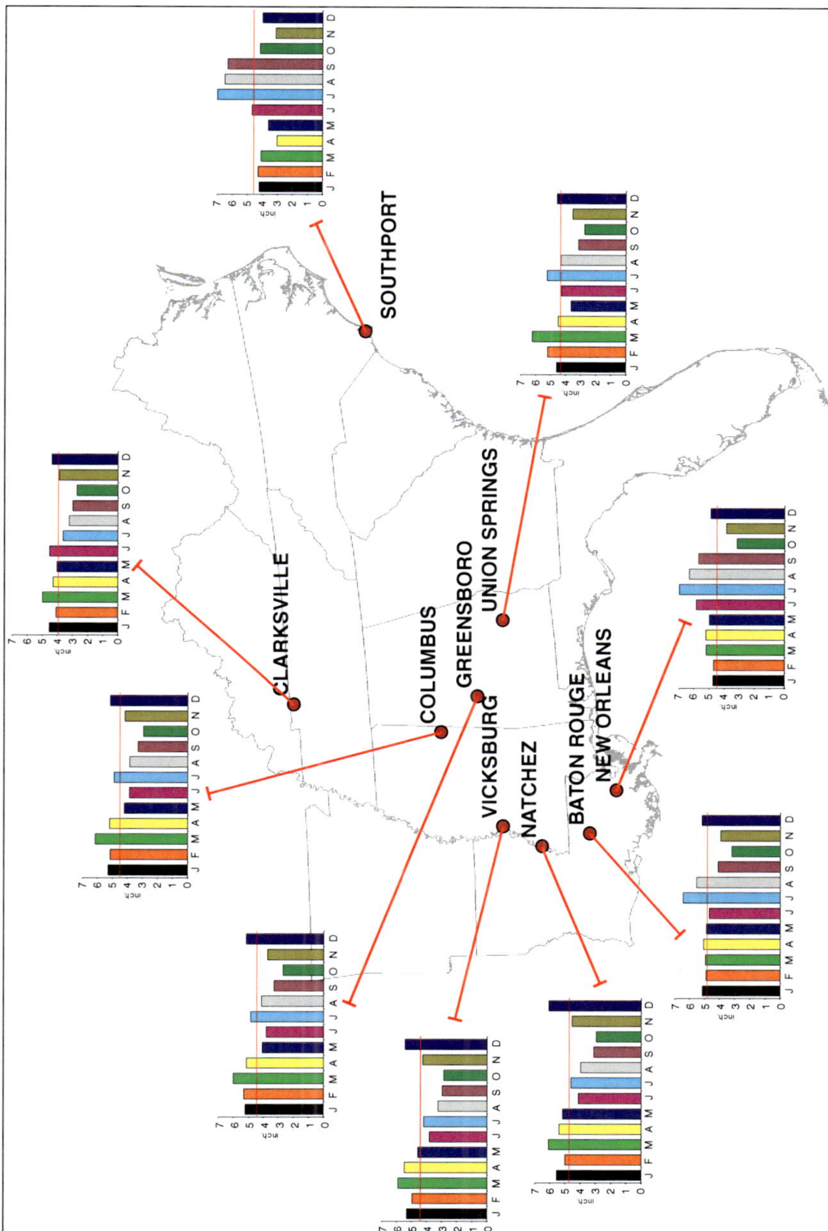

Fig. 2 Monthly distribution of precipitation for the historical stations in the Southeast US with mean annual precipitation (in inches). The monthly values are based on the entire record of a station used in this study

Table 1 Historical stations used with their first year of record and the length of continuous record

Station	First year of record	Length of record (Years)
Clarksville, TN	1855	143
Greensboro, AL	1855	143
Union Springs, AL	1868	130
New Orleans, LA	1871	127
Vicksburg, MS	1872	126
Columbus, MS	1876	122
Baton Rouge, LA	1889	109
Natchez, MS	1890	108
Southport, NC	1876	122

summaries, specifically monthly precipitation totals tabulated month-by-month, year-by-year. A continuous precipitation record exceeds 100 years for all nine stations and the length of the record varies from 108 years for Natchez, MS to 143 years for Clarksville, TN and Greensboro, AL (Table 1).

The source for historical precipitation records included data published by various branches of the US government (Bigelow, 1932–1936), the National Archives (Darter, 1942) as well as in newspapers (Sargent, 1814–1820) and journals (Sargent, 1821). Modern monthly precipitation data were extracted from the Historical Climate Network (HCN) data collection and the National Climate Data Center data set *Cooperative & NWS Sites – Monthly Precipitation Data.*

The historical precipitation records for four stations were combined with precipitation stations having close geographical proximity to obtain longer continuous records. The record for Natchez, MS was combined with Fayette, MS to extend the time series back to 1890. The missing record between November 1888 and April 1889 for Columbus, MS was fit together with nearby Carollton, AL. Five years (1881–1885) of a Nashville, TN record were used to fill the monthly precipitation dataset for Clarksville, TN back to 1855. The Greensboro, AL dataset was combined with overlapping Green Springs, AL data to obtain a continuous 123-year record. The overlapping sections of these records were analyzed to ensure comparability prior to data merging. Table 2 shows the length of overlapping periods between records used in merging procedure. No further, and more rigorous, homogenization procedures were used because of the scarcity of historical data within the area.

Table 2 Length of overlapping period for combined stations and their Pearson's correlation coefficients

Stations combined together	Overlapping timeframe	Pearson's r
Natchez, MS/Fayette, MS	1908–1932 (24 years)	0.832–0.973
Clarksville, TN/Nashville, TN	1872–1880, 1885–1892 (17 years)	0.704–0.954
Columbus, MS/Carrollton, AL	1884–1887, 1890,1892 (7 years)	0.807–0.971
Greensboro, AL/Green Springs, AL	1855–1859 (5 years)	0.798–0.847

Four of the nine historical stations – Natchez, MS, Vicksburg, MS, Columbus, MS and Greensboro, AL – belong to the high winter `South Appalachians and Mississippi Lowlands` precipitation regime (Fig. 1). Clarksville, TN is located just north of the high winter precipitation region, while Union Spring and Baton Rouge are found to the southeast and to the southwest, respectively. New Orleans, LA is located clearly outside the "South Appalachians and Mississippi Lowlands" region and was included in the analysis to illustrate the difference in a seasonal variability along the Gulf of Mexico. Southport, NC needs to be regarded as an extreme situation with an absolute opposite of precipitation variability in the Southeast US as opposed to the core winter region.

3 Results

3.1 Monthly and Seasonal Precipitation

Monthly precipitation climographs were constructed based on their long-term records. The average monthly precipitation for individual stations varied from 3.93 in. at Clarksville, TN to 5.12 in. at New Orleans, LA. The absolute winter precipitation ranged from 2.75 in. in 1890 at Southport, NC to 31.23 in. during the winter of 1974 at Vicksburg, MS. Climographs in Fig. 2 portray the monthly distribution of rainfall and the annual mean at selected stations. The December, January and February precipitation climatology at Natchez, MS, Vicksburg, MS, Greensboro, AL and Columbus, MS is consistent with their location within the core of the "South Appalachians and Mississippi Lowlands" high winter precipitation regime. March exhibits the highest monthly precipitation average of all four stations. It is consistent with the temporal peak in precipitation distribution occurring in March for the entire Southeast US, when all regions show average precipitation above the annual mean (Mojzisek, 2002). The rainfall amounts gradually decline from March to a low in October, when the total mean amount is less than 50% of the average precipitation during the high in March for these four stations.

At Union Springs, AL the annual variation in seasonal precipitation characteristics is similar to stations contained inside the core winter region. The yearly maximum occurs in March and minimum in October. There is, nevertheless, a strong drop in precipitation from March to April. Furthermore, the July local peak is stronger here than it is for stations located to the north because Union Springs comes under the influence of the Gulf of Mexico's warm moist air mass during summer. The annual mean precipitation of less than 4 in. in Clarksville is the smallest annual average among the historical stations. Clarksville, the northernmost station, has similar annual distribution with a high in March and a low in October.

The last three stations – New Orleans, LA, Baton Rouge, LA and Southport, NC – have entirely different characteristics of monthly precipitation, although the October minimum persists for New Orleans and Baton Rouge. The precipitation rises considerably above the annual mean only during the summer and early fall in

New Orleans and Southport. The high precipitation amounts of the summer months gradually decrease northward from the Gulf of Mexico. Climographs of individual stations also confirmed the lack of precipitation in the Southeast US during the fall.

3.2 Decadal Variability

The seasonality index was constructed for winter and spring by dividing seasonal totals by the annual amount. The analysis of variation between the spring-to-annual and winter-to-annual precipitation was conducted and is depicted in Fig. 3. The greatest difference between winter and spring precipitation corresponds to the earliest available record (Fig. 3d) in Greensboro Alabama. The winter-to-annual precipitation ratio exceeded that of spring by as much as 8%. Clarksville, TN also shows a positive spike, although it is much more moderate in winter-to-annual rainfall in the early 1860s. The level of the winter precipitation index decreased from the early 1860s to the early 1870s. At Greensboro, AL the winter index declined approximately 6%, accounting for a little over 26% of the annual precipitation around 1880, as compared to more than 32% in the early 1860s (Fig. 4d). The winter departure in the late 1870s is also pronounced in Vicksburg, MS, Union Springs, AL and New Orleans, LA where winter precipitation totals were as much as 7.5% below the spring amounts.

Values of the winter-to-spring ratio increased from 1880 to the beginning of 20th century at all stations, with the exception of Clarksville, TN. The end of the nineteenth century marked the largest disparity between winter and spring rainfall at three locations in or near the core of winter high regime. Over a third of the annual precipitation occurred during winter (Fig. 4a, b, e) and winter rainfall totals exceeded those of the spring season by more than 10% in Vicksburg, MS, Natchez, MS and Union Springs, AL (Fig. 3a, b, e). Clarksville, TN is a unique case to the late 1890s pattern in the Southeast. The winter index increased rapidly, reaching its maximum around 1882, and then weakened over the next decade. Winter accounted for about 5% less than the annual total for spring in the mid 1890s.

In the 20-year period from 1900 to 1920, winter precipitation declined for nearly all stations to reach the local minimum near 1910 before increasing to 1920. The trend is present for the entire region from Baton Rouge, LA in the southwest, through Vicksburg, MS to Southport, NC in the northeast (Fig. 4). At Baton Rouge, LA 1910 denotes the lowest ratio of winter precipitation during the entire length of the record. Two stations, Columbus, MS and Clarksville, TN do not reflect the low winter amounts around 1910, suggesting that different precipitation forcing mechanisms operate poleward of 33°N latitude.

No clear spatial pattern is evident during the next two decades from 1920 to 1940. In Clarksville, TN, Vicksburg, MS, Greensboro, AL and Baton Rouge, LA the proportion of winter precipitation remained above the spring precipitation until the local minimum in late 1930s, which is strongest along the Gulf of Mexico. For New

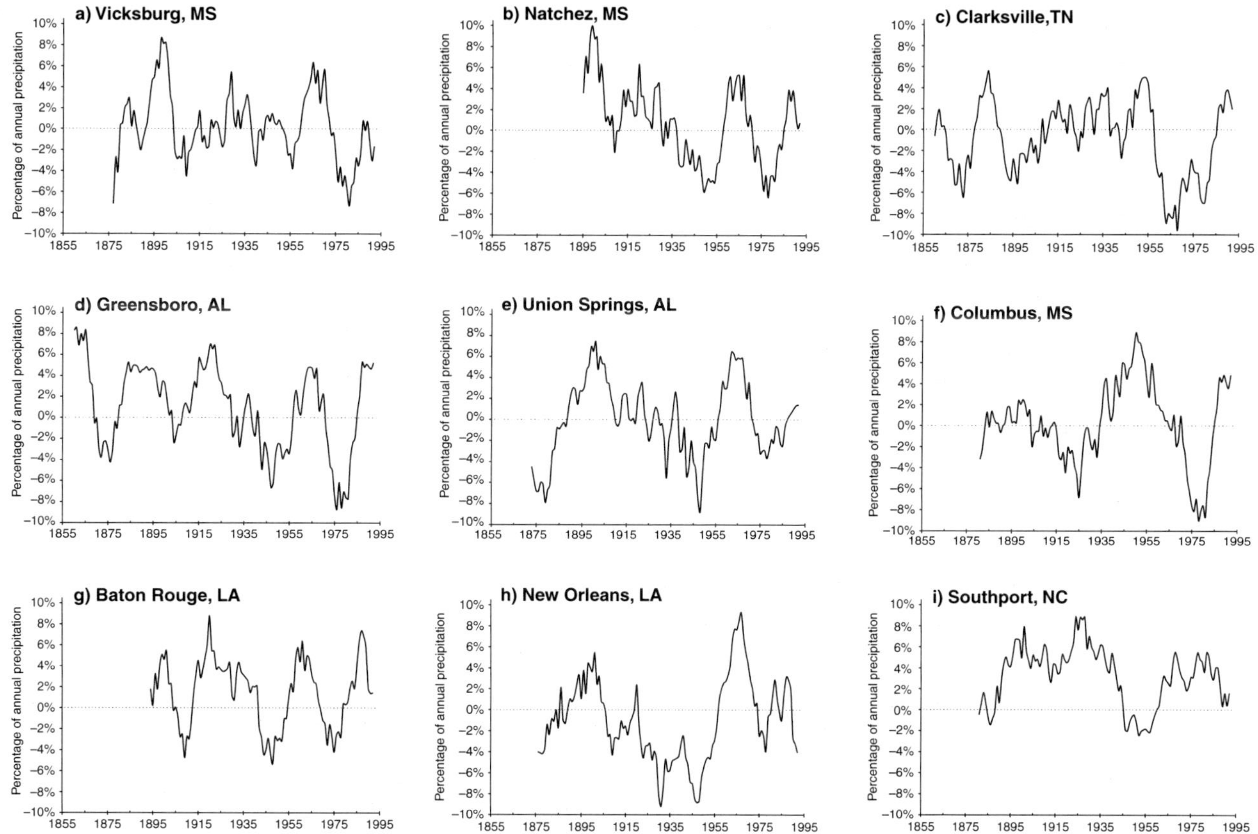

Fig. 3 Centered 11-year moving average of winter-to-spring ratio (positive values indicate the percentage of winter-to-annual precipitation exceeds that of spring-to-annual)

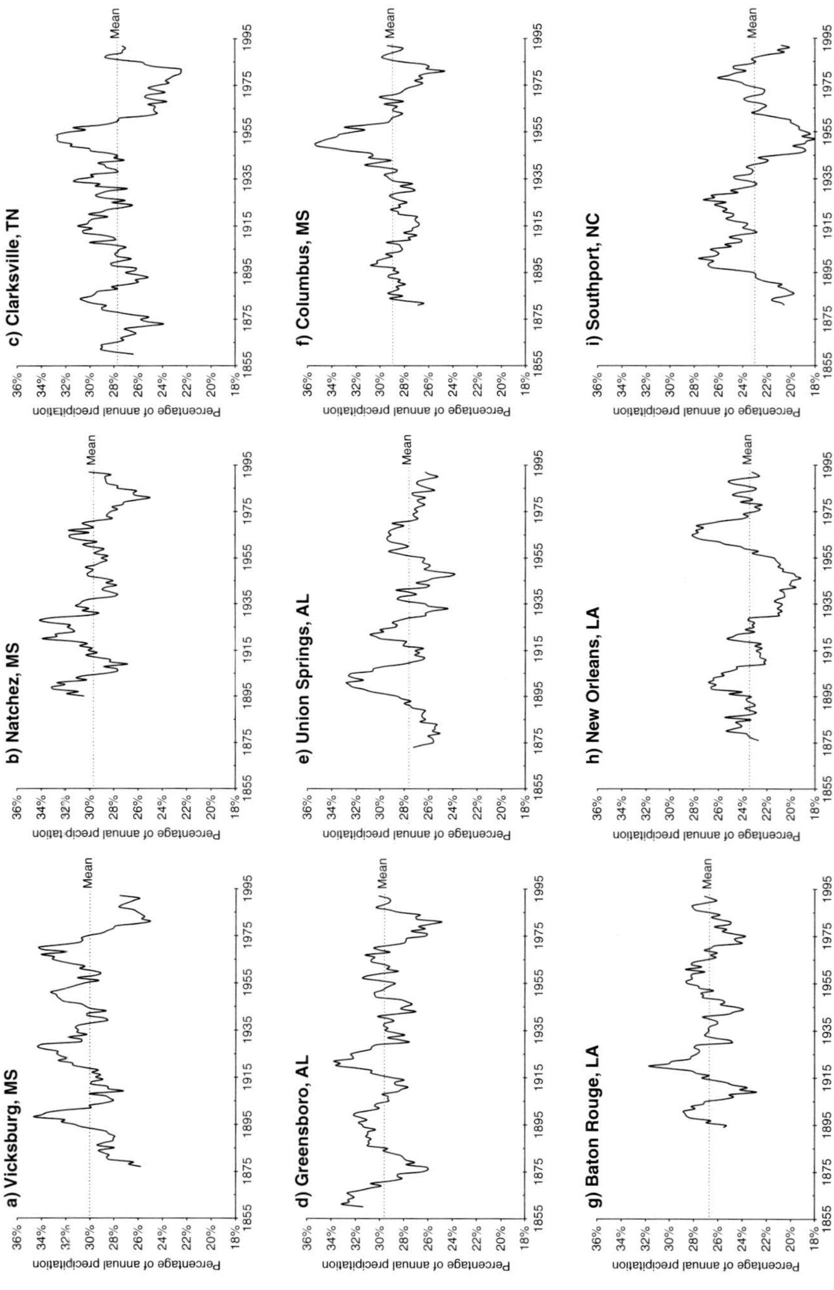

Fig. 4 Centered 11-year moving average of winter precipitation index

Orleans, LA (Fig. 3 h) the 1930 local minimum is also the absolute minimum in the ratio of winter-to-spring index. Stations in Louisiana and Alabama had a negative winter-to-spring ratio between 1940 and 1955. The trend cannot be attributed to the lower winter precipitation (Fig. 4c, d, g), with the exception of New Orleans, LA (Fig. 4 h), because the winter rainfall remained around the long-term mean. On the contrary, the deficiency in winter rainfall for all four stations exhibits the distinctive drop in winter-to-spring ratio in the late 1950s. Records from Columbus, MS (Fig. 3f) and Clarksville, TN (Fig. 3c) located to the north, indicate wet winter conditions from 1945 to 1960, with Columbus, MS reaching all-time high in winter precipitation (Fig. 4f) in 1950.

The 20-year period from 1960 to 1980 indicates a steady decline of the winter-to-spring index. Aside from Southport, NC the trend is clearly noticeable for the entire study area. The winter rainfall percentage reached its lowest point around 1978. All inland locations experienced an unprecedented minimum in winter precipitation that departed from the long-term mean by at least 5%. The 1978 winter minimum marked the highest disproportion between spring and winter rainfall at all four stations within the core "winter-high" regime. Spring rainfall totals topped winter amounts by 7% in Vicksburg, MS (Fig. 3a) and Natchez, MS (Fig. 3b), and by 9% in Columbus, MS (Fig. 3f) and Greensboro, AL (Fig. 3d).

The end of the study period, 1980-1997, experienced an increase in the winter-to-spring ratio throughout the core of the winter-high precipitation region as a result of increased winter precipitation (Fig. 4b, c, d, f). Stations to the south of the region show relatively flat trend, while the winter precipitation at the Southport, NC station to the east exhibits a decline in winter precipitation and its ratio to spring precipitation. The increase of winter-to-spring ratio during the last 20 years of the study period has two important characteristics: (a) the absolute or near-absolute lows of winter precipitation totals for the stations in the core of the high winter region in the 1980s, and (b) the decline in spring rainfall from 1985 to 1995 for the same stations and for Vicksburg to the north.

4 Interdecadal and Decadal Patterns

The El Niño-Southern Oscillation (ENSO) and Pacific/North American (PNA) teleconnections are two dominant signals of atmospheric variability on time scales from a few months to a few years (Wallace and Gutzler, 1981; Philander, 1990).The Pacific/North American teleconnection, a main mode of the Northern Hemisphere midtroposheric variability for the fall, winter, and spring months has a significant relationship with the variations in monthly totals of rainfall in the United States (Leathers et al., 1991). The ENSO events have been shown to be associated with the PNA. Several studies documented the teleconnection between the PNA pattern and ENSO events and its effect on circulation over North America (Horel and Wallace, 1981; Mo et al., 1998). The PNA index has a tendency to be positive during ENSO warm events, and to have negative index values during cold events (Yarnal and Diaz, 1986).

One has to be cautious in interpreting PNA-ENSO relationships. Two types of atmospheric circulation were found to be associated with high values of PNA (Rodionov and Assel, 2001). The authors suggested that first type is a true PNA with the ridge-trough pattern during the positive phase of the PNA index, while the second type is associated with strong warm ENSO events, which flatten the polar jet during the meridional type of upper-level flow. The second type of PNA circulation may have resulted in increased precipitation at several stations across the Southeast during the 1991/1992 El Niño event, despite the high values of the PNA index (Fig. 4).

Winter precipitation increases in the Southeast US during warm ENSO (El Niño) events and during years with zonal upper-level flows across the United States (high PNA index). Vicksburg, MS strongly reflects the response to PNA temporal variability. The raw precipitation data also reveal several extreme rainfall seasons during El Niño events, specifically for 1951, 1953 and 1965 events. The greatest winter precipitation occurred at Vicksburg, MS at the end of the 19th century, corresponding to the very strong 1899–1900 El Niño (Allan, 2000).

The Pacific Decadal Oscillation (PDO) is a long-lived El Niño-like pattern of Pacific climate variability. Both the PDO and ENSO modes have similar spatial signatures but they operate on distinctive temporal scales (Mantua et al., 1997). The warm and cold phases of PDO are defined by ocean temperature anomalies in the northern and tropical Pacific Oceans. The main PDO cycles have persisted for 20–30 years during 20th century, with cold PDO-regimes prevailing from 1890 to 1924 and 1947–1976, while the warm phase existed for 1925–1946 and 1977 to present periods (Fig. 4) (Mantua et al., 1997). The anomalies in climate variables during PDO events are similar to ENSO events, although they are not generally as extreme (Latif and Barnett, 1994; Mantua et al., 1997).

The monthly PDO index was correlated with standardized monthly precipitation totals for the historical stations. Only Clarksville, TN displayed persistent significant correlations with the PDO (at the 0.05 probability level). The relationship is detectable during winter months – December ($r = -0.229$), January ($r = -0.322$) and February ($r = -0.317$). The negative correlations signify the decrease of precipitation during the warm PDO phases. These results are contrary to general anomalies usually associated with PDO and can be attributed to the northerly location of Clarksville, TN.

The slower evolving PDO modulates ENSO-related predictability over North America (Mantua et al., 1997). During the high PDO phase, El Niño events tend to show stronger pattern of wetter winters in the southern portion of the United States, whereas drier winters are common during the low PDO – La Niña event coupling. The signal tends to be weaker and spatially inhomogeneous during El Niño-low PDO and La Niña-high PDO combinations (Gershunov and Barnett, 1998).

Several historical stations in the central and northern section of the Southeast show a reversed relationship between the winter precipitation and ENSO/PDO response. In Vicksburg, MS Natchez, MS Clarksville, TN Greensboro, AL and Columbus, MS the minimum winter precipitation occurred during high PDO phase coupled with the warm ENSO event (Fig. 3a, b, c, e, f, respectively). This PDO-ENSO combination would suggest an increase in precipitation in the Southeast. In addition, the La Niña cold event in 1954 stands out as the reverse pattern with relatively high winter rainfall amounts as opposed to the low totals suggested by the cold PDO – cold ENSO relationship.

The cumulative effect of PDO and ENSO phases is detectable in Southport, NC (Figs. 3i and 4i). The 1954 La Niña-low PDO corresponds with the absolute precipitation minimum at the location, while the strong 1982 El Niño-warm PDO connection resulted in high precipitation. The modulation of winter precipitation in Southport, NC by the PDO is noticeable also in the earlier part of 135-year record. The El Niño of 1932 occurred during warm PDO phase and there is a distinctive increase of precipitation at Southport, NC as well as the lower than annual mean precipitation during the 1886 La Niña cold event.

5 Conclusion

This study concentrating on the Southeast region of the USA adds to a large volume of evidence suggesting that regional precipitation exhibits distinct variability at the decadal scale. The analysis of the climographs, based on a true climatology extending over 100 years, shows a late winter (March) maximum peak and an October minimum in annual precipitation distribution. The core of winter high regime shows signs of precipitation above the annual mean during winter and early spring months.

The results of long-term time series analysis indicate wet winters during the earliest part of record prior to 1860. The abnormally wet winters are also apparent at the end of the nineteenth century for the majority of stations inside the "South Appalachians and Mississippi Lowland" region. The result is another piece of evidence that long continuous time series of climate observations are needed in interpreting natural regional and temporal cycles.

The results of analyzing the decadal and multidecadal oscillation indicate that historical stations located inland and away from the influence of maritime air, have a reversed relationship to the ENSO-PDO coupling. The geographical location seems to be the only plausible explanation. The effect of PDO diminishes inland and the interaction between PDO and ENSO becomes non-existent. The PDO-ENSO relationship was detected for major El Niño and La Niña episodes during last 140 years in Southport, NC.

In summary, this study provides a historical perspective of precipitation variability in the Southeast United States. The low spatial resolution of long-term historical data suggests that some of the results have to be interpreted carefully and that further research is needed to provide high-resolution paleoclimatic records from historical documents and instrumental records. A greater availability of long-term climate data will allow detailed investigation of the effect of decadal and multidecadal large-scale oscillations on the temporal and spatial variability of the climate.

References

Allan RJ (2000) ENSO and climatic variability in the last 150 years. In Diaz HF, Markgraf V (eds.) El Niño and the Southern Oscillation: Multiscale Variability, Global and Regional Impacts. Cambridge University Press, Cambridge, UK.

Bigelow FH (1932–1936) Summary of the climatological data for the United States, by section. US Department of Agriculture, Weather Bulletin W, Washington DC.

Bradley RS (1976) Precipitation history of the Rocky Mountain states. Westview Press, Boulder, Colorado.

Darter LJ (1942) List of climatological records in the National Archives. National Archives, Washington D.C.

Gershunov A, Barnett TP (1998) Interdecadal Modulation of ENSO Teleconnections. Bulletin of the American Meteorological Society 80: 2715–2725.

Henderson KG, Vega AJ (1996) Regional precipitation variability in the southern United States. Physical Geography 17: 93–112.

Henry AJ (1897). Rainfall of the United States, with annual, seasonal, and other charts. U.S. Weather Bureau Bulletin D, Washington, DC, 11–13.

Hirschboeck KK (1991) Climate and Floods. In National Water Summary 1988–89 – Floods and Droughts: Hydrological Perspectives on Water Issues, U.S, Geological Survey Water-Supply Paper 2375, 67–88.

Horel JD, Wallace JM (1981) Planetary-scale atmospheric phenomena associated with the Southern Oscillation. Monthly Weather Review 109: 813–829.

Latif M, Barnett TP (1994) Causes of decadal climate variability over the North Pacific and North America. Science 266: 634–637.

Leathers DJ, Yarnal B, Palecki MA (1991) The Pacific/North American teleconnection pattern and the United States climate. Part I: Regional temperature and precipitation associations, Journal of Climate 4: 517–528.

Lydolph PE (1985) The Climate of the Earth. Rowman & Allanheld, Totowa, New Jersey.

Mantua NJ, Hare SR, Zhang Y, Wallace JM, Francis RC (1997) A Pacific decadal climate oscillation with impacts on salmon production. Bulletin of the American Meteorological Society 78: 1069–1079.

Mo R, Fyfe J, Derome J (1998) Phase-locked and asymmetric correlations of the wintertime atmospheric patterns with the ENSO. Atmosphere-Ocean 36: 213–239.

Mock CJ (2000) Rainfall in the garden of the United States Great Plains, 1870–1889. Climatic Change 44: 173–195.

Mojzisek J (2002) Synoptic Characteristics of Precipitation Variability in the Southeast USA. M.Sc. Thesis, Geography, University of South Carolina.

Philander SG (1990) El Nino, La Nina, and the Southern Oscillation. Academic Press, San Diego, CA.

Rodionov S, Assel R (2001) A new look at the Pacific/North American index. Geophysical Research Letters 28: 1519–1522.

Sargent W (1814–1820) Summary of Meteorological Observations. Natchez Intelligencer.

Sargent W (1821) Summary of Meteorological Observations. Philadelphia Journal of the Medical and Physical Sciences 3: 7.

Soule PT (1998) Some spatial aspects of southeastern United States climatology. Journal of Geography 97: 142–150.

Trewartha GT (1961) The Earth's Problem Climates. University of Wisconsin Press, Madison.

Wallace JM, Gutzler DS (1981) Teleconnections in the 500 mb geopotential height field during the Northern Hemisphere winter. Monthly Weather Review 109: 784–812.

Yarnal BM, Diaz HF (1986) Relationship between extremes of the Southern Oscillation and the winter climate of the Anglo-American Pacific coast. Journal of Climatology 6: 197–219.

"Don't Want to See No More. . . Like That!": Climate Change As a Factor in the Collapse of Lowcountry Rice Culture, 1893–1920

James H. Tuten

Abstract This piece of historical analysis examines the role that climate events, especially hurricanes, played in hastening the failure of the Atlantic Coast rice industry more seriously than has earlier scholarship. In addition to problems with labor and market competition, a period of greater intensity and frequency of hurricanes, droughts and freshets exerted a financial, physical and psychological toll on the Lowcountry from 1893 to 1920 that convinced rice planters to abandon the industry.

Keywords South Carolina · Rice industry · Hurricanes · Droughts · Freshets

1 Introduction

In his 1937 book *The Seed From Madagascar*, former South Carolina Governor and rice planter Duncan Clinch Heyward described the factors which he thought brought about an end to the rice culture in the Lowcountry: foreign and domestic market competition, declines in labor supply and skills, declining soil fertility, and a series of particularly ill-timed hurricanes. Historians have largely agreed with his assessment while recently adding to our knowledge the macroeconomic forces and examining the issue of labor. However, historians have not delved into the climatic aspects with any depth (Heyward, 1993). This chapter examines the hurricanes that so damaged crops and claimed lives, while also analyzing hydrometeorological events such as freshets and droughts, that proved significant stressors in the waning days of the industry. Together with the aforementioned causes, climatic issues were an important factor in forcing planters to abandon rice culture in the Lowcountry of South Carolina between 1893 and 1920 (Clifton, 1978; Coclanis, 1989).

J.H. Tuten (✉)
History Department, Juniata College, Huntingdon, PA 16652, USA
e-mail: tutenj@juniata.edu

L.-A. Dupigny-Giroux, C.J. Mock (eds.), *Historical Climate Variability and Impacts in North America*, DOI 10.1007/978-90-481-2828-0_3,

2 Hurricanes

2.1 Hurricanes of 1893

Heyward (1993) knew whereof he spoke when it came to the collapse of rice culture. He had a front row seat, though he did miss the eye of one of the greatest storms affecting the Lowcountry. Heyward arrived in Charleston, South Carolina two days after the great hurricane of 1893 (Table 1). When he reached his train station on the Combahee River, near Beaufort, SC (Fig. 1) two days later, immediately he asked a trunk-minder who met him where he had parked the horse and buggy. The laborer, who had endured the driving wind, rain and storm surge of the hurricane, was taken aback by this question, and replied incredulously, "Buggy! I fetch de boat fur you" (Heyward, 1993).

At first, the August 27, 1893 hurricane did not seem as damaging to Charleston, SC as the storm of 1885 (*News and Courier*, August 28, 1893 and Mayes, 2006). However, the 1893 storm completely disrupted communication by telegraph, rail and river and interrupted life to the south of the city. Several days would pass before the extent of damage in the Ashepoo-Combahee-Edisto River Basin (hereafter ACE Basin) became clear (Fig. 1). No hurricane had ever killed more people in South Carolina, with some estimates claiming that over 2,000 people perished in the storm. As usual the black population suffered disproportionately (Marscher and Marscher, 2004). The majority of deaths occurred on the Sea Islands around Beaufort, SC

Table 1 Climate events damaging to rice culture, 1893–1920

Year	Date	Event
1893	August 27/28	Hurricane known as "Flagg Storm" Cat. 3
–	October 12	Hurricane, 60 mph
1894	September 26	Storm, 49 mph 3.6 foot tide considerable damage
1896	September 29	Storm, 62 mph
1897	September 21	Storm, 50 mph
1898	June-July	Drought
–	August 30–31	Storm, 52 mph
–	September 29	Storm on Combahee River
–	October 2	Hurricane, (severe on Altamaha) 62 mph 3.9 tide
1899	August 15	Storm, 57 mph
–	October 31	Storm, 58 mph pretty damaging
1903	May 9	Storm, 53 mph 3.9 ft tide
–	September 23	Freshet on Combahee
1906	September 17	Storm, 48 mph
1909	August 16	Storm, 50 mph
1910	October 19	Storm caused massive floods on Combahee
1911	August 27	Hurricane, 106 mph, 6 ft tide very destructive

Source: Heyward, *Seed From Madagascar*; West Point Mill Papers; Mitchell and Smith Papers, South Carolina Historical Society; Fraser, *Charleston! Charleston!*; Edgar, South Carolina. Mayes, *Reanalysis of Five 19th Century South Carolina Major Hurricanes Using Local Data Sources*.

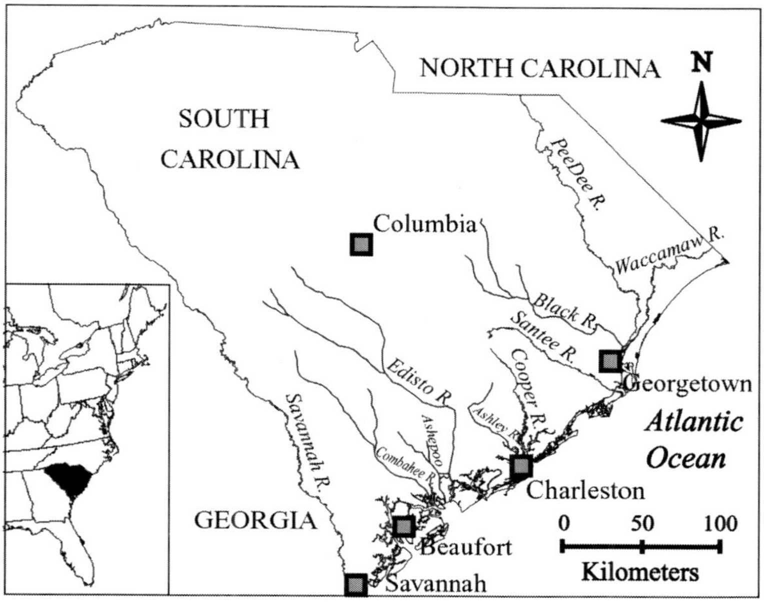

Fig. 1 Location map of South Carolina cities and rivers discussed in the text

where the population mostly farmed long-staple cotton and truck crops or mined phosphates (*News and Courier*, August 29, 30 & 31, 1893).

Stories of horror came back from the rice rivers. Eighty men and women died on the Clay Hall plantation of the Combahee River. The rest of the Clay Hall community survived by sitting on the roofs of their houses. On W. B. Bischoff's Pon Pon plantation, seven people drowned, including Laura Hamilton, who left safety to retrieve her baby only to be washed away in the deluge (*News and Courier*, September 1, 1893; Stoney, 1964; Vlach, 1993).

Plantation laborers had cut only a small amount of the year's crop when the storm hit. The storm destroyed nearly everything south of Charleston, SC. What the gale force winds did not blow down, floods drowned or salt water ruined. Heyward (1993) testified, "I did not make a pint of rice, nor did scarcely anyone on the river." On the other side on the Combahee River, Oliver Middleton Read suffered the greatest financial loss of his career, over eight thousand dollars. His neighbor, H. E. Bissell, told a reporter that his fields lay under five or more feet of water days after the storm, and "besides the loss of the greater part of the crop there has been a considerable loss of animals and houses" (*News and Courier*, September 1, 1893).

While the Combahee and Ashepoo plantations bore the brunt of the August 1893 storm, damage was widespread. In the Georgetown area too (Fig. 1), the destruction spread over a large area and left its mark on plantation residents. This category three cyclone battered the barrier islands along the coastline, where these thin strips of land boasted no cotton, only beach houses. The rice planter Dr. J. Ward Flagg and his family perished when the storm surge took away their home on Magnolia Beach near Pawley's Island (Pharo, 1937).

Ben Horry, born a slave on a rice plantation, worked for the Ravenel and Holmes Steamboat Company that plied the Lowcountry rivers in 1893. The day after the storm hit he left Charleston, SC on one of the steamboats bound for Georgetown at five a.m. The currents proved so strong that they did not reach Georgetown until nine that evening. Along the way, he saved a family floating on a homemade life raft. The following day Horry undertook a grisly duty: "After Flagg storm... [we] search for all them who drowned. Find dead horse, cow, ox, turkey, fowl... Gracious God! [I] don't want to see no more thing like that!" Eventually they did find the remains of the Flagg family and "fetch Doctor body to shore and watch still aticking." He recalled that the black servants living with the family, "Betsey, Kit and Adele" died too (Rawick, 1972).

Environmental hazards came in rapid succession upon the Lowcountry. Following the August 1893 Flagg Storm or Great Sea Island Storm as it is also called, tropical storms and hurricanes damaged rice crops in eleven of the next twenty years. In some of those years multiple storms struck. Later in 1893, a second hurricane struck South Carolina, while three storms blasted parts of the rice coast over a 32-day stretch in 1898 (Mayes, 2006). Such a collection of storms in one season affected all rivers, but in some years the effects of storms were felt only on a single river or a small area. Occasionally tropical cyclones appeared in an atypical manner, not blowing or raining with great ferocity, but dealing a harsh economic blow just the same. Duncan C. Heyward remembered the cyclone of 1910 as "the most peculiar storm which ever visited the coast." The US Weather Bureau predicted the October gale, but instead of lashing the coast with wind and rain, the hurricane turned seaward. This change in direction did not spare the planters of the ACE Basin, however, for as Heyward noted, "the prevailing wind for more than a month had been from the east backing much water up the river... Full moon and perigee came on the same day." This combination of factors resulted in one of the highest tides anyone had ever seen and all the plantations flooded. The water carried away the partially harvested crop, stopped all work, eroded banks and in Heyward's words "a scene of activity and prosperity was changed into one of stillness and desolation" (Heyward, 1993).

The 1893 hurricanes marked the beginning of a period of natural disasters during the last phase of rice culture that included hurricanes, freshets, and droughts. These natural hazards fell upon an industry already precariously positioned in the world market and facing a large array of agricultural challenges. The cataclysmic events delivered knockout blows for many planters. Importantly, this period of more frequent tropical cyclones came after a generation of quiescence from 1840 to 1870 (Mock et al., 2004). After 1893 everyone recognized that the survival of Atlantic Coast rice culture was in doubt.

2.2 Hurricanes and Tropical Storms After 1898

Between 1899 and 1909, rice planters endured five more notable storms and several freshets. Not only do dramatic hazards such as hurricanes, tropical storms, or floods affect both the natural and human landscapes, but additionally they often exact a psychological price on people. We see this today after major hurricanes when symptoms associated with post-traumatic stress disorder (PTSD) emerge. While historians have yet to undertake a detailed study of manuscript or oral history sources for PTSD in the aftermath of historical hurricanes, the contemporary psychological literature is telling. Symptoms displayed by affected individuals include: depression, loss of concentration, and the "disengagement from parts of life that were previously rewarding" (Harvey et al., 2007).

Several aspects of the PTSD literature suggest the effect that catastrophic climate events could have on the ability of individuals to cope with the aftermath. Most importantly, the incidence of PTSD's effects seem to vary by age, ethnicity, gender and how many hurricanes an individual has experienced. In recent studies, children exhibit far higher incidences of PTSD symptoms with as many as 71% showing at least moderate symptoms after Hurricane Floyd in 1999 (Russoniello et al, 2002; Weems et al., 2007). Adults examined after Hurricane Andrew in 1992 displayed PTSD symptoms in 15% of the non-Latino white population and 23% in the non-Latino African-American population. It should be noted that most of the studies on these topics only date back to Hurricane Hugo in 1989, leaving much to be learned about the psychological ramifications of major climate events. Nevertheless, these results must make historians and climatologists consider the role of PTSD in post-hurricane decisions making and recovery efforts. As the intensive research in Psychology develops our understanding of the human mind, historians and other scholars would be remiss not to take this knowledge into account as we research the past (Perilla et al., 2002).

The historical record does suggest that the demoralizing effect of cyclones frequently provided the final straw in a litany of problems faced by those engaged in rice culture. This is seen very clearly in the case of Duncan Clinch Heyward and other rice growers. For example, Elizabeth Allston Pringle, the famed "Woman Rice Planter" decided to give up planting when faced with the destruction following the 1906 hurricane as she dramatically wrote: "The loss is so widespread and complete. I fear the storm drops a dramatic, I may say tragic, curtain on my career as a rice planter" (Pringle, 1992).

2.2.1 The 1911 Hurricane

Following the storms of 1893, 1898, and 1910 another truly devastating hurricane struck in 1911 again claiming several lives. This storm proved to be the final straw for some. As Duncan Heyward put it, when "I saw the ocean actually coming up Meeting Street [in Charleston, SC]. . .I knew. . .that the death-knell of rice planting in South Carolina was sounded" (Heyward, 1993). The effects of the 1911 storm covered the whole Lowcountry, striking not only the plantations, but also the entire infrastructure of rice culture, and in particular that of Charleston, SC.

The storm's eye came ashore south of Charleston, SC and the city itself endured heavy flooding and wind damage, with only a few fatalities. The winds and flooding destroyed shipping and the entire fleet of the yacht club. George Swan, a harbor pilot, said of the storm, "We had one in 1893, but I think this one worse; I think it 'took the bun'" (Mitchell and Smith Papers). The storm caused at least a million dollars' worth of damage and boasted winds over 100 miles an hour, though the exact speed is uncertain since the winds broke the anemometer. By the dawn of the 20th century, loss of life occurred less frequently due to improved forecasting and communications from the local office of the Weather Bureau of the Department of Agriculture. In the rural plantation districts, however, word did not always arrive in time (Mitchell and Smith Papers).

The 1911 tidal surge flooded the storerooms of the West Point Rice Mill in Charleston, resulting in damage to the remaining rice stored from the crop of the previous year. Thus, one storm could affect two years' crops at once. Brokers and the mills themselves tended to own rice from the previous crop, and thus they bore the financial effects as well (Carolina Rice v West Point Mill).

The storm in 1911 placed Duncan Heyward on the verge of quitting, too. He scaled back his large planting operation, planting only two of the five plantations under his management on the Combahee River (Tuten, 2003). Most of the Combahee plantations remained active late into the 1910s. Heyward himself planted through the 1913 year, but his operations steadily declined. When he finally decided to give up rice planting altogether, he sold his plantations to "Northern capitalists," part of a trend followed by many planters or their heirs during those waning days of rice culture. Heyward moved away from the Combahee River and returned to Columbia, SC where he took the political appointment of collector of revenue and eventually became an insurance salesman (Heyward, 1993).

3 Freshets and Droughts

In addition to hurricanes and tropical storms, crops suffered from other environmental forces such as freshets and drought. Droughts mainly harmed growing rice in the spring and summer months. Drought could be dealt with through irrigation from the river unless conditions became very severe and the river too saline. Although rice lives as a semi-aquatic plant, it does not have a high tolerance for salinity in water. Hurricanes sometimes blew saline seawater far up into the rivers, thus damaging the crop. During periods of drought, the shortage of fresh water entering the rivers in the uplands allowed the salty water of the sea to encroach further upstream. If the trunk minder failed in his diligence, an open trunk allowed salt water on the fields. If, however, he planned well he could keep the trunks closed, preventing the salt water from hurting the rice. This standoff could not last because eventually the rice would require more water or a change of water and it might not be available. In the postbellum era, planters who operated plantations with freshwater springs or natural bays could establish "reserves" of fresh water for use under such circumstances. In a sense, the use of reserves represented a new use of the older rice cultivation method first established in Carolina in the eighteenth century (Cheves Papers).

In the seemingly cursed year of 1898 a summer drought proved to be the first of a series of terrible events. In July Henry Cheves worried to his brother that their Newport Plantation had already suffered so much from drought that "I fear. . .[it] is hanging on the ragged edge by this time" (Cheves Papers). It would hardly matter since the hurricanes and freshets of the harvest season would wipe out nearly all of the crop that year. The spatial extent of the most severe damage was seemingly capricious. The storms of that year left Duncan Heyward claiming no profit, although Oliver Middleton Read did clear a profit of several thousand dollars. This may have resulted from Read having harvested before the arrival of the storms and thus avoiding the damage (O.M. Read Papers).

Freshets – fresh water floods that came down river – occurred with more frequency than the cataclysmic hurricanes, and they also wrecked or damaged many crops. Planters no doubt preferred freshets to hurricanes, as the former were relatively predictable and rarely led to human fatalities. The expansion of the timber industry, along with the clearing of midland and piedmont forests for cotton planting, greatly increased the amount of impervious surface causing more water run-off and erosion in the watersheds of the rice rivers after the Civil War. As a result, freshets happened more often and with higher flood stages than in the antebellum period. Freshets occurred throughout the year, but inflicted the most harm during the fall harvest. To contend with the serious dangers of hurricanes and freshets, planters and laborers attempted to reinforce and raise the height of the rice-field banks. They also investigated means of using pumps to keep water out of the fields during the wrong stages of growth. These efforts met with limited success (Pringle (1992).

4 Human Changes to the Landscape

Some of the post-Reconstruction challenges involved in maintaining the rice plantations could be traced back to the unintended consequences of human changes to the landscape. Planters and rice farmers on the Cooper River near Charleston, SC endured particular hardships. The Santee Canal which connected the Cooper and Santee Rivers resulted in a more rapid water flow that eroded dikes and made plantations even more susceptible to freshets. Hardest hit was the Cooper River. Erosion, flooding and a new plant disease called rice blast combined to discourage all those involved with the industry on that river. The development of Charleston Harbor also created problems. At the turn of the century the jetties built by the federal government allowed for a deeper and wider ship channel and the creation of the Charleston Naval Shipyard in 1901. The land for the base came from four rice plantations that the government purchased. Expansion pressure from the city would make riverfront land more valuable for development than for agriculture, a pattern that was repeated on other rivers (Fraser, 1989).

The dredging of the Savannah River for shipping hurt rice planters there. The Army Corps of Engineers also constructed jetties and walls to keep the shipping channel clear. These activities altered the character of the river. A faster moving river eroded banks, allowed saltwater flooding higher up the river, and the new currents caused more freshets than before (Stewart, 1996). Colonel Elliott and Theodore Barker attempted to rouse other Savannah planters to file claims against the government for the resulting damage. Apparently they met with little success, for the

Corps of Engineers neither acted to undo their work nor paid the planters for their losses (Cheves Papers).

In addition, the disintegration of some plantations hurt others. The entire system of banks and canals was interdependent. Louis Haskell recognized the effects of the deterioration of adjacent plantations on his Delta Plantation on the Carolina side of the Savannah River as early as 1897. As neighboring plantations fell into disrepair, salt water traveled through them to challenge Delta's banks, which had never been vulnerable before (Cheves Papers).

5 Collapse of Commercial Rice Culture

Competition from the US Gulf Coast rice states and the loss of market share in Europe continued to deflate the demand for Carolina rice. Drought and freshets combined with severe losses from this active period of storms, creating conditions that drove planters out of the industry. By 1910 the board of directors of West Point Mills worried about the prospects of continuing their business under such circumstances. They had cause to be concerned as the supply outstripped demand in 1909 with the result that many planters did not even have their rough rice milled that year. Instead they stored it at the mill and waited for the market to improve. The mill's directors milled the smallest amount in their history except following the hurricanes of 1893 and 1906. The state of affairs became so precarious that planter and broker Henry Cheves remarked, "The rice market is in a dreadfully mixed up state and I fear all chances of holding up have been thrown away" (Cheves Papers).

In the face of mounting market and weather pressures, some planters continued after the storm of 1910. It remains remarkable that this generation of planters clung to rice planting so tenaciously. No matter how deep the commitment of a planter to rice culture only a tremendous supply of capital and a disregard for financial losses would allow an unsuccessful planter to continue production past 1910. Steadfastness was a major factor, as was location.

On the heels of such storms, planter after planter gave up commercial planting. Elizabeth Alston Pringle and the other Waccamaw planters quit planting commercially as a result of a 1910 storm (Rogers, 1970). However, many African-Americans continued to reside on the plantations. They no longer worked at growing rice commercially, but they continued to plant small plots of rice for their own food. On Sandy Island, for example, the all-black community grew rice for domestic consumption and as a trade commodity into the late 1940s. (Moore, 1997)

In 1860 there were at least 250 rice planters in the Lowcountry. By 1913 that number had dwindled to around two dozen. (Tuten, 2003) With a few exceptions, the Combahee Plantations remained active the longest. The estuary along which these plantations were located did suffer from storms, but less so than others from changes in the river itself. While canals and dredging altered flow and salinity in the Savannah, Ashley, Cooper, Santees, and Pee Dee Rivers, the murky swamps of the Big and the Little Salkahatchie Rivers that were the primary tributaries to the Combahee River remained mostly unpenetrated until after rice culture ceased altogether. Because the timber of the Salkahatchie River did not fall to lumbermen until after rice culture ended, freshets were less common on the Combahee than on other rice rivers (Elztroth and McClure, 1998). In addition, the ACE basin offered fewer employment opportunities than those available in the bigger port cities that were in close proximity, leading to less competition for laborers.

Even in light of the aforementioned environmental stressors there remains a further question about the management skills of the last Combahee planters. Did they remain active so long because of their talent and hard work or did their longevity depend on environmental elements beyond their control? Put another way, did luck make them appear to be more committed and skilled than they were? A circumstantial piece of evidence is that a number of these planters moved their operation to the Combahee from elsewhere, perhaps because they recognized its advantages (Linder, 1995).

Whether the Combahee planters deserve note for their resourcefulness or not, they too eventually gave up rice planting. Only a handful operated past 1920 and the final commercial planter, Theodore Ravenel harvested his last crop in 1927. All the Lowcountry rice planters eventually succumbed to the dire market conditions, weak labor supply, and the repercussions of a series of hurricanes, tropical storms and freshets (Doar, 1936).

6 Effects of the Collapse of Rice Culture

The end of rice planting brought changes on three levels: for individuals, plantation communities, and in land use. A planter's decision to quit not only affected his livelihood but also those connected to him through the industry. The first issue for the individual was finding a source of income; as Arthur M. Manigault revealed to a friend in 1906: "I have decided to stop rice planting and [am] looking around for something to do" (Cheves Papers).

Many planters, such as Langdon and Henry Cheves, had already tied themselves into a profession or business and secured a steady income from sources other than planting. For such persons, giving up planting proved less a financial hardship than a psychological one. The guilt over failing to perpetuate rice culture was one of the motivations that drove Duncan Clinch Heyward, Elizabeth Allston Pringle, and even Theodore D. Ravenel to write about their one-time callings.

For the African-Americans who worked in rice culture, either through labor alone or by virtue of both employment and abode, the cessation of planting hastened the process of migration into other employments. Over time this migration eroded many of the plantation communities in places such as Hobonny Plantation on the Combahee River and Weehaw Plantation on the Black River. Even so, some plantation communities lasted until after the Second World War. After 1900, with little employment left in phosphates, most African-American men who remained in the rural areas were fully devoted to timber, turpentine or railroad work, employments previously pursued in conjunction with rice. Often these jobs still occurred in conjunction with planting some cotton and efforts to raise most of their own food. Others joined the ranks of the cotton tenants and sharecroppers, in some cases on the highlands of the same plantations where they had grown rice (Tuten, 1992).

Since rice plantations represented a dominant form of land use in the coastal belt, the end of commercial planting led to major changes in land use, although not always major changes to the land itself. Most of the former rice plantations have either been maintained as hunting or nature preserves, broken up for development, or preserved as tourist attractions. While many are owned by the descendants of planters, a large number are not. Hunting clubs or elites from outside the region own some plantations and use them as hunting preserves. As part of the national trend to preserve nature and the environment, owners and non-profit organizations have since the 1980s been using rice plantations as nature or habitat preserves. Some plantations, especially those near the cities of Savannah, GA Charleston, SC and Georgetown, SC were divided into neighbourhoods; others on the Cooper River became part of a Navy base; while others still near Georgetown, SC became part of a lumber mill (Tuten, 2003). Beginning during the 1970s some plantations near Georgetown, Beaufort, and especially those located near Charleston, SC the hub of heritage tourism in the region, began to be opened to the public as historic sites or resorts (Brockington, 2006).

7 Conclusion

Planters and laborers on rice plantations faced challenges within the global rice market in 1892. The lost crops, ruined growing seasons, undermined levees, PTSD and even lost lives that resulted from the intense period of storms, cyclones, drought, and freshets from 1893 to 1911 drove almost everyone out of the rice industry. Atlantic Coast rice growers, dependent on a tidal method of cultivation, remained far more susceptible to extreme climate conditions than their main competitors in Arkansas, Texas and Louisiana. The vulnerability of the industry to natural disasters is a potent reminder of the role that climate plays in economic and social life. The legacy of rice culture lives on in the changed landscape, persistent Lowcountry food ways, and a pattern of fraught race relations. Today, a token amount of rice grows in South Carolina for sale to tourists and gourmets (Schulze, 2005).

Historical Papers

Cheves Family Papers, South Carolina Historical Society
O. M. Read Papers, In author's possession
Mitchell and Smith Papers, South Carolina Historical Society
Charleston, S. C. News and Courier

References

Brockington, L. G. (2006) *Plantation Between the Waters: A Brief History of Hobcaw Barony*. Charleston: History Press.

Clifton, J. (1978) Twilight Comes to the Rice Kingdom: Postbellum Rice Culture on the South Atlantic Coast. *Georgia Historical Quarterly* LXII: 146–154

Coclanis, P. (1989) *The Shadow of a Dream*. New York: Oxford

Doar, D. (1936) *Rice and Rice Planting in the South Carolina Low Country*. Charleston: Charleston Museum.

Elztroth, W. R., McClure, F. (1998) *Railroads and Sawmills: Varnville, S. C. 1872–1997*. Varnville Community Council, Varnville.

Fraser, W. J. (1989) *Charleston! Charleston!: The History of a Southern City*. Columbia: University of South Carolina Press.

Harvey, R., Smith, M. et al. (2007) The Hurricane Choir: Remote Mental Health Monitoring of Participants in a Community-based Intervention in the post-Katrina Period. *Journal of Health Care for the Poor and Underserved* 18: 356–361.

Heyward, D. C. (1993) *Seed from Madagascar*. Columbia: University of South Carolina Press.

Linder, S. (1995) *Historical Atlas of the Rice Plantation of the ACE Basin 1860*. Columbia: SC Department of Archives.

Marscher, W., Marscher, F. (2004) The *Great Sea Island Storm of 1893*. Macon: Mercer.

Mayes, D. O. (2006) "A Reanalysis of Five 19th Century South Carolina Major Hurricanes Using Local Data Sources." MA Thesis, University of South Carolina.

Mock, C. J., Mayes, D. O. et al. (2004) "Reconstructing South Carolina Tropical Cyclones Back to the Mid Eighteenth Century." *Preprints, Twenty Sixth Conference on Hurricanes and Tropical Meteorology*, Miami Beach, FL. American Meteorological Society, Boston, 671–672.

Moore, V. D. (1997) *Sandy Island: Nothing Is The Same*. Columbia, SC: Diachronic Research Foundation.

Perilla, J. L., Norris, F. H., Lavizzo, E. A. (2002) Ethnicity, Culture, and Disaster Response: Identifying and Explaining Ethnic Differences in PTSD Six Months After Hurricane Andrew. *Journal of Social and Clinical Psychology* 21: 20–45.

Pharo, E. (1937) *Reminiscences of William Hasell Williams 1811–1902*. Philadelphia: Patterson & White Company.

Pringle, E. (1992) *A Woman Rice Planter*. Columbia: University of South Carolina Press.

Rawick, G. (ed.) (1972) South Carolina Narratives. Vol. 2, *The American Slave: A Composite Autobiography*. Westport: Greenwood.

Rogers, G. C. (1970) *The History of Georgetown County, South Carolina*. Columbia: University of South Carolina Press.

Russoniello, C. V., Skalko, T. K. et al. (2002) Childhood Posttraumatic Stress Disorder and Efforts to Cope After Hurricane Floyd. *Behavioral Medicine* 28: 61–71.

Schulze, R. (2005) *Carolina Gold Rice: The Ebb and Flow of a Lowcountry Cash Crop*. Charleston: History Press.

Stewart, M. (1996) *What Nature suffers to Groe*. Athens: U Georgia P.

Stoney,S. (1964) *Plantations of the Carolina Low Country*. Charleston: Carolina Art Ass.

Tuten, J. (1992) "Live and Die on Hobonny": The Rise, Decline and Legacy of Rice Culture on Hobonny Plantation 1733–1980. MA Thesis, Wake Forest University.

Tuten, J. (2003) "Time and Tide: Cultural Change and Continuity Among the Rice Plantations of the Lowcountry, 1860–1930." PhD Thesis, Emory University.

Vlach, J. (1993) *Back of the Big House*. Chapel Hill: U North Carolina P.

Weems, C. F., Pina, A. A., Costa, N. M. et al. (2007) Predisaster Trait Anxiety and Negative Affect Predict Posttraumatic Stres in Youths After Hurricane Katrina. *Journal of Consulting and Clinical Psychology* 75: 154–159.

Climate in the Historical Record of Sixteenth Century Spanish Florida: The Case of Santa Elena Re-examined

Karen L. Paar

Abstract This chapter examines the challenges of re-creating and interpreting the role of climate in history for a time period with limited surviving written documentation. By drawing on data from dendroclimatology and archaeology, it explores the contributions these studies have made to our knowledge of the Spanish settlement of Santa Elena, located on present-day Parris Island, South Carolina from 1566 to 1587. This paper argues that only by pooling clues from surviving documents, archaeological investigations, and dendroclimatological data and analyzing these clues with a solid understanding of the historical context in which such colonization efforts took place, can we know the role that climate – or any other factor – played in determining this settlement's fate.

Keywords Santa Elena · Spanish Florida · Precipitation · Tree-rings

1 Introduction

Climate played a central role in some of the most dramatic events of Florida's early history, particularly those that involved the storms and hurricanes that are part of life along the southeastern United States coast. For example, a hurricane struck at a decisive moment in early September of 1565, while Pedro Menéndez de Avilés and his Spanish troops waited onshore at St. Augustine for a French attack led by Jean Ribault, who had anchored his ships outside the harbor there. Menéndez faced a shortage of supplies and had fewer troops than the French, but before Ribault and his men could land, a hurricane struck and forced the French ships away from St. Augustine to avoid being battered against the coast. Pedro Menéndez and his soldiers took the opportunity created by the storm to march north over land to the French Fort Caroline, which they took by surprise and defeated. The Spaniards

K.L. Paar (✉)
Director, Liston B. Ramsey Center for Regional Studies and Archivist,
Southern Appalachian Archives", Box 6706 Mars Hill College, Mars Hill, NC 28754, USA
e-mail: kpaar@mhc.edu

L.-A. Dupigny-Giroux, C.J. Mock (eds.), *Historical Climate Variability and Impacts in North America*, DOI 10.1007/978-90-481-2828-0_4,

saw evidence of divine favor in this storm that allowed them to expel the French Protestants, whom they considered heretics with an unjust claim to these lands (Solís de Merás, 1567; Menéndez de Avilés, 1565; McGrath, 2000).

Such stories are tantalizing, but by themselves, they offer few details for those seeking to recreate the historical climate of the United States Southeast. Spanish accounts of the hurricane striking at this crucial moment in the struggle between France and Spain over control of La Florida say far more about the determination of Pedro Menéndez de Avilés and his ability to inspire his followers than they do about the storm and its duration, course, and intensity. According to Gonzalo Solís de Merás (1567), Menéndez's brother-in-law who accompanied him on the Florida expedition, the hurricane struck with a north wind. References in his account of the preparations and journey to Fort Caroline suggest that the strong wind and heavy rains lasted at least ten days. In this document, however, the hurricane is not the subject as much as the backdrop for the events described. Far from making a dispassionate report of this storm, Solís de Merás calls its arrival at that particular moment a "miracle."

Such accounts of climatic events can be difficult to place in context, or, sometimes, to take literally. When climate does appear in the early records of Spanish Florida, it is often as an extreme event – a violent storm or a long drought. Governmental reports, as well as the letters that officials and clergy sent back to Spain, offer glimpses into the climate of this time that are difficult to piece together into a cohesive whole. In sixteenth-century Spanish America, mariners were the ones who kept the most systematic records of weather conditions, both on the mail ships that traveled back and forth between Spain and the American colonies, as well as on the fleets that made journeys twice a year with trade goods and treasure (García et al., 1999, 2000). Long before Pedro Menéndez de Avilés sailed to La Florida in 1565, the Spaniards had figured out the span of hurricane season and planned the sailing of the royal fleets around these times (Pérez-Mallaína, 1998; Hoffman, 1980).

Re-creating the climate of the United States Southeast in the sixteenth-century would be impossible by using just the documents that survive from this period, but fortunately, other sources of information exist for the study of the historical climate. These sources, when examined with the documents, provide valuable clues, not only about environmental conditions of the past, but also about the context in which historical events unfolded. The Spanish settlement of Santa Elena, located on Parris Island, South Carolina from 1566 to 1587, offers an excellent example of the value of a multidisciplinary approach to the re-creation of the contemporary climate and understanding of the role climate played in the course of the colony's history. Historians, archaeologists, and dendroclimatologists have all examined aspects of Santa Elena's climate and the settlement's history. In this chapter, I will explore both the contributions these studies have made to our knowledge of Santa Elena, as well as the limits of interpretations based solely on the results from one discipline. I will argue that only by pooling clues from surviving documents, archaeological investigations, and dendroclimatological data and analyzing these clues with a solid understanding of the historical context in which these colonization efforts took place, can we know what role climate – or any other factor – played in determining this settlement's fate.

2 Santa Elena's Climate: 1566–1587

Pedro Menéndez de Avilés founded the settlement of Santa Elena in 1566, after his mission to drive the French from their fort in the present-day Jacksonville, Florida area led him to land first to the south of there and establish the city of St. Augustine in 1565. In the mid-sixteenth century, Spain's colony of La Florida extended from the present-day US state of Florida to the Canadian province of Newfoundland. Spanish monarchs and their representatives fiercely opposed other European powers' efforts to gain a foothold in these lands.[1] The Santa Elena site was important to the Spanish for strategic reasons. Spain wanted to maintain control of Santa Elena's excellent harbor and keep the French from establishing a base for corsairs there. During the mid-sixteenth century, French corsairs preyed on the Spanish treasure fleets as they traveled up the Florida coast in the Bahama Channel, before heading across the Atlantic to Spain (Hoffman, 1980). French occupation of present-day Parris Island was not an idle concern, for in 1562, the French Captain Jean Ribault had founded the short-lived Charlesfort there and named the sound "Port Royal," the name it bears today (Ribaut, 1927; Laudonnière, 1975).

The Spanish occupied Santa Elena for two consecutive ten-year periods. The first ended in 1576, when a concerted attack by the region's native groups drove some 250 men, women, and children onto their ships and away from their flaming fort and town. When the Spanish returned to the present-day Parris Island site one year later, a series of storms thwarted them at every turn. Pedro Menéndez Marqués, the colony's acting governor, prepared for the hostile conditions at Santa Elena by ordering that the new fort be partially prefabricated in St. Augustine. The ships were loaded with the timbers and ready to leave, when a hurricane struck and wrecked all of the ships in St. Augustine's closed harbor. The Spaniards had to unload the ships and repair them. When they finally set out for Santa Elena, a storm forced the crew of one of the ships to throw many of the pieces for the fort overboard in order to save the vessel (Menéndez Marqués, 1577). Still, the Spanish managed to return to Santa Elena and build their fort, and the settlement endured until 1587, when government officials dismantled the fort and town by royal command (Lyon, 1984; Paar, 1999).

Santa Elena received the attention of paleoclimatologists in a study by Anderson et al. (1995) that analyzed tree-ring samples to create precipitation reconstructions for the years from 1005 to 1600 A.D. in the Savannah River basin of the present-day United States southeast. These reconstructions represent only broad fluctuations in growing season rainfall by documenting whether each year had an average amount of precipitation and the degree to which that year experienced either abundant rain or drought. These data do not provide information about specific weather events but create a general picture of the historical climate. For example, 1565 – the year of the hurricane described above – was in fact a year of above-average rainfall in the dry decade of the 1560s, according to the tree-ring evidence.

Anderson et al. (1995) compared their tree-ring analysis to the historical record on the Spanish settlement of Santa Elena in order to test their theories about food reserves for Indian chiefdoms in the Savannah River Valley, with the ultimate goal of understanding what effects cyclical extremes of drought and rainfall excess might have had on the development and decline of these chiefdom societies. Archaeological dating in the southeastern United States coastal plain was not accurate enough for these scholars to make meaningful comparisons with the dendrochronological data, so they used Santa Elena as a test case for the effects of the dry periods that they discovered. The Santa Elena site is approximately half way between the two locations where Anderson et al. (1995) collected the tree-ring samples they used to create their food reserve reconstruction.

The precipitation estimates from the Anderson et al. (1995) dendrochronological research and the clues from the historical documents basically correspond. For the years of the Spanish occupation of Santa Elena, the tree-ring samples show below-normal rainfall for the years 1566–1569. Anderson et al. (1995: 267) call the year 1566 "one of the driest of the decade, with only the years 1567 and 1569 drier." The tree-ring data indicate that rainfall was above average for most of the 1570s but that drought returned in the early 1580s and lasted at least until 1587, the year the Spaniards abandoned Santa Elena (Anderson et al., 1995).

Indeed, when Pedro Menéndez traveled north from St. Augustine to found the settlement of Santa Elena in April 1566, he learned from members of the Guale and Orista chiefdoms on the present-day Georgia and South Carolina coasts that they were experiencing a time of extended drought – eight months without rain, according to one witness in Guale and "many months" with no rain in Orista (Solís de Merás, 1567). The Solís de Merás (1567) account describes a climatic event of the sort that does not appear in the dendroclimatological data, and the author most certainly mentions it for purposes other than reporting the weather. Gonzalo Solís de Merás (1567) tells that the Guale received at least temporary relief from drought when their *mico*, or leader, knelt and kissed the cross and said he accepted the Catholic faith. He says that only half an hour later, a storm swept across the land and brought the Guale much rain. Events may have happened this way, or this may have been a dramatic rendering written in order to demonstrate divine approval for Spanish actions in those lands. If nothing else, this account shows the Guale in some distress due to drought and the Spanish taking advantage of the situation to bring these Native Americans to their faith. Solís de Merás (1567) also reports that before Menéndez intervened, a couple of young Spanish catechists were extorting corn, fish, and well-tanned deerskins from the Guale leader in exchange for their prayers for rain.

No further mention of dry weather in the late 1560s appears in the documentary record, but this period was clearly one of great food shortages at Santa Elena. The original ninety soldiers stationed at Santa Elena soon dwindled to approximately twenty, due to escapes inland, as well as a mutiny in which some forty men seized a ship and supplies and sailed away (*Archivo General de Indias Justicia* 999, No. 2, R. 9). When Captain Juan Pardo brought nearly 250 men as reinforcements in July of 1566, soon there was not enough food to support them all. Menéndez dispatched

Pardo and 150 of his men inland to find an overland route to Mexico and to bring the Native American towns they encountered into obedience to the Catholic God and Spanish King. Part of Menéndez's goal certainly was for the native population to support these troops as well, and in what would be two expeditions, Captain Pardo departed from Santa Elena and headed inland through present-day South Carolina, North Carolina, and into eastern Tennessee, visiting towns along the way. On the return leg of his second journey, Pardo distributed his men among six forts built in Native American communities across this region with the expectation that these communities would feed the soldiers stationed there (Hudson, 1990).

The arrival of nearly two hundred Spanish settlers – men, women, and children – by May of 1569 caused further strain on the food stores at Santa Elena. The settlers brought supplies with them, but they faced severe shortages by the fall. The Jesuit Father Juan Rogel described Spanish children crying for bread, when there were not even acorns to give them. Father Rogel said that these farmers should cultivate the land but that they were too weak from a lack of food. Provisions were so scarce that the settlers begged the priests to hold religious processions and special Masses to petition God for relief (Rogel, 1569; Paar, 1999). As Anderson et al. (1995) found, even if the settlers had been able to farm in 1569, this growing season was one of the driest in a decade of drought.

2.1 Role of Climate in Santa Elena's History: 1566–1576

The larger purpose of Anderson et al. (1995) in conducting this tree-ring analysis and examination of Santa Elena's history was to understand the effects of drought on Native Americans' food reserves. These researchers use references to Indian unrest that they find in the historical documents on Santa Elena as evidence that the droughts that appear in the dendrochronological data caused food shortages among the Native American population. They present the following incidents from the late 1560s and 1570. Indians destroyed all of the forts built by Juan Pardo in the interior of the Carolinas and eastern Tennessee within months of their construction in 1568. In July of 1570 – when rainfall still had not reached average levels – Juan de la Bandera, the lieutenant governor of the Santa Elena fort, interrupted a feast at the town of Escamazu and demanded corn from the leaders of Escamazu, Orista, and Ahoya. Bandera also instructed these leaders to provide food and shelter for forty Spanish soldiers until a supply ship arrived. When the soldiers appeared in these towns, the Indians fought against them until the Spanish colony's leaders managed to appease them (Rogel, 1570).

In attributing these uprisings to food shortages, Anderson et al. (1995) cite a long tradition where, in other places and other contexts, unrest was linked to crop failure. These researchers make the important point that the demands of the Spaniards would have further stressed an agricultural system already in serious trouble because of difficult climatic conditions. If anything, the Spaniards' demands on the Indians were even greater than Anderson et al. (1995) describe. On Captain Juan Pardo's first expedition into the interior, he and his men stopped at the towns on his way and asked the people there to become vassals of the Spanish king. When they agreed – at least according to the Spanish – by replying "Yaa" to the formulaic speech the Spaniards made on these occasions, Juan Pardo directed them to construct special houses for the Spaniards and to grow corn to fill them. Pardo's notary reported a remarkably high degree of compliance with these orders on the second expedition, when the Spaniards returned to these towns. Besides leaving men in the interior forts to be supported by native communities, Captain Pardo also ordered bearers to carry as much corn as possible back to Santa Elena, when he returned from his second expedition (Bandera, 1569). Spanish settlers as well as soldiers took food from the Native American population during this time of drought. By the spring of 1570, the settler men began going out with the soldiers to the native towns in the region to take food – and to "pacify" the Indians, when they responded with anger (Paar, 1999).

As we seek to understand the role of climate in history, we must avoid making simple correlations between even extreme events such as severe drought and the actions of historical actors. Food shortages alone did not lead to rebellion by the Native American population, and drought did not necessarily result in severe food shortages, nor was drought the only cause of food shortages in Santa Elena and the surrounding region.

There is a link between Spanish demands for food and a military response by the Indians in a number of instances, but the documentary record shows that many other factors were involved when Native Americans took such action. In some cases, the Indians were clearly angered by the Spanish demands, and the seizing of food appears to be an extension of a broader pattern of abuse by the Spaniards. The Jesuit Father Juan Rogel repeatedly complained about the Spanish soldiers' violent treatment of the Indians (Rogel, 1570). He believed that such behavior was the reason Indians destroyed the inland forts and killed the Spanish soldiers living in them. Father Rogel told of visiting Escamazu with Juan Pardo to placate the town following some offense by Spanish soldiers. While he and Captain Pardo were talking with the Escamazu, they overheard soldiers at Orista – these towns were apparently very close together – mistreating people there. Father Rogel asked if these soldiers would behave this way within earshot of their captain, then what would soldiers one or two hundred leagues away do (Rogel, 1568)? Beyond physical abuse, Spaniards also interfered in the Indians' social and political structures. Spanish demands for food likely forced Native American men to take a greater role in agriculture, which their cultures mostly assigned to women. The murder of several Guale leaders and disruption of succession in that chiefdom was one of the direct causes for the 1576 uprising, in which the Guale, Orista, and Escamazu allied to kill many Spanish soldiers at Santa Elena and to destroy the fort and town there (Martínez, 1577).

The periods of drought that appear in the dendrochronological record undoubtedly affected both the Spanish and Native American populations, but factors not considered in Anderson et al. (1995) must have strongly mediated the relationship between climate and the availability of food. After all, both the Indians and the Spaniards had adapted in various ways to the conditions they experienced in La Florida. The Indians had developed responses to a range of environmental factors. In this coastal region, they faced variations in climate as well as land that featured both sandy soil for agriculture and an estuarine habitat for plants and animals. Even during times of abundant rainfall, the peoples of these coastal islands likely did not rely on agriculture as much as those inland. Instead, they drew on a range of wild plants and animals, whose abundance was also affected by precipitation extremes, but not to the same degree as cultivated crops.

The account by Gonzalo Solís de Merás (1567) of the Orista Indians' first meeting with Pedro Menéndez de Avilés in April 1566 shows them celebrating their new friendship with feasts, even during a period of drought. Hospitality was an important value for these cultures, so the Orista likely would have shared their food with the Spaniards even if they had very little. Still, the descriptions of these feasts show the variety of foods available in the early spring – not only fish, shellfish, and oysters, but also corn and acorns that must have come from the previous year's crop (Solís de Merás, 1567). Redistribution of tribute through feasts and other means played a role in the political, social, and economic life of these chiefdoms and likely provided at least some protection for their members during periods of food shortages (Widmer, 1994).

Spaniards depended heavily on the Native Americans for food, but this was by no means their only source of sustenance. Likewise, drought was not the only reason that Spanish soldiers and settlers lacked food. More than one governor or lieutenant governor withheld supplies from their intended recipients in order to sell them at a profit to others. Even a far-flung colony like Santa Elena was part of an imperial system that coordinated the distribution of food and other goods. The shipments from Spain, Mexico, Cuba, and other parts of Spanish America to La Florida were unreliable, but supplies did flow into this colony, sometimes in abundance. Royal officials wanted the settlers to farm and provide food for the Spanish population, but they recognized that this was not always possible.[2] After all, the Santa Elena settlement existed because of its strategic, rather than its economic importance, although Spaniards always hoped to discover the mineral resources that other parts of the Americas had yielded. Philip II even began to provide an annual subsidy for the Florida colony in 1570, after the difficult years of the late 1560s. He increased this subsidy in 1578 and 1580, and other monarchs continued it into the eighteenth century (Paar, 1999; Sluiter, 1985).

Spaniards imported food into Santa Elena not only out of necessity, but also to preserve the culture of their home country. Although the Spanish diet in La Florida included indigenous foods such as locally grown corn, Spaniards preferred imported foods like wheat flour. The Spanish reliance on wheat, olive oil, and wine seems to have continued in La Florida, especially among the elite, although soldiers' rations also centered on flour and wine, when they were available. Efforts to grow grains like wheat and barley at Santa Elena failed, and all these items had to be imported (Settlers, 1576–1577).

2.2 Role of Climate in Santa Elena's History: 1576–1587

Knowledge of the historical climate is invaluable to understanding contemporary events, but climate's effects are often mediated by many other factors, as the examples above show. The 1576 uprising provides a good example of how environmental factors may have directly influenced historical events, although here too, interpretations can vary. Anderson et al. (1995) observe that this uprising took place during a period of abundant rain, and indeed, the dendroclimatological evidence shows that rainfall in the Santa Elena region was above average for most of the 1570s. Anderson et al. (1995: 268) assume that this precipitation resulted in a food shortage, and they conclude that "too much rainfall may have been as bad as too little" for crops. Based on this evidence from the tree-ring data, as well as information from the historical documents, I argue, however, that this abundant rainfall was not the cause, but rather the fuel, that made it possible for the native groups of the Guale, Orista, and Escamazu to launch this sustained campaign to drive the Spaniards from their lands. The written record suggests that the 1576 uprising was not an isolated event, but rather the beginning of a period of resistance that ended around 1583 (Paar, 1999).

The long duration of this rebellion is particularly striking because Pedro Menéndez Marqués, the Florida governor who assumed office in 1577, fought what he saw as the Indians' insubordination by attacking their very ability to subsist. In 1579, Governor Menéndez launched "wars of fire and blood" against many towns in the Guale and Orista chiefdoms. He not only burned their houses and killed their people, but he targeted their food supply, burning storage houses and cutting off corn plants in the fields. Except for brief periods of peace, the Native Americans continued to fight the Spaniards during these years. This resistance included both large-scale actions, such as the surrounding of the fort at Santa Elena by one thousand men in October 1580, as well as smaller skirmishes that made it difficult for Spaniards to grow crops or to hunt or fish for food.

The abundance of food supplies necessary to withstand such sustained Spanish attacks on the Native Americans' means of subsistence could only have occurred during a period of ample rainfall, when the crops or food reserves the Spaniards destroyed could be replenished. The recurrence of drought, which the dendroclimatological data suggest took place in the early 1580s, may well have ended the Native Americans' efforts to fend off Pedro Menéndez Marqués' attacks (Anderson et al., 1995). Indeed, in 1583, Governor Menéndez Marqués wrote the King that the leaders of towns in the area around Santa Elena had pledged their allegiance to Spanish rule during a season of drought, in which no rain had fallen in three months and there had been no corn harvest. He attributed this course of events to divine intervention (Menéndez Marqués, 1583). Whatever the political or social causes for the Indians' capitulation, the drought would have taken the last of the food reserves that fueled their struggle.

According to Anderson et al. (1995: 269), the drought continued at least until 1587, with that year being the driest of Santa Elena's twenty-year existence and "three of the four years before this (1583–1585) were characterized by below-average rainfall."[3] An explanation for why 1586 did not continue this pattern of drought may come from archaeology. Although no known document mentions this, archaeologists excavating the Santa Elena site have found evidence of a major storm there in 1586, one that left numerous tree holes that residents later filled with refuse, as well as damage to structures excavated there. The archaeologists propose that this may have been the same storm that Sir Francis Drake encountered on the present-day North Carolina coast in June of that year (South and DePratter, 1996; Ludlum, 1963).

Anderson et al. (1995) conclude that climate may have played an important role in Spain's decision to dissolve the Santa Elena settlement in 1587, which they learned was not only the driest of Santa Elena's existence, but the third driest of the entire sixteenth century. These dendroclimatological data are impressive and may well provide evidence of an important, though apparently unarticulated, reason for Santa Elena's demise. The documents show, however, that following Sir Francis Drake's raid on Spanish America – a raid that reached La Florida with the destruction of St. Augustine in 1586 – Spain reassessed the defense of its colonies. Within La Florida, the debate began about the consolidation of the Spanish presence there, rather than maintaining the far-flung forts at St. Augustine and Santa Elena. Family rivalry even figured in this heated discussion, as brothers-in-law in charge of the different settlements made their cases to officials in Spain. In the end, the King decided in favor of St. Augustine, and the soldiers and settlers at Santa Elena destroyed their fort and town and withdrew to the south (Paar, 1999).

3 Conclusions

As we seek to learn more about climate in periods of history with sparse documentation, the case of Santa Elena shows the benefit of sharing results across disciplines. Each discipline supplies a different piece of the puzzle in the re-creation of Santa Elena's climate. Dendrochronological data provide an overview of the fluctuations in rainfall over time and indicate when the extremes occurred. While the historical and archaeological records on climate in this period are much more uneven, these sources offer clues about specific events, such as hurricanes, that appear in the tree-ring data only as part of the overall rainfall for the year. Not only gathering the clues from different sources, but also "reading" them against one another, helps us to gauge the sources' reliability. For example, Governor Pedro Menéndez Marqués' report of drought in 1583 – a report in which he attributed the drought to divine intervention that forced the Native Americans' surrender to him – seems more reliable in light of the tree-ring data that report drought for that year (Menéndez Marqués, 1583; Anderson et al., 1995).

Collaboration among disciplines offers many challenges, particularly in interpreting the significance of results. As discussed above, dendroclimatological analysis provides information about the environmental context in which Santa Elena existed that we would never be able to recreate from the surviving documents alone.

Additional tree-ring analysis has proven useful for interpretation of other early European colonization efforts in the present-day United States, including the fate of the Lost Colony and the disastrous early days of the Jamestown settlement (Stahle et al., 1998). While the results such studies yield are dramatic, these data must be examined together with evidence from the surviving written documents and archaeological investigations in order to uncover the whole story in a historical climate case study. With careful attention to the wider context, such collaboration can offer greater knowledge and richer understanding of many aspects of the past – whether the history of climate, the size of chiefdoms' food reserves, or the motivations behind human actions.

Notes

1. Spain's definition of "La Florida" evolved over time. When Pedro Menéndez de Avilés signed his contract to conquer and settle La Florida in 1565, the territory as defined in his contract extended from the Bay of St. Joseph on the Gulf Coast, around the tip of the Florida peninsula, and north as far as Newfoundland.
2. One example of this is the King's chiding of the governor of Cuba in December 10, 1578 royal order, or *cédula*, that tells him that it is necessary to supply the people of Florida with the things they need for their sustenance – see the Royal Orders dated December 10, 1578, El Pardo, in "Cedulario de la Florida," 1570–1604, n.p., AGI Santo Domingo 2528 (Stetson Collection).
3. 1587 was a year of precipitation extremes in other regions. See, for example the discussion of this phenomenon also taking place in New England in LA Dupigny-Giroux (2008) Using bioindicators to explore climate variability. In: Dupigny-Giroux LA, Mock C (eds.) Historical climate variability and impacts in the United States. Springer, Berlin Heidelberg New York.

References

Anderson DG, Stahle DW, Cleaveland MK (1995) Paleoclimate and the potential food reserves of Mississippian societies: a case study from the Savannah River Valley. American Antiquity 60(2): 258–286

Bandera J (1569) The "long" Bandera relation. In: Hudson C, Hoffman P (trans) (1990) The Juan Pardo expeditions: exploration of the Carolinas and Tennessee, 1566-1568, with documents relating to the Pardo expeditions transcribed, translated, and annotated by Paul E. Hoffman. Smithsonian Institution Press, Washington London, pp. 205–296

Dupigny-Giroux LA (2009) Backward seasons, droughts and other bioclimatic indicators of variability. In: Dupigny-Giroux L-A, Mock C (eds.) Historical climate variability and impacts in North America. Springer, Berlin Heidelberg New York

García R, Prieto MR, López A et al. (1999) The Archivo General de Indias (AGI): a source of surface marine climatological information. Marine Meteorology and Related Oceanographic Activities 43: 61–68

García R, Gimeno L, Hernández E et al. (2000) Reconstructing the North Atlantic atmospheric circulation in the 16th, 17th, and 18th centuries from historical sources. Climate Research 14: 147–151

Hoffman PE (1980) The Spanish crown and the defense of the Caribbean, 1535–1585: precedent, patrimonialism, and royal parsimony. Louisiana State University Press, Baton Rouge

Hudson C (1990) The Juan Pardo expeditions: exploration of the Carolinas and Tennessee, 1566-1568, with documents relating to the Pardo expeditions transcribed, translated, and annotated by Paul E. Hoffman. Smithsonian Institution Press, Washington, DC

Laudonnière R (1975) Three voyages. Bennett CE (trans). University Presses of Florida, Gainesville

Ludlum DM (1963) Early American hurricanes, 1492–1870. American Meteorological Society, Boston

Lyon E (1984) Santa Elena: A Brief History of the Colony, 1566–1587. In: Lyon E (ed.) Pedro Menéndez de Avilés. Spanish Borderlands sourcebooks, vol. 24. Garland Publishing, New York, pp. 481–508

Martínez B (1577) Bartolomé Martínez to the King, February 17, 1577, Havana. In: Connor JT (ed.) (1925) Colonial Records of Spanish Florida, vol. 1. Florida State Historical Society, Deland, pp. 236–247

McGrath JT (2000) The French in early Florida: in the eye of the hurricane. University Press of Florida, Gainesville

Menéndez de Avilés P (1565) Carta del adelantado Pedro Menéndez de Avilés al Rey, Ribera de San Pelayo, fuerte de San Agustín, 15 de Octubre de 1565. In: Ruidíaz y Caravia E (ed.) (1893) La Florida: su conquista y colonización por Pedro Menéndez de Avilés. Colegio Universitario de Ediciones Istmo, Madrid (version reprinted in 1989), pp. 311–322

Menéndez Marqués P (1577) Pedro Menéndez Marqués to the King, October 21, 1577, Santa Elena. In: Connor JT (ed.) (1925) Colonial Records of Spanish Florida, vol. 1. Florida State Historical Society, Deland, pp. 262–277

Menéndez Marqués P (1583) Pedro Menéndez Marqués to Crown, December 27, 1583, St. Augustine, AGI Santo Domingo 224 (Stetson Collection)

Paar KL (1999) "To settle is to conquer": Spaniards, native Americans, and the colonization of Santa Elena in sixteenth-century Florida. Dissertation, University of North Carolina at Chapel Hill

Pérez-Mallaína PE (1998) Spain's men of the sea: daily life on the Indies fleets in the sixteenth century. The John Hopkins Press, Baltimore

Ribaut J (1927) The whole & true discouerye of Terra Florida: a facsimile reprint of the London edition of 1563. Together with a transcript of an English version in the British Museum with notes by H.M. Biggar, and a biography by Jeannette Thurber Connor. Florida State Historical Society, Deland

Rogel J (1568) Father Juan Rogel to Father Francisco de Borja, July 25, 1568, Havana. In: Zubillaga FSJ (ed.) (1946) Monumenta antiquae Floridae, 1566–1572. Monumenta Historica Societatis Iesu, Rome, pp. 317–329

Rogel J (1569) Father Juan Rogel to Juan de Hinistrosa, December 11, 1569, Santa Elena. In: Zubillaga FSJ (ed.) (1946) Monumenta antiquae Floridae, 1566–1572. Monumenta Historica Societatis Iesu, Rome, pp. 398–405

Rogel J (1570) Father Juan Rogel to Pedro Menéndez," December 9, 1570, Havana. In: Zubillaga FSJ (ed.) (1946) Monumenta antiquae Floridae, 1566–1572. Monumenta Historica Societatis Iesu, Rome, pp. 471–480

Philip II (1578) Royal Orders dated December 10, 1578, El Pardo. In: Cedulario de la Florida, 1570–1604, n.p., AGI Santo Domingo 2528 (Stetson Collection)

Settlers (1576–1577) The Settlers of Florida, February-March, 1576, Santa Elena, and November–April, 1576–1577, Madrid. In: Connor JT (ed.) (1925) Colonial Records of Spanish Florida, vol. 1. Florida State Historical Society, Deland, pp. 144–185

Sluiter E (1985) The Florida situado: quantifying the first eighty years, 1571–1651. In: Research Publications of the P.K. Yonge Library of Florida History, University of Florida (1). University of Florida Libraries, Gainesville

Solís de Merás G (1567) Memorial que hizo el Dr. Gonzalo Solís de Merás de todas las jornadas y sucesos del adelantado Pedro Menéndez de Avilés, su cuñado, y de la conquista de la Florida y justicia que hizo en Juan Ribao y otros Franceses. In: Ruidíaz y Caravia E (ed.)

(1989) La Florida: su conquista y colonización por Pedro Menéndez de Avilés. Reprint, Colegio Universitario de Ediciones Istmo, Madrid, pp. 110–257

South S, DePratter CB (1996). Discovery at Santa Elena: block excavation 1993. Research Manuscript Series 222. South Carolina Institute of Archaeology and Anthropology, University of South Carolina, Columbia

Stahle DW, Cleaveland MK, Blanton DB et al. (1998) The Lost Colony and Jamestown droughts. Science 280: 564–567

Widmer RJ (1994) The structure of southeastern chiefdoms. In: Hudson C, Tesser CC (eds.) The forgotten centuries: Indians and Europeans in the American south, 1521–1704. University of Georgia Press, Athens London, pp. 125–155

Part II
Reconstructing Extreme Events and Parameters

Historical Accounts of the Drought and Hurricane Season of 1860

Stephanie F. Dodds, Dorian J. Burnette, and Cary J. Mock

Abstract Spatial patterns of the drought and hurricane tracks were reconstructed for 1860. These reveal a typical La Niña pattern over the conterminous United States that would be analogous to an active Atlantic hurricane season. Precipitation frequency for April through October 1860 was calculated for the conterminous United States from 252 instrumental stations and transformed into percentiles relative to the modern record. These data were supplemented with instrumental and documentary accounts from diarists, newspapers, and ship logbooks, which allow for the rigorous reconstruction and reanalysis of the drought and hurricane season and their impacts on society. Widespread deficits in precipitation were observed in 1860 across a large portion of the central and southern Plains and the southeastern US. A total of nine tropical cyclones were detected in this new analysis, and up to three of those nine were previously unknown. The drought parched many crops from Plains into the southeastern US, but the landfall of three hurricanes within 120 km of New Orleans, Louisiana brought beneficial rainfall to the southeastern US. However, this rainfall came too late in the growing season for crops to recover, and also came with an intense storm surge, inland flooding, and wind damage. Synoptic-scale patterns indicate that high pressure systems centered over the central Plains and northeastern Gulf of Mexico steered the three land-falling US hurricanes into Louisiana and Mississippi.

Keywords Drought · 1860 · United States · Hurricanes

S.F. Dodds (✉)
Department of Geography, University of South Carolina, Columbia, SC 29208, USA
e-mail: doddss@mailbox.sc.edu

D.J. Burnette (✉)
Department of Geosciences, University of Arkansas, Fayetteville, AR 72701, USA
e-mail: djburne@uark.edu

C.J. Mock (✉)
Department of Geography, University of South Carolina, Columbia, SC 29208, USA
e-mail: mockcj@sc.edu

L.-A. Dupigny-Giroux, C.J. Mock (eds.), *Historical Climate Variability and Impacts in North America*, DOI 10.1007/978-90-481-2828-0_5,

1 Introduction

One of the most severe, sustained droughts across the western United States (US) in the last 700 years occurred during the mid-19th century (Cook et al., 2004). Tree-ring reconstructions of the Palmer Drought Severity Index (PDSI) (Palmer, 1965) suggest that this mid-19th century or "Civil War" era drought was centered over the Great Plains, but also felt across the Midwest and the southeastern United States (Fye et al., 2003; Herweijer et al., 2007). Intense drought appears to have been one of the key negative environmental factors that decimated the herds of bison across the Great Plains (West, 1995). Conditions in 1860 were particularly harsh and "gravely affected nearly all the horticultural Indians from east central Nebraska southward through eastern Kansas into Oklahoma" (Wedel, 1953). That year also saw suffocating dust storms across Kansas, some of which may have been similar to those experienced during the 1930s Dust Bowl drought, and may have been responsible for the mass migration of early settlers out of Kansas (Malin, 1946; Bark, 1978).

Important insight into the dynamics responsible for severe, sustained drought has been provided by the analysis of computer models used to simulate sea-surface temperatures. These models suggest that anomalously cool or La Niña-like conditions in the Pacific Ocean explain much of the decadal drought variance across North America during the Dust Bowl (Schubert et al., 2004; Seager et al., 2005a; Seager et al., 2005b; Herweijer et al., 2007). Analysis of US hurricane counts and Pacific sea-surface temperatures performed by Elsner et al. (2001) showed that US hurricanes were more likely during La Niña events. This would imply increased tropical system development across the Atlantic basin during La Niña events, and interestingly, Schubert et al. (2004) hypothesize that the wet anomalies over Mexico during the 1930s could be a function of a higher number of tropical systems, particularly in 1933 and 1936.

Proxy climate data and modeling studies also suggest anomalously cool, La Niña-like conditions occurred in the Pacific Ocean during the mid-19th century (Cole et al., 2002; Herweijer et al., 2006, 2007). Mock (2008) observed increased hurricane impacts over Louisiana during this period based on a comprehensive analysis of the available instrumental and documentary records. This active period of tropical systems included 1860, when three hurricanes, including a major category three, made landfall within about 120 km of New Orleans, Louisiana (Mock, 2008). This finding raises the possibility that synoptic-scale circulation patterns related to the intense drought in 1860 may have influenced hurricane tracks. It is also plausible that hurricane-related rainfall could have alleviated drought in the South Central US. Tree-ring PDSI reconstructions for 1860 fail to resolve a signal of increased rainfall across this area (Fig. 1, Cook and Krusic, 2004), but these hurricanes occurred from August into early October and instrumental PDSI data correlate best with tree-ring data during the summer (i.e., June/July/August, Cook et al., 1999). This limitation of proxy climate data to specific seasons demonstrates that the recovery and reanalysis of available instrumental and documentary climate records are imperative for a more complete understanding of 19th century climate extremes.

Significant socioeconomic losses have been attributed to seasonal climatic extremes, and an increase in some extreme events has been documented regionally

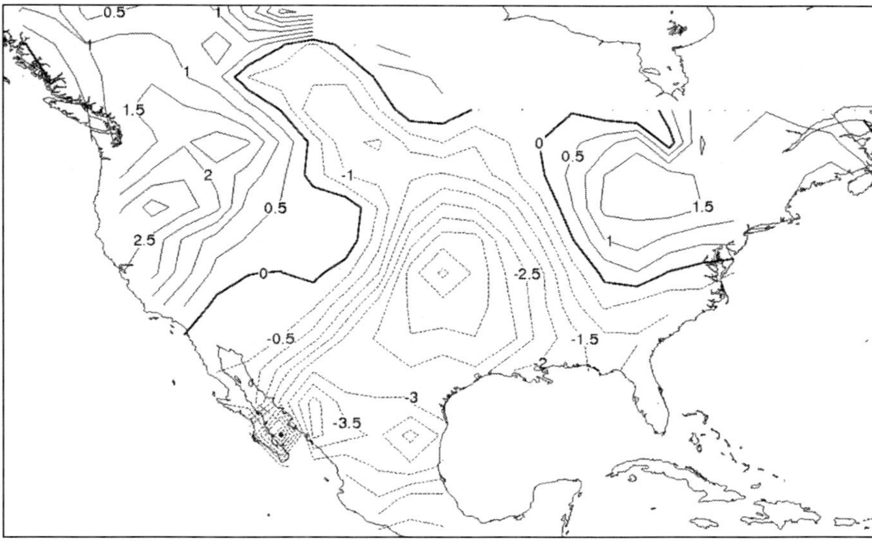

Fig. 1 The tree-ring PDSI reconstruction for 1860 (Cook and Krusic, 2004) showed intense drought conditions centered in the Plains, where dashed contours denote PDSI values less than zero. Zero is indicated by the solid bold contour line, and the other solid contours denote PDSI greater than zero. Note the low PDSI values across Texas and Louisiana despite the occurrence of beneficial rains from hurricanes in August, September, and early October

and globally (Trenberth et al., 2007). In fact, over the last four decades, a significant increase in precipitation totals across the United States has been accompanied by a significant increase in the number of consecutive days without precipitation across the eastern and southwestern portions of the country (Groisman and Knight, 2008), which suggests that more precipitation is falling in intense bursts. Instrumental and documentary weather records are available from earlier societies, who were sensitive to extreme events. These data are invaluable for studying the past because no other climate proxy can provide daily to sub-daily resolution. However, many of these records contain non-climatic biases because of the primitive recording practices used by these early observers. Nevertheless, exhaustive archival work and careful, selective analyses do allow for the recovery of unbiased weather observations (Chenoweth, 2007; Mock et al., 2007). In this chapter, we reconstruct the weather from April through October in 1860 to (1) identify the spatial extent of moisture anomalies, (2) identify the range of extremes, (3) reconstruct the hurricane season, and (4) assess the impacts of these extremes on society.

2 Data and Methods

2.1 Drought Reconstruction

Instrumental precipitation data used in this study were extracted from the handwritten documents of the US Army Surgeon General and the Smithsonian Institution

Fig. 2 Location map of instrumental and documentary data. Instrumental stations with at least one month of available data for April–October 1860 are displayed. These data were obtained from the National Archives (Darter, 1942), the Web Search Store Retrieve Display system of NOAA's Climate Database Modernization Program (Dupigny-Giroux et al., 2007), and from selected weather-sensitive diarists. Locations of cities and newspapers are also displayed on the map

(Darter, 1942; Dupigny-Giroux et al., 2007) and from selected weather-sensitive diarists (Fig. 2). In 1860, available instrumental precipitation records were sparse outside of the northeastern United States (Fig. 2). Thus, these records were supplemented with non-instrumental data (e.g., newspapers and diaries), which are critical in the reliable quantification of the number of rain days. All of these precipitation data can be biased due to the conversion of snowfall to liquid equivalent, the high placement of the rain gauges, infrequent observation of the precipitation gauge, changes in the physical environment around the gauge (e.g., tree growth, building construction), and changes in instrumentation (Mock, 1991, 2000; Daly et al., 2007). [See Conner, this volume for an account of station moves and the effect on data quality.] These biases can lead to undercounted precipitation, which is a common problem in historical precipitation records. Given that cold season precipitation is much more strongly susceptible to these biases (Mock, 1991, 2000), this study was limited to April through October 1860 to minimize the influence of cold season precipitation (e.g., Mock, 2002).

Warm season precipitation data are not bias-free, and therefore, two screening routines were executed to minimize non-climate variations in the precipitation data used in this study. Histograms of daily precipitation frequency were constructed

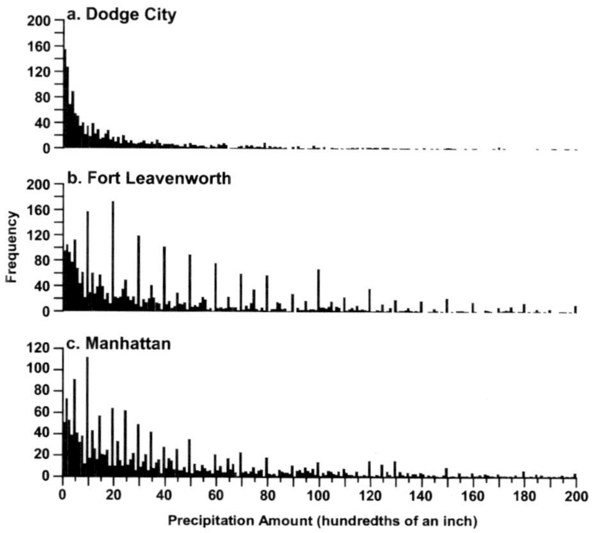

Fig. 3 Histograms of daily precipitation frequencies at Dodge City (**a**), Fort Leavenworth (**b**), and Manhattan (**c**, all in Kansas) vary in quality. Daily precipitation data at Dodge City, Kansas, appear to be of higher quality as indicated by the smooth, negative exponential curve. Undercount and 5/10 bias were observed in the Fort Leavenworth and Manhattan, Kansas stations. Inches are used for this study because those were the units recorded by the observers during the 19th century

to evaluate the quality of longer daily records in possession of and digitized by the authors (e.g., Fig. 3, Daly et al., 2007). A smooth negative exponential curve would be expected for higher quality precipitation data (i.e., a higher frequency of 0.01 in. amounts (0.0254 cm) than 0.25 in. amounts (0.635 cm) (Fig. 3a)). Spikes in precipitation totals ending in a zero or five or the so-called "5/10 bias" and truncation of the histogram for the smallest totals are indicative of estimated values, sporadic observations, and observer inattentiveness to lighter amounts (e.g., Fig. 3b, c; Daly et al., 2007). Most precipitation data used in this study were digitized only for the time period of interest (i.e., April through October 1860). Such short time series do not allow the daily precipitation to be screened reliably using histograms of daily precipitation frequency. Instead the spatial geography of dry and wet anomalies using precipitation day counts was examined, which minimizes the inadequate recording of small precipitation amounts when evaporation rates were high. Thus, a more important screening routine was to purge stations that were outliers compared to their surrounding regional network and clearly not homogeneous, and this routine was executed after monthly precipitation day counts at each station were transformed into percentiles (see below).

Counts of days with recorded precipitation at all instrumental and non-instrumental stations were summed on a monthly basis from April through October 1860 and then compared to their respective modern records. Modern precipitation records (post-1892) at or nearest to each study station were obtained from the U.S. National Climatic Data Center. These modern records varied in length from station to station, but all were at least thirty years long to capture the variations in precipitation frequency for each station. Monthly precipitation day counts were transformed into percentiles relative to the same month in the modern record (c.f., Mock and

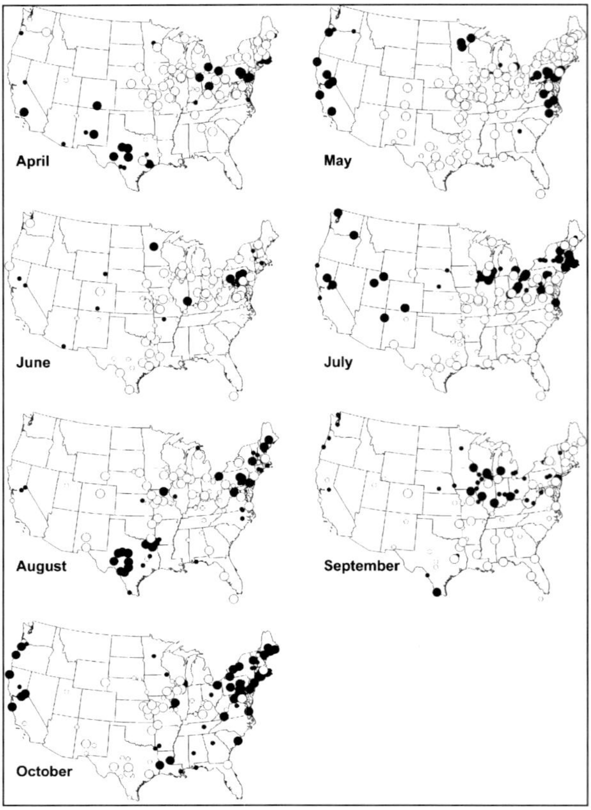

Fig. 4 Monthly precipitation percentiles relative to the modern (post-1892) instrumental record illustrate the spatial variability in precipitation from April through October 1860. Large open circles denote stations ≤10th percentile ("very dry"), small open circles denote stations < 25th percentile and >10th percentile ("dry"), small filled circles denote stations >75th percentile and <90th percentile ("wet"), and large filled circles denote stations ≥90th percentile ("very wet")

Bartlein, 1995; Mock and Brunelle-Daines, 1999). The percentiles were then classified into the following four categories: "very wet" for values ≥90, "wet" for values >75 and <90, "dry" for values <25 and >10, and "very dry" for values ≤10. These four categories were mapped on a monthly basis (Fig. 4), and then used along with general descriptions available from newspapers to illustrate impacts of the drought on settlement in 1860.

2.2 Hurricane Reconstruction

Using newspaper sources, Ludlum (1963) described the coastal impact of the 1860 hurricane season from three land-falling hurricanes along the northern Gulf of Mexico. This information was added later to the Atlantic HURDAT (Hurricane Database), archived by the NOAA National Hurricane Center (Landsea et al., 2004). However, analyses and reconstructions of all tropical cyclones in the mid-nineteenth century are not complete for most years, including 1860. Therefore, it is important to gather all available historical data before assuming track directions and intensities of historical tropical cyclones. This study significantly expanded the reconstruction of the 1860 hurricane season in the Western Atlantic Basin to include several newly found tropical systems.

The 1860 hurricane database includes information from approximately 25 coastal stations with instrumental data (e.g., Mount Vernon Barracks, Alabama; New Orleans, Louisiana), which are part of the subset of the national station network that was used to map precipitation frequency patterns. Data were taken from the US Army fort network, records from the US Coastal Survey and Smithsonian Institution, and miscellaneous voluntary weather observers. Some of the records include barometric pressure observations that enable detailed intensity analyses, following correction for elevation, latitude, and gravity (Mock, 2008). Additional detailed sub-daily instrumental records came from eleven ship logbooks taken from US and UK archives, located mostly close to land but also out in the open ocean. This study also utilized documentary information from over twenty different newspapers, with a heavy emphasis on the *New York Shipping and Commercial List, New York Herald, Charleston Courier, Charleston Mercury,* and *New Orleans Picayune*. Newspaper articles give a unique detailed perspective of the damages to the area as a first-hand account, which is an important aspect for accuracy given the frequency of repetitive stories in 19th century newspapers (Chenoweth, 2006). Other primary sources included approximately thirty personal diaries and other descriptive accounts, particularly those from plantations located in remote areas. These sources include ship protests from the New Orleans Notarial Archives, which provide some instrumental and verbal data on storms off the Louisiana coast (Mock, 2008).

All of the instrumental and documentary data were collated and mapped on a daily basis for the hurricane season of 1860. Storms were identified as tropical systems if they indicated a well-defined circulation, had winds or damage indicative of at least tropical storm force, and did not exhibit signs of extratropical behavior (e.g., no falling temperatures or frontal indications; c.f., Mock, 2004). Storms were also identified as those at hurricane and major hurricane strength when the data quality permitted. Tracks were drawn using typical methods in hurricane mapping (Landsea et al., 2004).

3 Results

3.1 Drought in 1860 and Its Impacts on the United States

Monthly precipitation percentiles from April through October are shown in Fig. 4, and indicate that the drought was most widespread from April through July 1860. This spatial pattern matches well with documentary data from newspapers and diaries and tree-ring PDSI reconstructions (Fig. 1, Cook and Krusic, 2004). The documentary data also reveal that occasional showers would renew hopes of better weather for the farmers, but hot and dry conditions returned quickly after much-needed rains. Across portions of Texas, the onset of drought was delayed in April

1860 when 8 out of 15 stations reported wet to very wet conditions (Fig. 4). This rainfall was thought to be the start of a prosperous growing season as reported by the *New Orleans Picayune* (1860) from Houston, Texas, dated April 17:

> Texas never enjoyed at this season of the year more flattering prospects of a full crop than at present. The season of frosts has been passed with impunity, and we have just been favored with a very general rain, rendering a plentiful corn crop almost a certainty. Corn, cotton, wheat and sugar cane promise well...

Weather observations from other parts of Texas were not as encouraging. G. Freeze (1860) from Boston, Texas wrote that the weather at the beginning of April 1860 was "very dry and dusty and sometimes strong wind." Conditions became worse in May, when dry to very dry conditions were most widespread across the central and eastern United States (Fig. 4). *The Arkansian* (Fayetteville, Arkansas, 1860) of May 18 states "Up to Monday night of this week, the air and earth were dry; vegetation began to look as though 'twas midsummer ... So dry spell in spring, for dryness and duration, never was before known in this latitude at this time of the year". By July, G. Freeze (1860) wrote:

> We got very hot and dry weather this month the Thermometer on the 7 8 and 9 got as high as 108° in the shade, and in the open air (in the garden in free current air) 120°. Caused by a hot wind from W and SW unknown here before. The common height of the Thermometer is generally at this place for the month of July between 90° and 100°. We got no rain in the last 3 months to wet the ground 1 inch deep, wich [sic] make our crops short.

A similar report from *The Navarro Express* (Corsicana, Texas, 1860) on July 14 indicated that "[d]uring the past week the thermometer has been standing at 100° and 112°. The weather continues very dry. The corn crops have been severely injured..."

Very dry conditions were observed across much of Kansas and Missouri in April and such conditions continued unimpeded through July (Fig. 4), but like the accounts from G. Freeze and others in Texas, dry weather was not the only story. High temperatures and strong winds were also common reports such as in the *Fort Scott Democrat* (Fort Scott, Kansas, 1860) on July 28th:

> Saturday, of last week, and Monday and Wednesday of this week, were three of the hottest days we ever experienced. The thermometer on each of those days, rose to 112 degrees IN THE SHADE! On Saturday afternoon, a strong gale of wind prevailed from the South west. The blast was hot and scorching as the breath of a furnace, and as withering and destructive in its effects, as the sirocco of the Arabian desert. In the sun, the thermometer went up to 132. Men and animals were alike prostrated, and it actually seemed as if some terrible calamity was impending. The effect on our blighted crops was truly alarming; and it is now evident that unless we have heavy rains within a week everything will be ruined.

These high temperatures, low rainfall, and high winds were the perfect conditions for dust storms, and 1860 was a particularly bad year for blowing dust in Kansas (Malin, 1946). Descriptions of these dust storms provided by early homesteaders reveal how "a person could scarcely be seen one hundred yards" and that they breathed the dust and swallowed it down (Malin, 1946). Crops in Kansas were desiccated as reported by *The State Record* at Topeka, Kansas on May 26th (*Chicago Tribune*, 1860):

> From a recent visit into the country during the past week, we are satisfied that the wheat crop of this section will this year be almost a total failure, not producing, on an average, as much seed as was sown. A large breadth of country was last winter sown to winter wheat, which is now headed out about six inches above the ground, and very thin at that, being literally worthless. Some of the spring wheat put in this spring may produce a partial crop, but a large portion of that is also a failure.

Crop conditions were just as grim to the east in Missouri, where on June 8th the *St. Joseph Free Democrat* (St. Joseph, Missouri) reported (*Chicago Tribune*, 1860):

> A general felling of gloom is beginning to pervade all classes in this usually most prosperous section of the state, and unless the parched earth is speedily visited with copious rains, it is evident that we are to have a repetition of the famous famine year of 1833. Much may undoubtedly be done to break the force of the blow which the loss of our crops will illicit.

Percentiles of precipitation days indicated that dry to very dry conditions were generally the rule from Iowa eastward into Indiana through June (Fig. 4), but documentary evidence showed that the weather was even more unsettled. In fact, intense thunderstorms were reported during the first week of June 1860 including a "Great Tornado ... literally destroying the towns of Camanche, Iowa, and Albany, Illinois" (*Chicago Tribune*, 1860). Meteorological data indicated wet to very wet conditions across northern Illinois and southern Wisconsin in July 1860, and newspapers confirm this wet period, especially on July 21st when 19 stations reported heavy rain across Iowa, Illinois, and Wisconsin. The heavy rainfall experienced locally across this region was beneficial for crops and reports such as "more acres of wheat sown the last Spring than ever before, and the yield is heavy per acre, and rain generally plump and good" were common (*Chicago Tribune*, 1860).

Instrumental observations indicated dry to very dry conditions were observed in the southeastern United States through July 1860 (Fig. 4), and the available documentary data support these observations. John Horry Dent (1860) of Barbour County, Alabama terms 1860 as the "fatal dry year" with only a few days of rain and no substantial amounts from February to July. His diary temporarily stops in August due to the tremendous drought conditions that were "ruinous" to the crops. According to Washington Cochran Smith (1860) of Abbeville, South Carolina, "dry" was used to describe conditions seven times in early April. Late April through May brought close to average rains. Conditions in June through July were hot and dry again, and Cochran (1860) described 10 days throughout these months as "dry", "very dry" or "very, very dry". On July 11 he states "the warmest day I ever knew. Thermometer 102°." Severe repercussions for the crops in South Carolina, Georgia, Alabama, and Mississippi were reported in late July (*Chicago Tribune*, 1860):

> Never in my life have I witnessed such injury to growing crops by drouth as I have in the above States, in the last three weeks. ...[farmers from the area] testified to the wide-spread destruction of corn and cotton, vegetables and fruits,...
>
> The corn in many places is entirely destroyed—even unfit for fodder. Planters waited, hoping for rain, and so the blades and stalks have dried from the root to tassel...

> Cotton is wilting—shedding leaves, forms, blooms, and bolls. Even copious and continuous showers henceforth will not only fail to arrest this destruction, but aggravate it; by causing more shedding and a second growth, which will be too late for maturity in the fall...

The landfall of three hurricanes along the Gulf Coast brought increased rainfall to southeastern United States from late summer into the fall (Fig. 4, see hurricane season results below). However, drought conditions remained entrenched across the central Plains of Kansas and Oklahoma and a large part of Missouri, where precipitation percentiles were persistently dry to very dry (Fig. 4). Remarkably, there was enough rainfall across Kansas that some reports mentioned "some localities here and there, have a very good crop," but for the most part the crops across Kansas were a failure in 1860 (*Chicago Tribune*, 1860; Malin, 1946).

3.2 Atlantic Tropical Cyclones of 1860 and Their Impacts on the United States

The current official Atlantic Hurricane Database (HURDAT) contains seven tropical cyclones for the 1860 hurricane season. This study's analysis identified up to three additional storms (Storms 1, 2, and possibly 7, Fig. 5), and it added more days to the track history for five storms (Storms 3, 4, 5, 8, and 9, Fig. 5). Storm 1 was likely a tropical or subtropical storm of a few days given the lower pressure revealed by several ship reports, particularly from the *Metropolis*. Storm 2 made landfall near Vera Cruz, Mexico around the end of July and the first day of August. Storm 3 lost tropical storm strength as it moved through the interior Southern US, probably south of Covington, Georgia. The storm brought about 3 in. (7.62 cm) of rain to coastal South Carolina, before re-emerging as a marginal tropical storm off North Carolina.

In terms of previously identified storms, the analysis of the track for Storm 4 remained similar to that of Fernandez Partagas and Diaz (1995), but shifted a bit westward to include impacts on eastern North Carolina and Massachusetts. The history of Storm 5 was much expanded from previous work, largely based on four ship reports and an instrumental record in Bermuda by the Royal Engineers that indicated a drop in barometric pressure and an increase in NE winds at tropical storm force. Storm 7 appears to have mostly been extratropical, but the possibility of subtropical or tropical status on the first day or so of its history off North Carolina cannot be ruled out, since severe non-tropical gale force winds at the peak of hurricane season are not common occurrences in the area.

A few extensions to the earlier track histories of Storms 8 and 9 were added by this current analysis. We found no clear evidence of the storm described by Fernandez-Partagas and Diaz (1995) from September 18–21 beginning at 22.20°N, 66.10°W and ending at 39.00°N, 68.50°W, since geographic locations from a few data points were interpolated and analysis was largely based on one report of questionable reliability. Our analysis suggests that some of these data could represent misdates of the October storm (Storm 9). There was also one potential tropical depression that we could not confirm which made landfall near Victoria, Texas (not

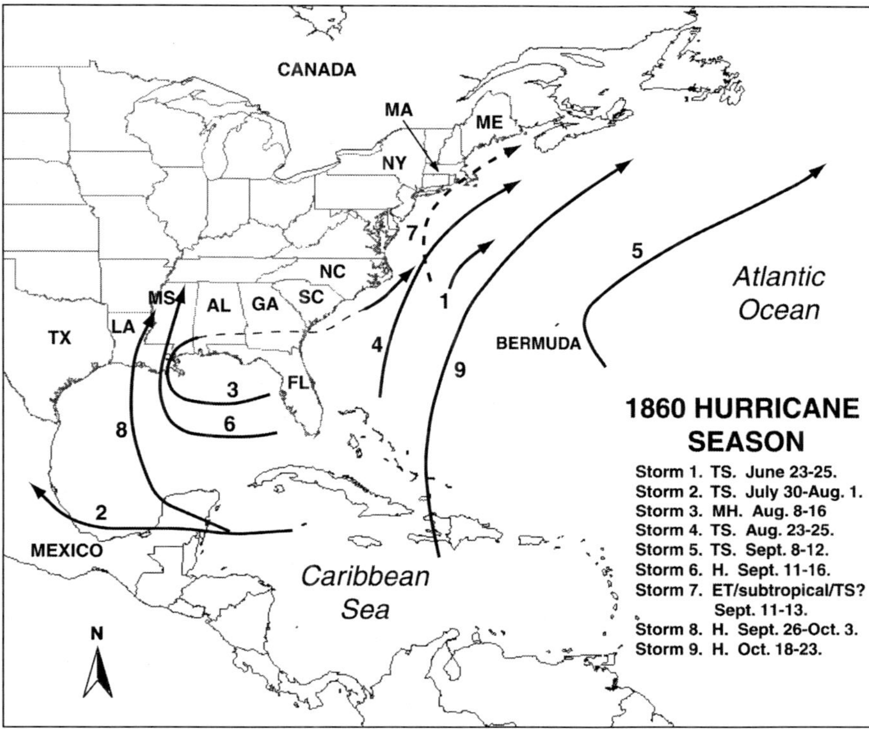

Fig. 5 Tracks of Atlantic tropical cyclones in 1860 (MH = major hurricane, H = hurricane, TS = tropical storm, and ET = extratropical cyclone). *Dashed lines* indicate periods of time when the tropical cyclone decreased below tropical storm strength or had extratropical characteristics (Storms 3 and 7, respectively). Up to three new tropical cyclones were identified in this study, which brought the total storm count in 1860 to nine

listed in Fig. 5). This disturbance brought beneficial rain to Texas during the latter half of August 1860 and by the end of the month, 14 out of 19 stations reported wet or very wet conditions (Fig. 4). The total number of nine tropical cyclones west of 55°W is above the long-term average of between seven and eight storms from 1851 to 2008.

Tropical rainfall was a welcome sight in the southeastern US. Unfortunately, most of it came from three hurricanes, which brought additional hazards to this drought-stricken region. The first of the three hurricanes made landfall as a category three southeast of New Orleans, Louisiana late on August 11th with maximum sustained winds of 190 kph (Storm 3, Fig. 5; Landsea et al., 2004; Mock, 2008). T. Harrison kept a weather record in New Orleans, Louisiana and reported: "Aug. 11- Began to rain at night; ended at 9 P.M.; quantity fallen 1.20 [inches]. Very stormy; very high wind during the evening, sometimes blowing with a force of 3 or 9" (*New Orleans Picayune*, 1860). Massive storm surge reports were common such as those in the *New Orleans Picayune* (1860), which mentioned water rising "some 3 ft in the Mississippi" and inundating "...the entire parish of Plaquemines, from Dr. Wederstrandt's down to the Quarantine Station. The water rose to a depth of 4 ft on the public roads at Pointe-a-la-Hache..."

After sweeping around New Orleans, Louisiana, the storm headed east toward Biloxi, Mississippi, and Mobile, Alabama, early on August 12th, where (*New Orleans Picayune*, 1860):

> ...The wind veered from N.E. to N.W. and N., and back to N.E., increasing in violence every hour, until sundown of Saturday, when a night of pitchy darkness set in after a day of incessant rain, and from that time until 4 o'clock A.M., of Sunday morning, the storm raged in all its fury, nothing being heard save the rushing of the mighty wind, the remorseless dash of the agitated billows, and the harsh howling of the Storm King, as amid torrents of rain, and Erebusean darkenss [sic], he ordered the elemental carnival. After 4 o'clock, A.M., Sunday, the wind shifted gradually to the S. W., decreasing in violence until it subsided into a fresh breeze.

Water Street in Mobile, Alabama was flooded in the storm surge. The storm was less destructive to personal property overall though, and newspapers speculated that this was because "[f]rom early morning it gave forewarnings of its approach, and the people had the whole day to provide against it" (*New Orleans Picayune*, 1860). However, the storm had also weakened to a category one hurricane with maximum winds of 130 kph by the time it arrived at Mobile, Alabama, on August 12.

Fewer instrumental stations across the southeastern United States reported dry to very dry conditions in August 1860 due to the increased tropical moisture associated with the major hurricane and additional possible disturbance over Texas (Fig. 4), but the additional rainfall was localized and too late in the growing season to allow for widespread recovery in crop yields (*Chicago Tribune*, 1860):

> It is now ascertained, beyond a doubt, that the prevalence of an unprecedented drouth [sic], concurring and running through nearly the whole length of the food crop making season, all though the States, of South Carolina, Georgia, Alabama, Mississippi, Louisiana, Texas, including a portion of Arkansas, has produced the most disastrous effect ever known. It is true through all this region, some locations have been blessed with rains to make fair crops, and some even to spare... But these oases in the desert of drouth has made, bear a proportion of the whole area ... cessation of the spring rains. All accounts concur, though, that these rains have come and are coming, too late to benefit to any noticeable extent, the great breadth of the corn crop.

A second hurricane made landfall in nearly the same location about a month later in the early morning hours on September 15th with maximum winds of 170 kph (Storm 6, Fig. 5). New Orleans, Louisiana experienced more damage along the lake, but this damage was "principally upon what had not been destroyed by the storm of the 11th of August" and the "repairs made since that time stood the storm perfectly" (*New Orleans Picayune*, 1860). This suggests that the storm was below major hurricane strength. The most severe damage was located farther to the east in the vicinity of Biloxi, Mississippi where the newspaper headlines read "The Town of Biloxi a Heap of Ruins" and "Nearly Every House at the Balize Blown Down." The storm's rage was also felt in Mobile, Alabama, where a warehouse fire destroyed three thousand bales of cotton, and headlines told of a storm loss around $500,000 (*New Orleans Picayune*, 1860).

Storm 6 moved farther inland across the southeastern United States than its predecessor and brought increased rainfall across Louisiana, Mississippi, Alabama, and Tennessee (Fig. 5). Interestingly, this sixth tropical cyclone of the season caused inundation where the first and stronger land-falling storm had not because "...in August the swamps were nearly dry, and the water from the lake found a natural outlet; whereas, yesterday, the swamps being full, the water rose in the streets of Milneburg [near New Orleans, Louisiana] and covered the railroad track in the distance" (*New Orleans Picayune*, 1860). This appears to be only a localized effect because five out of seven instrumental stations across Louisiana, Mississippi, Alabama, and Tennessee still reported dry to very dry conditions in September (Fig. 4). This included New Orleans, Louisiana, where precipitation in September 1860 was 3.59 in. (9.12 cm) below the September 1937–2007 average, and a similar deficit was observed at Columbus, Mississippi.

Approximately two weeks later, on October 2nd, a third hurricane made landfall to the southwest of Morgan City, Louisiana, with maximum winds to 170 kph (Storm 8, Fig. 5, Landsea et al., 2004). The track of this storm dealt a more severe blow to New Orleans, Louisiana, as reported in the *New Orleans Bee* on October 4th (*New York Herald*, 1860):

> One of the most terrific hurricanes that ever visited New Orleans raged during the whole of yesterday. Following closely upon two destructive predecessors within the space of seven weeks, it has been more violent than either of the others, and as intelligence reaches us from the country around we fear to learn a sad story of its direct effects. In this city we already know of great destruction and a melancholy loss of life.

Across the surrounding Louisiana Parishes the "open cotton was scattered, so as to make the ground in some places white, as if covered with snow" and the sugar cane "greatly suffered; thus blasting the last hopes of our sugar planters for this season" (*New Orleans Picayune*, 1860). The remnants of this hurricane impacted Louisiana, Mississippi, Alabama, and Tennessee through the first week of October 1860, and five out of seven stations reported wet to very wet conditions (Fig. 4).

4 Discussion and Conclusions

Drought conditions were observed across much of the central and southern Plains, portions of the Midwest, and in the southeastern United States. The core of the drought appeared to be centered in the central Plains of Kansas, Oklahoma, and Missouri, where conditions were persistently dry to very dry from April through October 1860, and impacts on the crops were most severe. The tree-ring PDSI reconstructions for 1860 also identify a peak intensity of the drought across this region (Fig. 1, Cook and Krusic, 2004), but analysis of the historical climate data revealed a more complex story including areas that suffered from drought and intense storms. The spatial pattern of the precipitation percentiles from April through October 1860 and the tracks of the three land-falling US hurricanes (and also perhaps related to the above-normal tropical activity in general) suggest one strong high pressure system was centered in the central Plains and another in the northeastern Gulf of Mexico with "rings of fire" around their peripheries (i.e., thunderstorms and hurricanes would tend to move around the high pressure centers). This pattern would allow Storms 3, 6, and 8 to be steered into the Gulf of Mexico and then into Louisiana and Mississippi through a weakness between the two high pressure centers.

Intense heat and wind that accompanied this drought allowed for the development of dust storms across Kansas, and some may have been as intense as those during the 1930s Dust Bowl drought (Malin, 1946). Cunfer (2005) suggested that the intense heat, low precipitation, and low soil moisture better explained dust storm geography during the Dust Bowl than the overuse of the land due to economic depression as suggested by Worster (2004). The descriptions of the dust storms in 1860 would seem to support such a hypothesis. A recent study by Cook et al. (2008) showed that model simulations of the Dust Bowl drought could be improved by including dust-emission into the forcing, and previous modeling studies have shown that dust may be a key land-surface feedback that exacerbates ongoing drought conditions (Miller and Tegen, 1998; Rosenfeld et al., 2001). Thus, dust storms may have been involved in prolonging other long-term droughts shown in the paleoclimate record, and studies have shown a correlation between aeolian movement in the Great Plains and the occurrence of drought prior to western settlement (e.g., Muhs and Holliday, 1995).

The instrumental and documentary historical climate data appear to corroborate the widespread "Civil War" drought conditions that prevailed throughout the early 1860s according to paleoclimatic and modeling studies (Fye et al., 2003; Herweijer et al., 2006, 2007). These historical climate data also show that high temperatures and erosional aeolian activity accompanied this drought in 1860, which would indicate a high rate of evapotranspiration and may have been a key land-surface feedback responsible for prolonging the drought conditions during the mid-19th century (Miller and Tegen, 1998; Rosenfeld et al., 2001; Cook et al., 2008). Most of the annual precipitation that falls across the Plains, Midwest, and the southeast occurs during the warm season and many of the agricultural products consumed by the United States population are also grown at this time, which cause prolonged dry spells during this period to be more detrimental (Groisman and Knight, 2008).

The occurrence of three strong hurricanes near New Orleans, Louisiana, from August through October 1860 is a particularly noteworthy event that is not evident in the modern record. If such a pattern were repeated today, then the impacts on New Orleans, Louisiana, and surrounding coastal cities may be much more severe than those of Hurricane Katrina in 2005, especially given the increase in population along the coasts. This demonstrates how the modern climate record can give an incomplete assessment of the full range of climatic extremes. In 1860, the subsequent hurricanes proved disastrous to the waterways, but in the absence of the intense drought lowering the water tables and swamps in the low country of Louisiana, Mississippi, and Alabama, damage from flooding could have been greater. Unfortunately, the rainfall brought to inland areas by these hurricanes was insufficient and occurred too late in the growing season to improve the yield of the parched crops.

Previous modeling and tree-ring studies indicate La Niña conditions were responsible for the weather patterns prevalent in North America during 1860 (Cole et al., 2002; Herweijer et al., 2006, 2007). This pattern also influences the paths of tropical cyclones and could possibly deter impacts along the Atlantic coast (Elsner et al., 2001). Understanding past analogs of tropical cyclone tracks and the teleconnections that influence these tracks are vital to identify a range of storm tracks that could occur in the future. This study illustrates the value of using instrumental and documentary historical climate data to provide important analyses of past climatic extremes. Many other valuable historical climate datasets await careful recovery and reanalysis from archives across the United States, and will give further clues about the full range of climatic extremes that can occur today.

Acknowledgements We gratefully acknowledge help from the staff at various historical libraries and repositories. Some funding for archival research was supported by NSF Grants ATM-0502105, ATM-9904383, BCS-0622894 and SBR-0349986.

References

Bark LD (1978) History of American droughts. In: Rosenberg NJ (ed.) North American Droughts. Westview Press, Boulder.

Chenoweth M (2006) A reassessment of historical Atlantic basin tropical cyclone activity, 1700–1855. Climatic Change 76:169–240.

Chenoweth M (2007) Historical climatology. In: Elias S (ed.) Encyclopedia of Quaternary Science. Elsevier, UK.

Chicago Tribune (1860) Chicago, Illinois, June 4, June 5, June 12, July 23, August 3, August 11, August 13, September 5, September 12.

Cole JE, Overpeck JT, Cook ER (2002) Multiyear La Niña events and persistent drought in the contiguous United States. Geophys Res Lett 29:1647, doi:10.1029/2001GL013561.

Cook ER, Krusic PJ (2004) North American drought atlas. Lamont-Doherty Earth Observatory and the National Science Foundation. http://iridl.ldeo.columbia.edu/SOURCES/.LDEO/.TRL/.NADA2004/.pdsi-atlas.html. Cited 29 December 2007.

Cook ER, Meko DM, Stahle DW, Cleaveland MK (1999) Drought reconstructions for the continental United States. J Climate 12:1145–1162.

Cook BI, Miller RL, Seager R (2008) Dust and sea surface temperature forcing of the 1930s "Dust Bowl" drought. Geophys Res Lett 35:L08710, doi:10.1029/2008GL033486.

Cook ER, Woodhouse CA, Eakin CM, Meko DM, Stahle DW (2004) Long-term aridity changes in the western United States. Science 306:1015–1018.

Cunfer G (2005) On the Great Plains agriculture and environment. Texas A&M University Press, College Station.

Daly C, Gibson WP, Taylor GH, Doggett MK, Smith JI (2007) Observer bias in daily precipitation measurements at United States Cooperative Network Stations. Bull Am Meteorol Soc 88: 899–912.

Darter LJ (1942) List of climatological records in the National Archives. National Archives, Washington, DC.

Dent, JH (1860) Plantation Diary. Manuscript Collection, University of Alabama, Tuscaloosa, Alabama, and in the Troy State University Library, Troy, Alabama.

Dupigny-Giroux L-A, Ross TF, Elms JD, Truesdell R, Doty SR (2007) NOAA's Climate Database Modernization Program: rescuing, archiving, and digitizing history. Bull Amer Meteor Soc 88:1015–1017.

Elsner JB, Bossak BH, Niu X-F (2001) Secular changes to the ENSO-U.S. hurricane relationship. Geophys Res Lett 28:4123–4126.

Fort Scott Democrat (1860) Fort Scott, Kansas, July 14.

Freeze G (1860) Smithsonian Institute Weather Observations. Available at National Archives and Records Administration, Washington, DC.

Fye FK, Stahle DW, Cook ER (2003) Paleoclimatic analogs to twentieth-century moisture regimes across the United States. Bull Am Meteorol Soc 84:901–909.

Groisman PY, Knight RW (2008) Prolonged dry episodes over the conterminous United States: new tendencies emerging during the last 40 years. J Climate 21:1850–1862.

Herweijer C, Seager R, Cook ER (2006) North American droughts of the mid to late nineteenth century: a history, simulation, and implication for Medieval drought. Holocene 16:159–171.

Herweijer C, Seager R, Cook E, Emile-Geay J (2007) North American droughts of the last millennium from a gridded network of tree-ring data. J Climate 20:1353–1376.

Landsea CW, Anderson C, Charles N, Clark G, Dunion J, Fernandez-Partagas J, Hungerford P, Neumann C, Zimmer M (2004) The Atlantic hurricane database re-analysis project. In: Murnane RJ, Liu KB (eds.) Hurricanes and Typhoons: Past, Present, and Future. Columbia University Press, New York.

Ludlum DM (1963) Early American hurricanes 1492–1870. American Meteorological Society, Boston.

Malin JC (1946) Dust storms part one 1850–1860. Kans Hist Q 14:129–133.

Miller RL, Tegen I (1998) Climate response to soil dust aerosols. J Climate 11:3247–3267.

Mock CJ (1991) Drought and precipitation fluctuations in the Great Plains during the late nineteenth century. Great Plains Res 1:26–57.

Mock CJ (2000) Rainfall in the garden of the United States Great Plains, 1870–1889. Climatic Change 44:173–195.

Mock CJ (2002) Documentary records of past climate and tropical cyclones from the southeastern United States. PAGES News 10:20–21.

Mock CJ (2004) Tropical cyclone reconstructions from documentary records; examples from South Carolina. In: Murnane RJ, Liu KB (eds.) Hurricanes and Typhoons: Past, Present, and Future. Columbia University Press, New York, pp. 121–148.

Mock CJ (2008) Tropical cyclone variations in Louisiana, USA, since the late eighteenth century. Geochem Geophy Geosy 9:Q05V02, doi:10.1029/2007GC001846.

Mock CJ, Bartlein PJ (1995) Spatial variability of late-Quaternary paleoclimates in the western United States. Quaternary Res 44:425–433.

Mock CJ, Brunelle-Daines AR (1999) A modern analogue of western United States summer paleoclimate at 6000 years before present. Holocene 9:541–545.

Mock CJ, Mojzisek J, McWaters M, Chenoweth M, Stahle D (2007) The winter of 1827–1828 over eastern North America: a season of extraordinary climatic anomalies, societal impacts, and false spring. Climatic Change 83:87–115.

Muhs DR, Holliday VT (1995) Evidence of active dune sands on the Great Plains in the 19th century from accounts of early explores. Quaternary Res 43:198–208.

New Orleans Picayune (1860) New Orleans, Louisiana, April 24, August 13, August 14, August 16, August 19, September 16, September 17, October 9.

New York Herald (1860) New York, October 6.

Palmer WC (1965) Meteorological drought.Weather Bureau Research Paper 45, U.S. Department of Commerce, Washington, D.C.

Partagas JF, Diaz HF (1995) A Reconstruction of Historical Tropical Cyclone Frequency in the Atlantic from Documentary and other Historical Sources: 1851–1880 Part I: 1851–1870. Climate Diagnostics Center, NOAA, Boulder, CO

Rosenfeld D, Rudich Y, Lahav R (2001) Desert dust suppressing precipitation: a possible desertification feedback loop. Proc Natl Acad Sci USA 98:5975–5980.

Schubert SD, Suarez MJ, Pegion PJ, Koster RD, Bacmeister JT (2004) On the cause of the 1930s Dust Bowl. Science 303:1855–1859.

Seager R, Harnik N, Robinson WA, Kushnir Y, Ting M, Huang H-P, Velez J (2005a) Mechanisms of ENSO-forcing of hemispherically symmetric precipitation variability. Quart J Roy Meteor Soc 131:1501–1527.

Seager R, Kushnir Y, Herweijer C, Naik N, Velez J (2005b) Modeling of tropical forcing of persistent droughts and pluvials over western North America: 1856–2000. J Climate 18:4065–4088.

The Arkansian (1860) Fayetteville, Arkansas, May 18.

The Navarro Express (1860) Corsicana, Texas, July 14.

Trenberth KE, Jones PD, Ambenje P, Bojariu R, Easterling D, Klein Tank A, Parker D, Rahimzadeh F, Renwick JA, Rusticucci M, Soden B, Zhai P (2007) Observations: surface and atmospheric climate change. In: Solomon S, Qin D, Manning M, Chen Z, Marquis M, Averyt KB, Tignor M, Miller HL (eds.) Climate Change 2007: The Physical Science Basis. Cambridge University Press, Cambridge, pp. 235–336.

Washington Cochran Smith (1860) Manuscript Diary. South Caroliniana Library, University of South Carolina.

Wedel WR (1953) Some aspects of human ecology in the central Plains. Am Anthropol 55: 499–514.

West E (1995) The Way to the West. University of New Mexico Press, Albuquerque.

Worster D (2004) Dust Bowl: The Southern Plains in the 1930s. Oxford University Press, New York.

Reconstructing 19th Century Atlantic Basin Hurricanes at Differing Spatial Scales

David A. Glenn and Douglas O. Mayes

Abstract This chapter introduces concepts of hurricane reanalysis and hurricane reconstruction at multiple spatial scales. Two case studies are used to illustrate these respective processes: (1) reconstruction of the 1850 hurricane season and (2) reanalysis of the 1854 Great Carolina Hurricane. Historical documentary evidence from newspapers, ship logs, personal correspondence, and instrumental weather reports were compiled and utilized during the analysis. The storms in 1850 were objectively classified by strength as tropical storm, hurricane, or major hurricane intensity and the storm track was reconstructed using all available locational data across ocean and inland areas. Hurricane intensity for the 1854 Great Carolina Hurricane was determined through several analyses: wind damage, storm surge, and local impact analysis. Results suggest that the 1850 hurricane season contained at least four tropical cyclones and two suspect storms (e.g. potential tropical cyclones requiring additional substantiating evidence). Results also indicate that although the Great Carolina Hurricane of 1854 made landfall near Savannah, Georgia the impacts of the hurricane were probably less than a Category 2 storm in southern South Carolina and around a Category 1 in Charleston, SC. The methodologies and examples presented in this chapter can be used to reassess and expand the North Atlantic Basin hurricane database, as well as to improve and extend historical hurricane chronologies of other ocean basins around the world.

Keywords 1850 · Hurricane reconstruction · Charleston · Documentary evidence

D.A. Glenn (✉)
Department of Geography, University of South Carolina, Columbia, SC 29208, USA
e-mail: David.Glenn@noaa.gov

D.O. Mayes (✉)
Department of Geography, University of South Carolina, Columbia, SC 29208, USA
e-mail: dougandmichele@earthlink.net

L.-A. Dupigny-Giroux, C.J. Mock (eds.), *Historical Climate Variability and Impacts in North America*, DOI 10.1007/978-90-481-2828-0_6,

1 Introduction

Hurricanes are one of the most extreme climatic events affecting communities, landscapes, and ecosystems around the globe. Recently, there has been tremendous scientific debate about the magnitude of global hurricane variability during the last century, especially prior to satellites (1960s) and aircraft reconnaissance observations (mid 1940s) (Landsea, 2007; Mann and Emanuel, 2006; Holland and Webster, 2007; Webster et al., 2005; Hoyos et al., 2006). In this study we demonstrate the utility of revisiting and reanalyzing historical weather observations to reconstruct the occurrence of hurricanes more than 150 years ago. We narrow our approach to hurricanes of the North Atlantic Ocean and analyze all available data at multiple spatial scales: synoptic, regional, and local. This chapter uses two case studies to illustrate the methodology for: (1) reconstructing the 1850 hurricane season and (2) reanalyzing the physical and localized impacts of the Great Carolina Hurricane that struck South Carolina in 1854.

Hurricane reconstruction and reanalysis utilize documentary evidence of hurricane damage as well as descriptive and instrumental accounts of meteorological conditions from a variety of historical sources. Each meteorological scale of analysis (i.e. local, regional, or synoptic) is conducted once all available observations are obtained. Synoptic scale analyses were utilized in reconstructing the 1850 hurricane season while both local and regional scale analyses were used to reanalyze the 1854 Great Carolina Hurricane. It is important to note that although larger scale analysis is used in reconstructing the entire track of a hurricane, mesoscale and local scale analyses are also used in determining a storm's intensity at any given point in time.

This chapter addresses two main objectives meant to illustrate the distinct methodology for revisiting and reconstructing historical hurricanes. The first objective is to demonstrate the processes involved in reconstructing the track and intensity of undocumented historical storms. The 1850 hurricane season is utilized as an example of these processes and specifically focuses on recreating large scale (synoptic) daily weather maps from available surface observations. The second objective is to demonstrate the processes involved in reanalyzing the track and intensity of previously documented hurricanes, specifically storms that are in need of greater local scale damage and intensity analysis. As an example, the 1854 Great Carolina Hurricane is utilized to give researchers a step by step methodology for revisiting historical hurricanes. Both objectives represent the methodological techniques needed to compile the most complete and current record of historical hurricanes.

2 Data

2.1 Newspapers

Newspapers are an extremely valuable source of information pertaining to historical weather events, especially hurricanes (Mock, 2004; Fernandez-Partagas and Diaz, 1995; Chenoweth and Landsea, 2004; Landsea et al., 2004) and the resulting hurricane damage. Newspapers accounts contain detailed descriptions of the events that transpired during and after the passage of a hurricane, and typically varied from general reports to very specific information, such as damage sustained by a particular building or plantation. The data used within this chapter were collected from newspapers published in cities from New Orleans, Louisiana, to London, England, and many others in between. All newspapers with appropriate spatial and temporal coverage were examined from sources that range from local archives to major university and municipal libraries. Accounts published in newspapers came from a wide variety of sources including ships logs, letters from private citizens, official meteorological observations, interviews and first hand accounts of reporters.

Newspaper accounts can vary greatly in detail and locational specificity. In larger urban areas such as Charleston, SC and New York, NY newspapers often contain specific descriptions of both the type of damage and location, including block-by-block descriptions with specific address information. As an example, the *Charleston Daily Courier* (9 September 1854, p. 2) reported that, "Bethel M. E. Church, South corner of Pitt and Calhoun streets, was badly damaged, the edifice being nearly completely unroofed. Damage $2,500." However, not all accounts are highly specific; especially those pertaining to less densely populated areas or locations distant from the newspaper's home city. For example, this description came from the *Charleston Courier* (29 August 1850c, p. 3) which gave the weather observations from Smithville, North Carolina, "Schr. H. Wescott, Jeremiah Foster, master, from Charleston, S.C., bound to this port, in ballast, went ashore in a gale on the night of the 24th inst., four miles to the northward of Ball Head Light House, all hands saved."

Newspaper accounts are especially valuable because the volume of data published in them provides a context for analyzing the severity of damage at a regional scale, and for analyzing atypical accounts documented at specific locations. Newspapers occasionally published side-by-side comparisons of a current hurricane with those of previous storms believed to be of a similar strength. For instance, the *New York Herald* (3 September 1850c, p. 3) quoted a report from the Montgomery (Ala.) Journal, Aug. 26 ". . .the storm, from its violence, reminded us of that October 6th, 1837; though this, occurring earlier in the season, when there is much less open cotton in the fields, will not likely prove so disastrous." Apart from setting the historical context, these side-by-side comparisons enable a more accurate interpretation of damage that occurred in the immediate area.

Newspaper reports can have their drawbacks. They catered to their audience, focusing on areas within the city or reports from the wealthier citizens scattered throughout the country and along the coast. Newspapers were not immune to inaccurate or highly suspect reports. One such example came from Fayetteville, North Carolina in August of 1850 in which it was reported that "A violent storm occurred in North Carolina on the 7th inst. The tide rose higher at Fayetteville than has been known for fifty years" (*New Orleans Picayune*, 17 August 1850c, p. 3). While this account may be entirely accurate, it is highly improbable that this resulted from a hurricane. When taken in the context of other regional damage and meteorological reports, it is likely that this damage resulted from a separate severe weather event, such as a tornado or straight line winds, unassociated with a hurricane and thus could not reasonably be interpreted as a hurricane observation.

2.2 Diaries and Journals

Descriptive accounts are also found in unpublished sources including personal correspondence, diaries, and journals housed in local archives such as, the South Carolina Historical Society and South Caroliniana library. These data were often daily journals of farmers and plantation owners who lived in rural areas along the coast and throughout the state of South Carolina. Accounts from these sources help to fill in the geographic gaps between larger urban areas that were covered by newspaper reports. When these unpublished source materials originated in more highly populated areas such as Charleston, South Carolina they are very valuable as an independent source that can be used to verify or call into question information found in other published materials such as newspapers. These descriptions generally went into great detail about the damage that occurred in the immediate vicinity of the writer's property. References to previous hurricanes were also found in these sources, which aid in understanding the difference in storms' impacts at varying locations.

2.3 Meteorological Data

Meteorological data came from both instrumental and documentary records of weather conditions. Instrumental records were kept by a wide range of observers including: the United States Army Signal Service, which recorded weather observations (Fleming, 1990) at fixed times according to standard procedures; private citizens, who recorded weather observations by standard and non-standard methods at varying times; and ships at sea, whose observations were generally published in newspapers only in the case of extreme weather events.

Official government weather records were recorded at fixed times during a 24 h period. The number of observations varied from three to five times daily, and the hours of observation varied between records, but generally began in the morning around sunrise and ended between 9:00 PM and 11:00 PM. These instrumental records are of great value because of the regularity with which they were recorded, and the general reliability of the record. One drawback of these records is the lack of coverage of some potentially important times during the height of individual storms. Occasionally the records would include extreme values as the last observation, but not in most cases. As a result, key barometric pressure values or wind speeds may not be included in the record (e.g., Chenoweth and Landsea, 2004).

Newspapers occasionally published detailed meteorological records taken by the official weather observer during the storm. The published records typically included wind speed and direction, barometric pressure and occasionally temperature. These reports are a valuable supplement to official records kept by the Signal Service or Weather Bureau recorders because newspapers were not restricted to publishing information from the standard observation times of the official records.

Specific meteorological descriptions of weather conditions were also found in archival materials such as journals, diaries, and correspondence (Sandrik and

Landsea, 2003; Mock, 2004). These sources are very valuable for reconstructing the general path and intensity of hurricanes as they made landfall (e.g., Chenoweth, 2003) and traversed the landscape. These accounts could be as general as indicating wind direction, or as detailed as including information about temperature, precipitation, wind speed and wind direction. The major value in these records is to fill in the geographic gaps in the official and published records and to supplement standard meteorological observations. Accounts that included wind speeds and directions were valuable in reassessing current track and landfall location estimates.

3 Methods

3.1 Track Reconstruction at Sea

The first task in reconstructing or reanalyzing the track of historical hurricanes is to compile all the pertinent weather records into files that can be used to create daily weather maps. These maps are plotted by hand and are crucial in estimating the location of a hurricane from all available observations. In order to reconstruct a reliable track chronology of a hurricane one must have several meteorological observations surrounding the hurricane over time. Ideally, every observation would have the following: latitude/longitude, barometric pressure, air temperature, date, time of day, wind direction, verbal wind intensity estimate, storm duration, damage description of the ships merchandise or goods, and a structural ship damage description. Often times, however, the available ship reports contained only a few of the aforementioned storm observations such as location, date, wind direction and estimated wind strength. Given that a hurricane's wind field is cyclonic and symmetric, the approximate center is located at a 20° angle to the inflow of the hurricane (e.g. a southeast wind suggests a storm center to the west of the observation) (Jelesnianski, 1993; Landsea et al., 2004). The central location of a hurricane can be found by analyzing the surface wind direction of multiple observations. Multiple wind direction observations are also valuable in understanding the direction of storm movement. In the absence of wind direction, the storm track is estimated to be near the location of multiple hurricane observations.

3.2 Intensity Estimation at Sea

Intensity estimation is more difficult to quantify than track estimation, especially for hurricanes in the open ocean. Most weather accounts, especially from ships at sea, were not required to make instrumental observations during the early half of the 19th century. During the 1850 hurricane season only a handful of the weather records contained barometric pressure, thus most intensity estimations were conducted through analysis of verbal wind classifications (i.e. gale, violent storm, hurricane, tremendous hurricane, violent hurricane, etc.). Typically, verbal wind intensity estimates of "gale" equated to tropical storm intensity, "hurricane" equated to minor hurricane intensity (Saffir-Simpson Category 1–2), and "tremendous" or "violent" hurricanes equated to major hurricanes (Saffir-Simpson Category 3–5). It should be noted, however, that the designation of major hurricane status is difficult unless there are substantial accounts to indicate major hurricane intensity (Landsea et al., 2004). Several of the weather accounts, mainly those in the direct path of a hurricane and especially on land, used the terminology "hurricane", "tremendous hurricane", or "violent hurricane". The most frequently occurring verbal wind intensity estimates was the term "gale", and because of this, multiple gale observations were needed to confirm tropical cyclone occurrence.

3.3 Landfall Reanalysis Data Classification

The two primary indicators of the intensity of historical tropical cyclones in South Carolina are (1) damage caused by the force of the wind, and (2) coastal flooding associated with storm surge. All of the damage accounts were classified and standardized before being mapped and analyzed. Damage accounts ranged from very broad descriptions covering large geographic areas to very specific accounts of damage reported for individual structures or trees. The first step was to classify each account into categories relating to storm surge flooding, wind damage, or general damage.

The first category was limited to accounts of storm surge flooding along the coast or coastal creeks and rivers. All accounts of flooding were included in this category regardless of the scale or level of detail. This was done because there were generally fewer accounts of storm surge, and even general descriptions were useful in understanding the size, strength and track of the tropical cyclone as it moved along the coast. In highly populated areas such as Charleston, South Carolina, the depth of storm surge was often described in great detail and could be coupled with published reports of the high and low tide occurrences.

A second category of damage was created for general damage descriptions. These descriptions were considered general due to the lack of detailed damage information, for the broad geographic area that they described, or for damage accounts that could not be clearly attributed to either the wind or storm surge. As an example, the following account comes from the *Charleston Mercury* (11 September 1854b, p. 2) "On the Back Beach the damage is universal, all of the houses having suffered more or less." This could not be categorized as either wind or flood damage due to the vague language used in the description and because the houses were located on Sullivan's Island which experienced storm surge flooding. Thus, the damage could have been caused by either storm surge or the wind.

The third wind damage category included all accounts of damage that could be directly attributed to the force of the wind on a particular object. For example, the following account was a typical wind damage report that clearly highlights the force of the wind, "A large portion of the tin roof of the Charleston Hotel was ripped off" (*Charleston Mercury*, 9 September 1854a, p. 2). Wind damage accounts were analyzed and classified by the severity of damage. Three classifications were utilized: Slight, Moderate, and Severe. These classifications are based on the Saffir-Simpson hurricane scale (Simpson, 1974) and the modified Fujita scale developed by Emory Boose and colleagues (Boose et al., 2001; Boose et al., 2004). Broader classifications of damage severity were used because the Saffir-Simpson scale accounts for the geographic extent of damage within its categories. Individual damage reports cannot directly be assigned to a Saffir-Simpson category because they represent a specific instance of damage at an instantaneous point in time. The process of estimating the Saffir-Simpson intensity classification can only be performed after multiple reports have been classified and mapped.

3.4 Damage Mapping

The data used in these case studies were collected throughout the last 150 years. In that expanse of time, geographic names have changed, communities have moved, and addressing systems (e.g. methods for identifying specific locations) have undergone numerous changes. Some of this has been well documented and some has not. Mapping a discrete damage account often involved multiple sources and a great deal of time. Most regional damage accounts were located using a standard modern atlas. Additionally, the United States Geological Survey (USGS) Geographic Names Information System (GNIS) (http://geonames.usgs.gov/) website is an invaluable tool in locating some of the lesser known locations or historic areas. Other historical sources of geographic information were utilized for locations that were not found in conventional channels. Using these sources, regional wind damage and storm surge flooding maps were produced for the 1854 Great Carolina Hurricane. Accounts of flooding were mapped with confidence at the local scale even in the absence of specific addressing information because of the way flooding accounts were reported in newspapers and other hurricane accounts (e.g. water depth and street name or intersection).

3.5 Storm Surge Analysis

The regional scale aspect of this study focuses on the geographic extent and description of coastal flooding reports. Specific descriptions of surge heights were rarely reported, with the exception of highly populated areas such as Charleston, SC. For this reason, only local scale comparisons of modeled storm surge heights and observed storm surge heights were performed. Comparisons were made between the descriptive flooding accounts and the SLOSH (SLOSH, 1999) (Sea, Lake, and Overland Surges from Hurricanes) model output. Many coastal barrier islands are subject to inundation during Category 1 or 2 hurricanes. Descriptive coastal accounts from newspapers, personal diaries, or personal correspondence often noted whether the island was completely or partially submerged. This information could then be used to make estimations of storm surge heights and the corresponding Saffir-Simpson intensity. In the city of Charleston, SC where higher resolution data were used to create more detailed maps, we were able to make direct comparisons between observed flooding and modeled flooding based on SLOSH. This comparison was then used to estimate the corresponding Saffir-Simpson intensity category.

3.6 Wind Damage Analysis

Regional wind damage reports were mapped throughout South Carolina and the United States Southeast for the 1854 Great Carolina Hurricane. Wind damage maps clearly illustrate the spatial distribution of different damage severity levels. Broad areas of similar damage severity were examined more closely to determine the approximate Saffir-Simpson intensity level of damage. This process included a re-examination of each damage report within the context of the original source material, as well as other damage reports in the area. For instance, widespread reports of severe damage were examined for both structural and vegetative impacts. Severe damage reports were also checked for individual versus a coincidence of multiple sources. Hurricane intensity was categorized as major (i.e. Category 3 or greater) in cases where damage maps showed widespread severity and context analysis revealed the reports to be accurate and reliable.

3.7 Hurricane Intensity Estimation over Land

During the 1854 Great Carolina Hurricane, intensity estimates were conducted using the Saffir-Simpson intensity scale for different geographic regions along the South Carolina Coast and Charleston, SC. The distinction between areas that experienced major hurricane intensity versus non-major hurricane intensity was generally unambiguous. However, it was often difficult to distinguish between strong Category 1 intensity and weak Category 2 intensity or Category 3 versus Category 4, even with multiple data types and sources. In Charleston, SC instrumental meteorological data and high-resolution storm surge flooding accounts improved the confidence of intensity estimates for the Charleston area and the surrounding islands.

4 Results

4.1 Reconstruction Case Study of the 1850 Hurricane Season

The 1850 hurricane season was previously analyzed by Chenoweth (2006) and Bossak and Elsner (2004), who expanded on the work of Ludlum, (1963) by identifying five and three tropical cyclones, respectively. Our reconstruction results

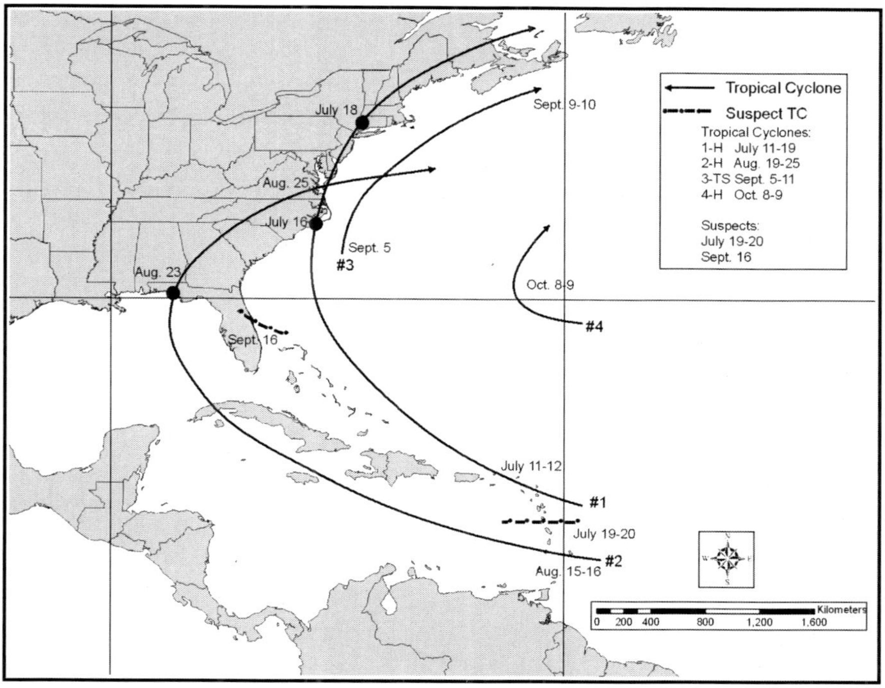

Fig. 1 Track map of the 1850 hurricane season with estimated storm strength and potential suspect storms noted. *Solid (dotted) lines* indicate confirmed (suspect) tropical cyclones

indicate that at least four tropical cyclones occurred in 1850, with other two systems being potential or "suspect" storms (Fig. 1). A brief description of each storm's track and intensity will now be presented.

4.2 Individual Tropical Cyclone Descriptions

Storm 1, July 11–19: The first storm originated in the northeast Caribbean Sea on the 11th and 12th of July, being observed on the islands of St. Kitts, Antigua, and St. Martins with gales and a wind shift from north to southwest (*New Orleans Picayune*, 13 August 1850a, p. 4). The change in wind direction suggests that the storm traversed the Lesser Antilles in a west to west-northwest direction. On the 14th and 15th, verbal indications of "heavy gales" and "hurricane" were observed near 29.5°N, 69.5°W (*Charleston Courier*, 29 July 1850a, p. 3). By July 16th and 17th, gales were experienced near Cape Hatteras, North Carolina with a report to the west of this location noting, "At Newbern and Washington (N.C.) the effects of the late Gale were very disastrous. The tides in Newbern rose 2, 3, and 4 ft in the stores and dwellings on and near the wharves, sweeping away and injuring naval stores, groceries, lumber, salt, &c..." (*Wilmington Chronicle*, 31 July 1850a, p. 2).

Storm data suggest that after landfall near Cape Hatteras, the storm continued on a northward track, striking the New England states on the 17th and 18th. On the 17th, at Fort Monroe near Norfolk, VA a peripheral barometric pressure of 29.067 in. (29.338 inches corrected) was observed. A peripheral pressure of 29.34 in. or 993 millibars equates to surface winds of 59 knots based on the Atlantic pressure-wind relationship (Landsea et al., 2004).

Fort McHenry, MD observed wind shifts from southeast on the 17th to north-northwest with 1.30 in. of rain by early morning on the 18th. Baltimore, MD experienced southeast gales and an unusual high tide on the 17th (*New York Herald*, 19 July 1850a, p. 7). On the 18th Philadelphia, PA experienced gales with a wind shift from southeast to northeast and in Brooklyn, New York "prostrated trees" were observed in the area (*New York Herald*, 20 July 1850b p. 7). On the 19th Fort Adams, RI observed winds shifting from southeast to southwest with a minimum peripheral barometric pressure of 29.732 in. (29.636 in. corrected). A peripheral pressure of 29.64 in. or 1003 millibars is equivalent to surface winds of 44 knots, based on the Atlantic pressure-wind relationship (Landsea et al., 2004). Additional instrumental weather records from the 16th–19th July exhibited temperatures above 70°F (21°C) from South Carolina to New Hampshire and an absence of any significant temperature gradients at the local (storm) scale which maintained the barotropic nature of this storm. Nearly 40 newspaper articles and 15 instrumental records depict the progression of this hurricane from landfall in North Carolina to traversing the Northeastern United States. Available weather accounts suggest that this tropical cyclone was a minor hurricane when it struck the United States and is tentatively assigned Category 1 Saffir-Simpson hurricane intensity.

Storm 2, August 19–25: The second tropical cyclone of 1850 was first observed near the Windward Islands on 15–16 August. Newspaper reports on the 19th and 20th noted this very intense storm as a "very violent hurricane" (*New Orleans Picayune*, 15 August 1850d, p. 4). However, these data occur too late to match with gale and hurricane observations from two days later in the Gulf of Mexico and in Florida. Winds shifted from northeast to southeast during the afternoon on the 21st at Fort Brooke (Tampa) Florida. On the next day, cooler daytime temperatures (minimum of 77°F (25°C) at 3 p.m.) were observed at this location with rainfall of 0.35 in. (8.89 mm), and "heavy squalls from South about 12 h" as the storm tracked northward just a few hundred miles west of Tampa, Florida. The tropical cyclone was observed at sea again on the 22nd as gales were experienced near Tortola (*New York Herald*, 29 September 1850d, p. 4). On the 23rd, near 26°N, 85.8°W, a ship observed the storm as a "hurricane" with a wind shift from east to south.

Data surrounding landfall indicated that the storm struck the panhandle of Florida, just east of Pensacola late on the 23rd when hurricane conditions were observed from Pensacola to St. Marks and Apalachicola (*Mobile Commercial Register*, 30 August 1850, p. 3). Mount Vernon Barracks, near Mobile, AL, observed temperatures in the 70–80°s F (21–27°C) and a wind shift from the east-northeast to northwest throughout the 23rd and early on the 24th. At Fort Pascagoula in Pascagoula, Mississippi, barometric pressures fell steadily throughout the day on the 23rd reaching an observed minimum of 29.35 in. or 994 millibars (corrected).

A peripheral pressure of 994 millibars equates to 60 knot surface winds based on the Atlantic pressure-wind relationship (Landsea et al., 2004). Given that Pascagoula is roughly 100 miles west of the estimated landfall location, storm intensity at landfall is assigned at least Category 2 Saffir-Simpson hurricane intensity (85–90 knots).

Southeasterly gales were observed on the 24th and 25th across the Southeastern US states of Georgia, South Carolina and North Carolina, before the storm exited out to sea near the Virginia/North Carolina border. On the night of the 24th, Wilmington, North Carolina reported "...a heavy gale from the southwest, accompanied with heavy rain" (*Wilmington Chronicle*, 28 August 1850b, p. 2). On the 26th at Fort McHenry in Maryland, daytime temperatures stayed in the 70°s F (21°C) and 2.50 in. (63.5 mm) of precipitation occurred with remarks of "N.E. storm from 2 a.m. to 9 a.m.". Also early on the 26th, Fort Monroe near Norfolk, VA observed wind shifts from northeast to northwest and a minimum pressure of 28.88 in. or 978 millibars (corrected). A peripheral pressure observation of 978 millibars is equivalent to 75 knots based on the Atlantic pressure-wind relationship (Landsea et al., 2004). Instrumental weather records across the Southeast on the 24th–26th indicate that the storm maintained tropical characteristics during and after landfall, based on time-series analysis of barometric pressure falls and warm air temperatures devoid of significant temperature gradients. Over forty newspaper articles and ten instrumental records describe the life cycle of this hurricane across the Southeast. All available weather accounts suggest that this tropical cyclone was a strong minor hurricane and is tentatively assigned Category 2 Saffir-Simpson hurricane intensity.

Storm 3, September 5–11: The third storm of 1850 originated southeast of Cape Hatteras, North Carolina, on September 5th. Several gales were observed off Cape Hatteras on the 6th, the Capes of Delaware on the 7th, Cape Cod, Massachusetts on the 8th, and the coast of Newfoundland on the 9th and 10th (*Charleston Courier*, 12 September 1850, p. 3). Newspaper reports of this system contained several gales with wind shifts from east to north as well as southeast to northwest, indicating that the tropical cyclone stayed off shore for the duration of its existence. On 6–7 September, Fort Monroe, VA observed a wind shift from southwest to east without any substantial pressure drops below 1010 millibars. Fort Mifflin, near Philadelphia, PA observed an identical wind shift with remarks of "heavy squall from the west at 3 p.m. accompanied by wind, rain, and thunder". It is hypothesized through analysis of instrumental weather observations over land, that a frontal system approached the East Coast of the US on the 5th which steered the storm away from the coastline and out to sea. Over 40 newspaper articles illustrate the track of this storm at sea, highlighting the storm's impact on shipping traffic in the northwestern Atlantic Ocean. All available weather observations suggest that this storm was at least tropical storm in strength, with the possibility of being upgraded to a minor hurricane should future archive data indicate a stronger tropical cyclone.

Storm 4, October 8th–9th: The fourth tropical cyclone of 1850 was observed in the open ocean east of Bermuda on the 8th and 9th of October. First observations noted this storm as a "severe hurricane" on the 8th near 30°N, 60°W (*The London Times*, 12 November 1850, p. 8). On the 9th the storm was observed near 31.5°N, 63.5°W as a very heavy gale with a wind shift from south-southeast to northwest.

This storm tracked very close to Bermuda without making landfall there. A total of nine newspaper articles observed this open ocean storm. All available weather accounts suggest that this tropical cyclone was a minor hurricane and is assigned Category 1 Saffir-Simpson hurricane intensity.

4.3 Tropical Cyclone Suspects in 1850

The next group of storms is considered to be potential or "suspect" storms because there is either currently insufficient substantiating evidence to determine the tropical nature of the weather system or too few data points to indicate full tropical cyclone classification. Frequent early and late hurricane season weather systems contain gales, but with low resolution data coverage it is unclear whether the gales are tropical or non-tropical. Thus, these storm systems are considered potential tropical cyclones and are included on the suspect list. The baroclinic or barotropic nature of each suspect storm was also determined. There were two tropical cyclone suspects observed by ships at sea during the 1850 hurricane season. Additional weather reports off South Carolina in late September suggest another suspect storm, but time-series analysis of land-based air temperatures and atmospheric pressures objectively establish it as non-tropical in nature. This storm is included to illustrate the process of differentiating between tropical and non-tropical storms. Future archival research may reveal undiscovered weather observations of these suspect storms, which may lead to their removal (addition) due to non-tropical (tropical) characteristics.

The first tropical cyclone suspect was observed in the Lesser Antilles on the 19th of July. Two newspaper accounts (i.e. *Charleston Courier*, 23 August 1850b, p. 3) indicate that the island of Dominica experienced gales on this date; however, this account cannot be substantiated by any additional data sources or linked to any existing storms. Due to the low latitude of the island and frequent occurrence of tropical cyclones in this region, this potential tropical cyclone will remain a tropical suspect. The second suspect storm was documented in three newspaper accounts and observed off Cape Canaveral, Florida as a heavy gale on the 16th of September (*Charleston Courier*, 20 September 1850, p. 3). Fort Brooke in Tampa, FL observed a slight decrease in barometric pressure and air temperature, a wind shift from southeast early on the 15th to northwest late on the 16th, and rainfall of 0.40 in. (10.2 mm) with remarks of "heavy squalls during the evening". Due to a lack of substantiating evidence, this system will also remain a tropical suspect, as tropical and non-tropical gales are potentially observed off the Florida coast during the month of September.

At first glance, a third potential tropical cyclone occurred about 125 miles southeast of Charleston, SC on September 30th. Three newspaper accounts record the effects of this storm. Newspaper reports document significant northeast winds with the system that stayed off shore. There is a lack of wind observations from each quadrant (i.e. lack of any winds with a southerly component) and analyses of a ship logbook and land-based temperature records note that temperatures cooled well below 70°F (21°C) during the time of highest winds and closest proximity to the South Carolina coastline. These data are indicative of a frontal system passage, therefore, non-tropical characteristics. Moreover, the barometric pressure observations greater than 30.00 in. or 1015–1020 millibars observed on the Bark Fenelon (National Archives and Records Administration, 2008), bound from New York to Apalachicola, were higher than normally observed within tropical cyclones. Wind observations aboard the Fenelon during the storm showed southwesterly flow early on the 28th, becoming northerly by late on the 29th. The Fenelon's wind direction remarks also suggest the passage of a frontal system just before the peak wind observations southeast of Charleston, SC. This storm is not classified as a tropical suspect but is instead a likely extratropical cyclone moving off the Georgia/Carolina coastline.

4.4 Reanalysis Case Study of the Great Carolina Hurricane – September 8, 1854

4.4.1 Meteorological Observations and Storm Track

This storm is known as the Great Carolina Hurricane (Tannehill, 1938). The current official HURDAT track of this storm shows landfall in the vicinity of St. Catherine Sound, Georgia (Jarvinen et al., 1984; Landsea et al., 2004). It then traversed inland crossing into South Carolina between Allendale, South Carolina and Augusta, Georgia. Upon entering South Carolina the storm turned to the northeast and weakened as it passed over Columbia, South Carolina. It continued to move northeastward through North Carolina and emerged back into the Atlantic Ocean near the Virginia border. This track is supported by numerous instrumental and documentary weather observations. The following is a newspaper account of the winds experienced in Savannah, Georgia from September 7–9:

> It began to blow on the evening of the 7th and continued throughout the night, but from 10 o'clock on the morning of the 8th until 4 o'clock pm it blew a perfect hurricane. In the evening it changed to the southeast, blew heavily in the quarter for some hours, then toward morning it hauled to the south and southwest where it continues (*Charleston Mercury*, 12 September 1854c, p. 2).

Thomas Chaplin (Tombee Plantation journal, 1845–1886), a resident of St. Helena, SC recorded in his journal that the winds commenced blowing from the east on September 7th, with increased intensity through the 8th, and did not begin to subside until around noon on Saturday, September 9th. A similar account from the Beaufort district (Gignilliat Family Papers, 1828–1901) indicated high northeast winds on the 7th, increasing to a northeast gale in the 8th that shifted suddenly to the southwest with clearing skies on the following day.

At St John's, SC the Black Oak Agricultural Society (Black Oak Agricultural Society records, 1842–1925) reported winds from the northeast on September 7th, with clouds, wind and rain, with the same flow direction on September 8th winds. The remarks read, "Stormy. Gale from NE." Four tenths of an inch of rain was recorded. On September 9th the southeast wind, shifted to the west with one inch of rain recorded. In Charleston, South Carolina Samuel Wilson recorded his experience (Samuel Wilson, 1854). The rain and high winds from the east began in the forenoon on September 7th, increasing through the day blowing with great violence. The violent winds continued through the night and the next day finally subsiding on the 9th. On the 10th, Mr. Wilson reported a heavy rain with a gale of short duration. In Georgetown, SC, the *Charleston Mercury* (13 September 1854d, p. 2) reported the gale at northeast and southeast for 48 h and then at south-southwest for 12 h. This storm lasted for almost three days and was recorded as the longest duration storm in memory. These observations all support the current best track estimation in the HURDAT database.

4.4.2 Wind Damage

It is important to note that even as far south as Savannah, Georgia wind damage associated with this hurricane was generally not severe. There was very little structural damage; usually only the roofing material was stripped off buildings. However, one case of major structural damage was reported in the *Charleston Daily Courier* (9 September 1854, p. 2) when a two storey wooden building on King Street, Charleston, South Carolina was blown down. The remaining structural damage was classified as slight to moderate, limited to damage to roofs and the destruction of fences. Numerous accounts of damage to trees were reported from Savannah, Georgia to Charleston, South Carolina. The following is a description of the scene in Savannah:

> We passed through a portion of South Broad street about dusk. It presented a melancholy spectacle. From Abercorn to Bull street nearly every tree is blown down. The few that remain standing are limbless and leafless—nothing but naked trunks remain to tell the effects of the gale. It will take years to replace the beautiful trees which now lie prostrate in that street (*Savannah Daily Morning News*, 9 September 1854, p. 2).

In St. Helena, SC winds uprooted trees and blew down fences. In Charleston, South Carolina there were very few trees blown down, but the foliage was stripped from many trees. Inland there were no reports of significant wind damage. In Branchville, South Carolina the effects of the storm were classified as very minimal. Most locations throughout South Carolina indicated high winds, but no notable damage, thereby relegating the hurricane to Category 2 type impacts along the coast, close to the Georgia/South Carolina state border. One exception was observed in Charleston, South Carolina where, although the foliage was completely stripped from the trees (suggesting more intense winds), other features in the area were not similarly damaged. It is likely that the stripping of foliage from vegetation resulted from less intense winds of longer duration rather than more intense winds of shorter duration. Wind damage in Charleston was consistent with Category 1 impacts.

4.4.3 Flooding

Storm surge flooding associated with this storm was extensive. Flooding was reported from Savannah, Georgia, to Georgetown, South Carolina. Hutchinson's Island, in the Savannah River just north of Savannah was completely submerged with only the roofs of houses and trees visible from the city. In the eastern part of Savannah, Georgia the water was several feet deep in the cotton yards and along the wharves. Farther to the north in Saint Helena Sound, much of Warren Island was completely submerged and the Coosaw and Combahee rivers merged to cover Buzzards Island. Reports from Edisto, South Carolina call this hurricane one of the most terrific in the collective memory of the inhabitants. The surge made a clean sweep over the beachfront in five or six places inundating hundreds of acres of high ground. The tide is said to have risen higher than in any previous storm including that of 1804 (*Charleston Mercury*, 13 September 1854d, p. 2).

In Charleston, SC the water on East Bay Street (Fig. 2) was up to 4 ft deep (1.2 m) in many places. The water backed up through Atlantic and Water Streets into Meeting Street, parts of which were covered to a depth of 2–3 ft (0.61–0.9 m). The western portion of Sullivan's Island was covered with water on Friday morning, September 8th, 1854, when the storm was at its height. On the beach fronting the Atlantic Ocean the four cottages that made up "Tennessee Row" were completely swept away, while other buildings were swept away or damaged by the storm surge. The *Charleston Mercury* (9 September 1854a, p. 2) reported the appearance of Sullivan's island as "very dismal" following the storm.

The storm surge was still substantial all along the coast to the mouth of the Santee River in the north. Every bridge between South Island and the Santee ferry was carried away by the surge. There was also significant damage to the rice crop along the Santee, Pee Dee, and Waccamaw rivers. In Georgetown, South Carolina the surge was reported to be as high as that of the hurricane of 1822. Salt water was pushed upriver as far as anyone could remember, greatly damaging to the rice crop [see Tuten, this volume for a discussion of the demise of the rice industry in this region]. Storm surge around Savannah, Georgia was very likely in the Category 3 range. This hurricane is officially listed in HURDAT as a Category 3 hurricane at landfall in northern Georgia, and no evidence was found to refute that classification. Surge heights in Charleston, South Carolina were consistent with Category 1 impacts. If reports from Hutchinson's Island and Savannah, Georgia are accurate, it is very likely that this was a Category 3 hurricane with Category 2 impacts in southern coastal South Carolina and Category 1 impacts in Charleston, South Carolina further north.

4.4.4 Local Analysis

This section examines the extent of flooding in the city of Charleston, SC during the Great Carolina Hurricanes of 1854, using higher resolution data from the city of Charleston mapped at block level resolution. By working at this scale, highly accurate maps depicting observed flooding in Charleston were generated (Fig. 2). These

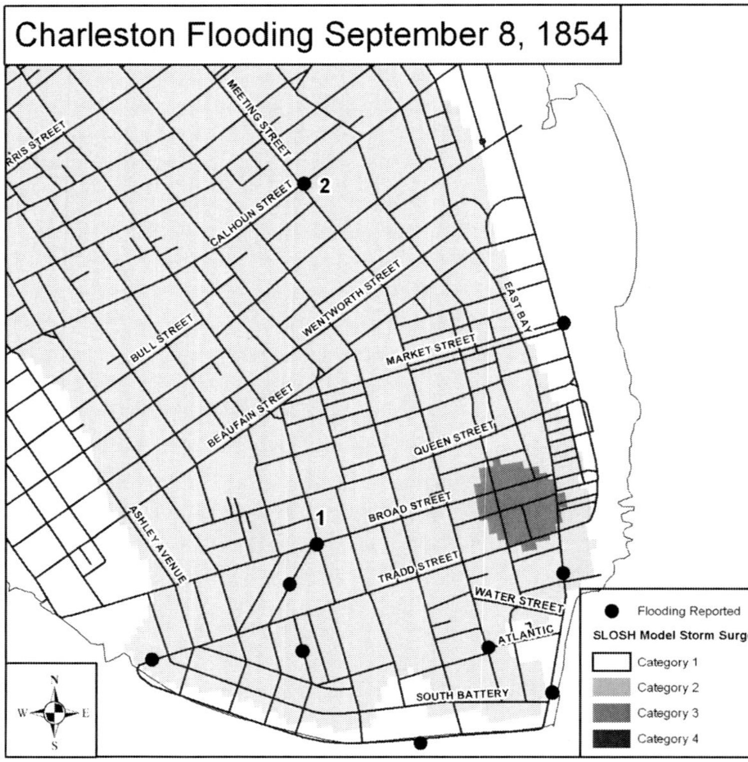

Fig. 2 Flooding map showing reported locations and extent of storm surge flooding within the city of Charleston, South Carolina compared to expected storm surge flooding based on modern SLOSH model. Location (1) shows the furthest extent of flood waters reported on the southern and western side of the city. Location (2) shows where the flood waters extended up Calhoun Street to Meeting Street

maps were then overlaid and compared to storm surge models for the Charleston area. Examining storm surge at the local scale also illustrates how hurricanes of different intensities and magnitudes that come ashore over a large geographic area can have similar and significant impacts at the local level. Flooding in the city of Charleston during the hurricane of September 8, 1854 was significant. Extensive damage was done to the wharfs, piers, and the accompanying store houses along East and South Bay streets. Much of the Battery sea wall was completely washed away, with floodwaters rising to about the top of the sea wall, and estimated to be 4 ft (1.2 m) deep in East Bay Street. The location of many of the reports of flooding came from areas that were near the border of expected Category 1 and 2 storm surge. Two reports, however, are strong evidence of Category 2 flooding within the city. The floodwaters were uninterrupted to the west and south of location 1 (Fig. 2). On the opposite side of Charleston neck, location 2 shows where the flood waters extended up Calhoun Street to Meeting Street. This comparison shows that the storm surge associated with this hurricane was clearly in the Category 2 range in the city of Charleston, South Carolina (Fig. 2).

5 Discussion

This research plays a valuable role in supplementing our understanding of tropical cyclones currently found in the HURDAT database, and their physical, social and economic impacts on coastal communities. By understanding the physical impacts of past hurricanes we can better understand the historical role of tropical cyclones on society. This knowledge can be incorporated into practical and theoretical planning applications including storm surge models such as the one developed by the CaroCOOPS (Carolinas Coastal Ocean Observing and Prediction System) project and coastal vulnerability studies (Cutter et al., 2000; Purvis and McNab, 1985). There are also implications for the hurricane history of a particular region. For example, the hurricane of September, 1854 has commonly been referred to as the "Great Carolina Hurricane." The evidence suggests, however, that the impacts of this hurricane were less than Category 3 in South Carolina. It can also contribute to better understanding and recognizing the susceptibility to coastal flooding in specific geographic locations.

This research focused on the 1850 hurricane season and the Great Carolina Hurricane that struck South Carolina in 1854, but a similar approach can be taken for numerous coastal regions of the Atlantic Ocean and Gulf of Mexico. The reconstruction example also demonstrates how the HURDAT database can be extended back in time, at least into the earlier decades of the 19th century. Reconstructing historical hurricane tracks and intensities is critical for understanding long term trends in hurricane variability. Longer temporal records of hurricanes can give a broader view of the vulnerability and risk of coastal communities to hurricanes. Moreover, the methodology used in this reconstruction can be applied to many other ocean basins and can allow researchers to create a more comprehensive and accurate global hurricane climatology. Reassessing weather and climate from documentary evidence does have some limiting factors that must be acknowledged. One factor involves the geographic coverage of the available data. The bulk of the reports often come from highly populated areas. Hurricanes typically come ashore in less populated areas resulting in uncertainty about exact location of landfall, as well as the intensity of the hurricane. Researchers can address these caveats by performing the analyses we have outlined, specifically noting the use of different types of information from many different sources.

Reconstructing climate often involves analyses at multiple scales. The methodology for reconstructing the 1850 hurricane season is a synoptic approach, centered on the utilization of large scale recreated daily weather maps. In this approach, one can discern the location of tropical cyclones and the extratropical systems that influence their movement. This type of analysis is invaluable in determining hurricane intensity at landfall and it can often provide more precise information about landfall location. Additionally, our research demonstrates the value of incorporating higher resolution regional and local level data into the analysis of historical hurricanes.

References

Black Oak Agricultural Society. Black Oak Agricultural Society Records, September 7–8, 1854. (1171.03.02) South Carolina Historical Society.

Boose, E. R., K. E. Chamberlin, and D. R. Foster. 2001. Landscape and Regional Impacts of Hurricanes in New England. *New England Ecological Monographs* 71: 27–48.

Boose, E. R., M. I. Serrano, and D. R. Foster. 2004. Landscape and Regional Impacts of Hurricanes in Puerto Rico. *New England Ecological Monographs* 74: 335–352.

Bossak, B. H. and J. B. Elsner. 2004. Plotting Early Nineteenth Century Hurricane Information. *Eos Transactions* 85: 199.

Chaplin, T. B. 1822–1890. Tombee Plantation Journal, September 7–9, 1854. (34/198) South Carolina Historical Society.

Charleston Courier. July 29, 1850a, p. 3. Charleston, South Carolina.

Charleston Courier. August 23, 1850b, p. 3. Charleston, South Carolina.

Charleston Courier. August 29, 1850c, p. 3. Charleston, South Carolina.

Charleston Courier, September 12, 1850d, p. 3. Charleston, South Carolina.

Charleston Courier, September 20, 1850, p. 3. Charleston, South Carolina.

Charleston Daily Courier. September 9, 1854, p. 2. Charleston, South Carolina.

Charleston Mercury. September 9, 1854a, p. 2. Charleston, South Carolina.

Charleston Mercury. September 11, 1854b, p. 2. Charleston, South Carolina.

Charleston Mercury. September 12, 1854c, p. 2. Charleston, South Carolina.

Charleston Mercury. September 13, 1854d, p. 2. Charleston, South Carolina.

Chenoweth, M. 2003. The Eighteenth Century Climate of Jamaica Derived form the Journals of Thomas Thistlewood, 1750–1786. Philadelphia, PA: American Philisophical Society. 153 pp.

Chenoweth, M. 2006. A Reassessment of Historical Atlantic Basin Tropical Cyclone Activity, 1700–1855. *Climatic Change* 76(1–2): 169–240.

Chenoweth, M. and C. W. Landsea. 2004. The San Diego Hurricane of October 2, 1858. *Bulletin of the American Meteorological Society* 85: 1689–1697.

Cutter, S. L., J. T. Mitchell, and M. S. Scott. 2000. Revealing the Vulnerability of People and Places: A Case Study of Georgetown County, South Carolina. *Annals of the Association of American Geographers* 90: 713–737.

Fleming, J. R. 1990. Meteorology in America 1800–1870. Baltimore, MD: The Johns Hopkins University Press. 288 pp.

Fernandez-Partagas, J. and H. F. Diaz. 1995. A Reconstruction of Historical Tropical Cyclone Frequency in the Atlantic from Documentary Evidence and Other Historical Sources. Part I: 1851–1870. Boulder, CO: Climate Diagnostics Center, Environmental Research Laboratories. NOAA.

Gignilliat Family Papers, September 7–9, 1854. (10866). South Caroliniana Library, University of South Carolina.

Holland, G. J. and P. J. Webster. 2007. Heightened Tropical Cyclone Activity in the North Atlantic: Natural Variability or Climate Trend? *Philosophical Transactions of the Royal Society – A* 365(1860): 2695–2716.

Hoyos, C. D., P. A. Agudelo, P. J. Webster, and J. A. Curry. 2006. Deconvolution of the Factors Contributing to the Increase in Global Hurricane Activity. *Science* 312: 94–97.

Jarvinen, B. R., C. J. Neumann, and M. A. S. Davis. 1984. A tropical cyclone data tape for the north Atlantic basin, 1886–1983: Contents, limitations, and uses. NOAA Tech. Memo. NWS NHC 22, Miami, FL, 21 pp.

Jelesnianski, C. P. 1993. The Habitation Layer. Global Guide to Tropical Cyclone Forecasting. WMO/TC-No. 560, Report No. TCP-31, World Meteorological Organization; Geneva, Switzerland.

Landsea, C. W. 2007. Counting Atlantic Tropical Cyclones Back to 1900. *EOS* 88(1): 197, 202.

Landsea, C. W., C. Anderson, N. Charles, G. Clark, J. Dunion, J. Fernandez-Partagas, P. Hungerford, C. Neumann, and M. Zimmer. 2004. The Atlantic Hurricane Database Re-analysis Project: Documentation for the 1851–1910 Alterations and Additions to the HURDAT Database. In: Hurricanes and Typhoons: Past, Present and Future, R. J. Murname and K.-B. Liu, eds. New York: Columbia. 464 pp.

Ludlum, D. M. 1963. Early American Hurricanes: 1492–1870. Boston, MA: American Meteorological Society. 198 pp.

Mann, M. and K. Emanuel. 2006. Atlantic hurricane trends linked to climate change. *EOS* 87(24): 233–241.

Mobile Commercial Register. August 30, 1850, p. 3. Mobile, Alabama.

Mock, C. J. 2004. Tropical Cyclone Reconstructions from Documentary Records: Examples for Charleston South Carolina, USA. In: Hurricanes and Typhoons: Past, Present, and Future, R. J. Murnane and K. B. Liu, eds. New York: Columbia. 464 pp.

National Archives and Records Administration. 2008. 1792–1965. Records of the Weather Bureau. Records Group 27.5.5, Records of the Marine Division (1796–1861). College Park, Maryland.

New Orleans Picayune, 13 August 1850a, p. 4. New Orleans, Louisiana.

New Orleans Picayune, 15 August 1850b, p. 3. New Orleans, Louisiana.

New Orleans Picayune, 15 September 1850d, p. 4. New Orleans, Louisiana.

New Orleans Picayune, 17 August 1850c, p. 3. New Orleans, Louisiana.

New York Herald, July 19, 1850a, p. 7. New York City, New York.

New York Herald, July 20, 1850b, p. 7. New York City, New York.

New York Herald, September 3, 1850c, p. 3. New York City, New York.

New York Herald, September 29, 1850d, p. 4. New York City, New York.

Purvis, J. C. and A. McNab, Jr. 1985. Hurricane Vulnerability for Charleston County. *Climate Report No. G14*. South Carolina Water Resources Commission, Columbia, SC. 17 pp.

Sandrik, A. and C. W. Landsea. 2003.Chronological Listing of Tropical Cyclones affecting North Florida and Coastal Georgia 1565–1899. *NOAA Technical Memorandum NWS SR-224*, Southern Region, National Weather Service.

Savannah Daily Morning News. September 9, 1854 p. 2. Savannah, Georgia.

Simpson, R. H. 1974. The hurricane disaster potential scale. *Weatherwise* 27: 169–186.

SLOSH. 1999. FEMA, and EPD, Coastal SC SLOSH Inundation. Available online at: http://www.cla.sc.edu/geog/hrl/scemd_datadown.htm

Tannehill, I. R. 1938. Hurricanes, their Nature and History. Princeton, NJ: Princeton University Press. 257 pp.

The London Times. November 12, 1850, p. 8. London, England.

Webster, P. J., G. J. Holland, J. A. Curry, and H.-R. Chang. 2005. Changes in tropical cyclone number, duration, and intensity in a warming environment. *Science* 309: 1844–1846.

Wilmington Chronicle. July 31, 1850a, p. 2. Wilmington, North Carolina.

Wilmington Chronicle. August 28, 1850b, p. 2. Wilmington, North Carolina.

Wilson, S. 1791–1861. Samuel Wilson journal, September 7–10, 1854. (43/298) South Carolina Historical Society.

The Sitka Hurricane of October 1880

Cary J. Mock and Stephanie F. Dodds

Abstract This study reconstructed a unique and previously poorly understood storm known as the October 1880 Sitka Alaska Hurricane. Data were comprised of daily instrumental records from military posts, exploration surveys, and ship logbooks, as well verbal descriptions from diaries. Hourly pressure data from the USS *Jamestown* were corrected for elevation, latitude, and gravity, and a meteogram was constructed to assess the specific storm impact at Sitka. Pressure records reveal that the Sitka hurricane is clearly a very abnormal weather event, likely one of the strongest storms ever to strike western North America in the historical period. The plotting of data indicates that the storm track originated off Eastern Siberia and had no associations with any possible typhoon from the western Pacific Ocean.

Keywords Sitka · Alaska · Logbooks · USS Jamestown

1 Introduction

Strong coastal storms are a commonly recurring natural hazard for Alaska, at times causing coastal erosion, storm surges, coastal flooding, blizzards, and heavy snowfall. This is particularly evident for the Bering Sea and mid-southern Alaskan coastline in the fall, winter, and spring months, with some storms approaching hurricane-strength. Several case studies of strong storms have been conducted (e.g., Larsen et al., 2006). Mason et al. (1996) demonstrated that chronologies of strong coastal storms can be reconstructed from documentary evidence that extends back to the late nineteenth century for the Nome, Alaska area. Conversely, very little research has been conducted on strong storms for the southeastern Alaskan coast.

C.J. Mock (✉)
Department of Geography, University of South Carolina, Columbia, SC 29208, USA
e-mail: mockcj@sc.edu

S.F. Dodds (✉)
Department of Geography, University of South Carolina, Columbia, SC 29208, USA
e-mail: doddss@mailbox.sc.edu

L.-A. Dupigny-Giroux, C.J. Mock (eds.), *Historical Climate Variability and Impacts in North America*, DOI 10.1007/978-90-481-2828-0_7,

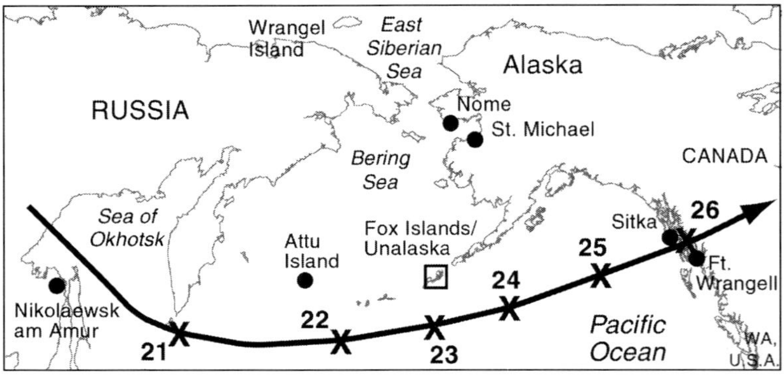

Fig. 1 Map of the study area, selected places mentioned in the text, and inferred track of the Sitka Hurricane. The numbers along the track indicate the beginning of specific dates in October 1880

This region is perhaps less susceptible to storm surges, and mid-latitude cyclones tend to be in the cyclolysis phase when traversing through the Northeastern Pacific Ocean (Martin et al., 2001). However, case studies of some strong non-winter storms that struck the United States Pacific Northwest (Lynott and Cramer, 1966) indicate that very strong storms occasionally occur in the region, comparable in strength with tropical hurricanes.

Of particular interest to southeast Alaska, is a storm that was observed on January 1, 1948 near Juneau, Alaska with a corrected sea-level pressure of around 959 mb (Burt, 2007) and somewhat lower at the Sitka Magnetic Observatory (*Daily Sitka Sentinel*, 1948, January 2, p. 2) – this is the record lowest barometric pressure known in the modern record (post 1900). Its storm impact, however, was of short duration and mostly from wind, with little precipitation and limited societal impacts (*Daily Sitka Sentinel*, 1948, January 2, p. 2).

This chapter describes the characteristics and storm history, as reconstructed from documentary and instrumental data, of a very unusually strong storm that made landfall near Sitka Alaska on October 26, 1880 (Fig. 1). This storm, termed the "Sitka Hurricane", is likely among the strongest known storm that ever impacted the region in the historical period.

2 Historical Data

Historical records have been extensively utilized to reconstruct case studies of major storms within the past several hundred years. These high resolution sources include ship logs, diaries, annals, and newspapers. Examples of reconstructions include the 1588 storm that hit the Spanish Armada (Douglas et al., 1978), the Storm of 1703 in southern England (Wheeler, 2003), and numerous case studies on hurricanes (e.g., Vaquero et al., 2008). In the late nineteenth century, some early instrumental records from Sitka, Alaska are available from the United States Navy, United

States Signal Service, and exploration surveys; mostly in terms of temperature, precipitation, wind speed, wind strength, cloudiness, written remarks, and pressure. Qualitative descriptions of wind direction, wind speed, rain, and snow provide clues of storm impact and characteristics. Wind strength followed the Beaufort scale and was recorded on a numerical scale from 0 to 12, with 0 indicating a calm condition and 12 indicating a violent storm. Corrected pressure data for gravity, latitude, and elevation (to sea-level) by the authors provide evidence of storm intensity and specific geographic location, including its "V-shaped bargraph" signature.

The historical data used to reconstruct the Sitka hurricane's characteristics and track were obtained from original archival sources where possible. These include several logbooks from the US National Archives (particularly the USS *Jamestown*), the Smithsonian Institution (the William Dall papers), several whaling logbooks from the New Bedford Whaling Museum Library, some published weather data from ships as given by the Monthly Weather Review (1882), weather data from an eastern Siberian location from Wild (1882), and a published Signal Service report (Turner, 1886).

3 The 1880 Sitka Hurricane in Alaska

Although reference has been made to a "hurricane" at Sitka, AK in 1880 in several Alaskan history books (e.g., Scidmore, 1893), details about the storm remained completely unknown until this study. The primary record documenting the specifics of the Sitka Hurricane in Alaska comes from the logbook of the USS *Jamestown*, commanded by Henry Glass. It was moored off the Sitka harbor during October 1880. The logbook recorded meteorological data on an hourly basis, enabling the formulation of a meteogram of weather conditions from October 22 to November 2, 1880 (Fig. 2). The authors carefully examined the hourly barometric data for the entire year of 1880 (the USS *Jamestown* rarely ventured out of port), finding no obvious discontinuities and the mean is close to standard sea-level pressure, clearly suggesting that the data are of high quality and standardized.

Corrected barometric data reveal relatively higher pressure, at or above 1010 mb, for October 22–24, followed by gradual decreasing pressure on October 25, and a sharp V-shaped bargraph on October 26. The V-shaped characteristics show the lowest pressure at 958 mb before midday. Winds generally were from the east at the beginning of October 25, changing to east/northeast and then to southeast/east-southeast following the lowest pressure reading. These wind shifts suggest that the storm center remained to the south of Sitka, and the 958 mb lowest pressure is likely a conservative estimate of the peak intensity. The pressure generally decreased from October 26 to the beginning of November, but remained below 1010 mb, indicating the instability perhaps typical behind strong cold fronts. Another V-shaped bargraph on October 31 represents the passage of another low pressure system.

Wind force was low from October 22 to 24, and it increased from October 25 to 26 as the Sitka Hurricane struck the region with a peak strength of force 12, the

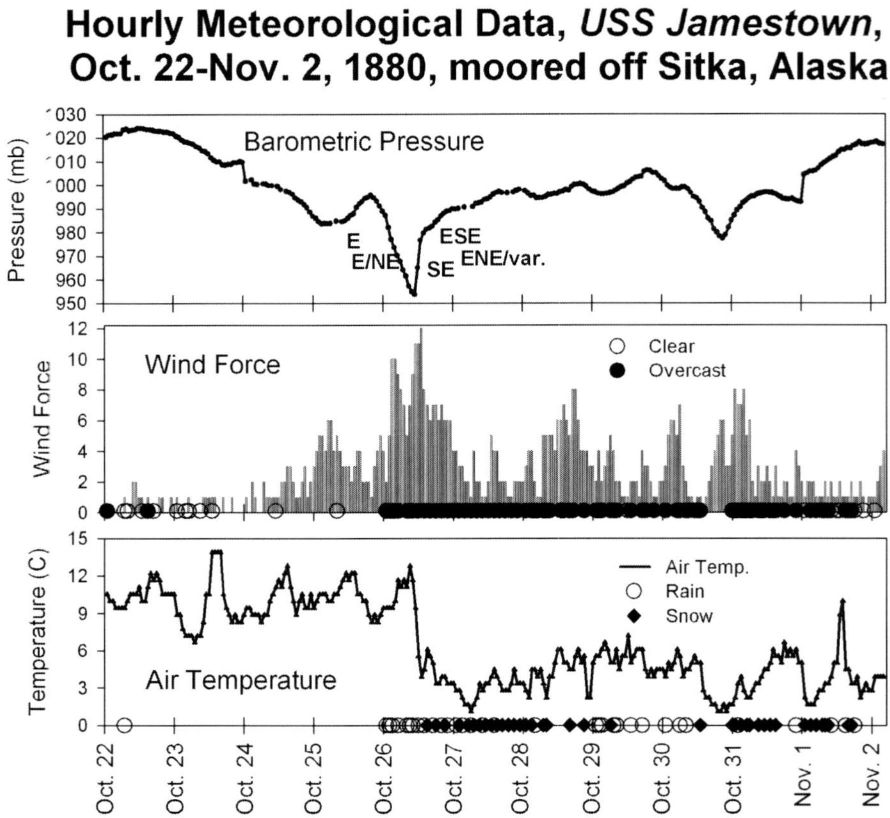

Fig. 2 Meteogram of USS Jamestown

highest possible. Air temperature remained relatively high prior to the likely cold frontal passage on midday of October 26. Precipitation started on October 26 as rain, eventually changing to snow by late afternoon. The following description of the weather from 8 am to noon was written in the logbook of the USS *Jamestown*:

> Weather first hour of watch partly clear. Wind in strong squalls from NE to ENE first two hours. At 9:15 wind increased in force to a strong gale, and continued until 11:15 at which time it shifted to S.E. increasing in force to very heavy gale. At 9:40 clouds and heavy mist arose to the NE and heavy nimbus clouds passed rapidly to the SW. At 10, became overcast and rain set in which continued until end of watch. Barometer falling rapidly until 10:40 at which time it commenced to rise and rose rapidly rest of watch. Wind during all the watch in squalls at short intervals. Starboard mizzen topsail sheet bitt [?] to which chain of quarter anchor is secured, twisted during a heavy squall. Ten [?] of air suddenly decreased when wind shifted to SE. Barometer reached its lowest 28.20 [uncorrected] at 10:35.

From October 27 to early November, relatively lower temperatures hovered around or slightly above the freezing mark (0°C), and were associated with mostly cloudy conditions, sporadic 4–6 h periods of gusty winds, and mostly alternating snow/rain conditions. These conditions are likely associated with a dominance of a cold air mass related to an extension of the polar jetstream southward of Sitka. In the days following the October 26 storm, the snowfall was "heavy" and "wet" at times. Like the pressure readings, the alternating snow and rain squalls were probably indicative of instability of mesoscale features commonly occurring behind cold fronts (Carleton, 1991).

The only other weather observation available for analysis was taken from the s.s. "Princess Alice", moored south of Sitka at Fort Wrangell, Alaska (Monthly Weather Review, 1882). Early on October 26, it reported an E.S.E. gale, followed by persistent strong southerly winds which suggest the low pressure center stayed to the west. Its lowest barometric pressure observation in the late morning was 967 mb. Given its distance from Sitka of about 215 km, and assuming that the storm center was likely very close to Sitka, a possible lowest pressure in the center within the 952–956 mb range is quite feasible.

4 Track and Genesis of the October 1880 Storm

Although North Pacific historical data are relatively sparser compared to other data networks in the world (e.g., the North Atlantic Basin), enough exist to construct the origin and track of the Sitka Hurricane (Fig. 1). Arnott (2005) implied that some typhoon remnants from the Western Pacific go through extratropical transition, become strong coastal non-tropical systems and strike Alaska. This situation was the case of the famous Columbus Day storm in 1962 for the Pacific Northwest (Lynott and Cramer, 1966) but it does not appear to be the case for the 1880 Sitka storm. The logbook of the USS *Palos*, based at Shanghai, China (not shown on the map) for a few weeks prior to October 26, reveal pressures consistently at mean sea level or higher, thus showing no indications of a tropical system. The logbook of the USS *Alert*, based within the same region, shows similar results. Both of these logbooks reveal evidence of tropical cyclones in late September and the first few days of October, but the timing of these storms is too far apart to be related to the Sitka hurricane. Garcia-Herrera et al. (2007) describe a typhoon striking the Philippines around October 18–22, but the track of this storm stayed near the southeast Asia mainland.

Daily weather data from the USS *Jeannette*, located east of Wrangel Island and northwest of the Herald Islands in the East Siberian Sea, show no distinctive trends in barometric pressure and weather that would suggest any potential for formation from the far north. A weather station in southeastern Siberia, (Nikolaewsk am Amur), recorded daily weather data with 7, 9, and 1 fixed hour times (Wild, 1882). Plots of daily temperature reveal that a distinctive push of cold air from the northwest and west occurred around October 16–20, likely reflecting the passage of the southern portion of a cold front (Fig. 3). This was followed by a decrease in surface barometric pressure on October 21–23, which may be related to baroclinic activity and cyclogenesis further east. This information suggests the possibility of the Sitka hurricane's origin being to the north of Nikolaewsk am Amur, but the formation was occurring well off Northeast Asia as a strong extratropical system and moving

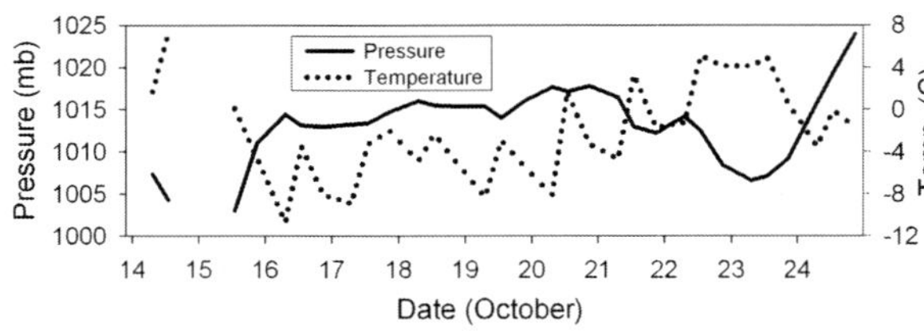

Fig. 3 Pressure (converted to mb, but not corrected to sea-level given the lack of metadata) and temperature (Celsius) at Nikolaewsk am Amur, Russia during October 1880. The fixed observation times each day were at 7, 1, and 9 h (Wild, 1882)

eastward. Snow was reported in the records on the 23rd, but this likely represents instability perhaps on the back side of a longwave trough well to the west of storm cyclogenesis.

Unfortunately, a spatial data gap exists in the northwest Pacific, with the closest evidence of a storm around Chichagof harbor, on Attu Island (52°51′ N, 173°11′ E). A weather abstract lists an October 1880 lowest pressure of 28.992 in. (uncorrected, about 980 mb), but no date or detailed metadata on the record is listed. However, it is assumed that the value is related to the storm, given its extreme event nature, and that it was likely undergoing strengthening.

Several weather observations describe specifics on the storm as it struck the central and eastern Aleutian Islands on October 22 and 23. An instrumental record, kept by William Dall on the USCS schooner *Yukon* south of Unalga Pass/Unalaska (near 54°N, 166° 32W), reveals storm impact most apparent on October 23 (Table 1). The lowest pressure suggests a value around 960 mb, indicating that the storm likely had intensified towards its peak strength in its life cycle. The wind shifts from SE to NE and to NW indicate that the low pressure center passed to the east and

Table 1 Meteorological data taken aboard the USCS schooner *Yukon* in October 1880. Barometric and temperature were converted from the original data which were in inches and degrees Fahrenheit respectively

Date Oct.	Time	Barometer (mb)	Temperature (Celsius)	Wind direction
23	4 A.M.	993.63	9.2	SE
	8 A.M.	990.92	10.0	SE
	12 M	985.50	10.6	E by N
	4 P.M.	975.34	11.1	E
	8 P.M.	968.56	11.1	NE by E
	12 P.M.	966.87	10.6	NNE
24	4 A.M.	969.92	11.1	NW
	8 A.M.	978.05	10.0	"
	12 M	985.16	10.6	"
	4 P.M.	989.56	9.4	"
	8 P.M.	993.63	9.4	"
	12 P.M.	993.63	9.4	"

south of the schooner *Yukon*. Dall wrote in his notebook that "The wind which has been rather light in the early A.M. increases, shifts to the eastward and blows very hard in squalls. Toward evening a stiff gale blowing… The continued rough and adverse weather is very wearing on all hands." A nearby logbook from the whaler bark *Helen Mar* near the Fox Islands (about 54°N, 166° 35′ W) also noted the wind shifts from SE to NE from October 23–24. The *Helen Mar* logbook also noted a "heavy gail from the N.W. with plenty of rain" for October 25, indicating a cold frontal passage and the low pressure center well to the east.

Some meteorological stations that are located well away from the proposed storm track provide some related synoptic information. A decrease in pressure on October 25 was noted by the weather observer at Fort St. Michael, located well to the north off western Alaska (Monthly Weather Review, 1882); however, no further information was provided and the original weather manuscripts have not been found. This change likely relates to the prominent buildup of a strong longwave trough, with a prominent cold air mass traversing southward, to enable steering of the storm well to the south into Sitka. The Monthly Weather Review report (1882) suggests the storm heading towards Hudson's Bay, but no further documentation of the storm in the interior of North America has been found, although population densities in this region were extremely sparse during this time. Examination of daily weather data from Washington state in the United States reveals no storm impact, and a slight increase of pressure around October 26. This increase may relate to some ridging off the West Coast that enabled the steering of Sitka Hurricane in a northeastward direction, potentially enhanced in strength by upper air divergence and increasing its forward speed, towards the Alaskan coast.

5 Conclusions

The Sitka hurricane of October 1880, while clearly possessing neither tropical origins nor any tropical characteristics in its storm history, likely ranks among the strongest documented meteorological storms that have struck the West Coast of North America. It is comparable in strength with strong major hurricanes that form and traverse off western Mexico. Being extratropical in nature, the Sitka hurricane most likely had its zone of strongest winds more spread out from the center, and was likely of larger size than typical tropical hurricanes, adding to its hazard potential. Sitka, Alaska has not documented such a storm in the modern (post 1900) record, and the storm struck the region during a time of early settlement when population was extremely sparse. Therefore, such events at that magnitude are not even considered as worst case scenarios in hazard planning, zoning, and management for the region today.

Long and continuous chronologies of severe coastal storms have been successfully reconstructed and the impacts well-documented from historical data for some regions such as New England and England (e.g., Wheeler, 2003). Historical data, particularly ship logbooks, shows the high feasibility of using historical evidence to reconstruct major storms for other regions, such as coastal Alaska which currently lag much of the research being conducted in other areas. A pressing need still exists to reconstruct pre-World War II twentieth century storms from unexploited newspapers and instrumental data. Such reconstructions back into the early nineteenth century can be extended by careful analysis of records from logbooks aboard US Navy vessels that routinely remained in port in late nineteenth century, whaling logbooks from busy activity during much of the mid nineteenth century, hourly Russian data from the Sitka Observatory as well as some logbooks from ships of exploration.

Acknowledgements We gratefully acknowledge help from the staff at the United States National Archives in Washington, DC, the Smithsonian Institution, and the New Bedford Whaling Museum Library. We also thank Mike Chenoweth for insightful discussions. Some funding for archival research was supported by NSF Grant ATM-0502105.

References

Arnott J (2005) Typhoons in Alaska? Extratropical Transition and its Implications for Northern Alaska. NWS Alaska Region Research Notes RN-05-0003. Fairbanks, Alaska

Burt C (2007) Extreme weather: A guide and record book. W.W. Norton & Company, New York

Carleton AM (1991) Satellite remote sensing in climatology. Belhaven Press, London

Daily Sitka Sentinel (1948) January 2, 1948. Sitka, Alaska

Douglas KS, Lamb HH, Loader, C (1978) A Meteorological Study of July–October 1588: The Spanish Armada Storms. Climatic Research Unit Publication No. 6. East Anglia, United Kingdom

García-Herrera R, Ribera P, Hernández E, Gimeno L (2007) Northwest Pacific typhoons documented by the Philippine Jesuits, 1566–1900. J. Geophys. Res. 112: D06108, DOI:10.1029/2006JD007370

Larsen C, Atkinson D, Walsh J, Lingaas J, Arnott J (2006) Dynamical Development of the Bering/Chukchi Sea Storm, October 2004. American Geophysical Union Fall Meeting. 11–15 December, 2006. San Francisco, CA

Lynott RE, Cramer OP (1966) Detailed analysis of the 1962 Columbus Day windstorm in Oregon and Washington. Mon. Wea. Rev. 94: 105–117

Martin JE, Grauman RD, Marsili N (2001) Surface cyclolysis in the North Pacific Ocean. Part I: A synoptic climatology. Mon. Wea. Rev. 129: 748–765

Mason OK, Salmon DK, Ludwig SL (1996) The periodicity of storm surges in the Bering Sea from 1898 to 1993, based on newspaper accounts. Clim. Change 34: 109–123

Monthly Weather Review (1882) International Meteorology. Mon. Wea. Rev. 10: 10–15

Scidmore ER (1893) Appleton's guide-book to Alaska and the northwest coast. Appleton and Company, New York

Turner LM (1886) Contributions to the natural history of Alaska. Govt. Print. Office, Washington, DC

Vaquero JM, García-Herrera R, Wheeler D, Chenoweth M, Mock C J (2008) An historical analogue of 2005 Hurricane Vince. Bull. Amer. Meteor. Soc. 89: 191–201

Wheeler D (2003) The Great Storm of November 1703: A new look at the seamen's records. Weather 58: 419–427

Wild H (1882) Annalen des physikalischen Central-Observatoriums. St. Petersburg, Russia

Daily Synoptic Weather Map Analysis of the New England Cold Wave and Snowstorms of 5 to 11 June 1816

Michael Chenoweth

Abstract Daily weather maps for 5–11 June 1816 depict the unprecedented June snowstorms and freezing weather events in the northeast US. New sources of data never previously used include ships' logbooks, newspaper extracts and weather journals. The highlight of this work is the discovery and documentation of a hurricane affecting Florida for a four-day period during the cold wave. In addition, newspaper accounts have added new instrumental temperature data including the lowest temperature yet found in the US during this remarkable cold wave. This chapter highlights the large amount of unused data still available for use in reconstructing historical climate variation.

Keywords Daily weather map · 1816 · Tambora · New England

1 Introduction

The so-called "Year without a summer", 1816, is one of the most well-known climate extremes of early United States history. It has been documented in general interest books (Perley, 1891; Stommel and Stommel, 1983), historical studies (Hoyt, 1958), for its meteorological aspects in the northeast US (Ludlum, 1966) and in Canada (Wilson, 1985), its economic and social impacts in Europe and North America (Post, 1977) and as the subject of an historical workshop on world climate in 1816 (Harrington, 1992). The cold summer was part of an extended period of global cooling following the April 1815 eruption of Tambora, Indonesia (Stothers, 1984). Weather reports from ships around the world were used to better describe the evolving state of the global climate before, during, and after 1816 (Chenoweth, 1996) and to provide the most accurate measurements of the post-Tambora surface cooling of the earth (Chenoweth, 2001).

In this chapter, I return to the famous June cold wave and snowstorms in New England that are the most well-known of the sequence of cold snaps, snows and

M. Chenoweth (✉)
Unaffiliated Independent Scholar, Elkridge, Maryland, USA
e-mail: m.chenoweth@att.net

L.-A. Dupigny-Giroux, C.J. Mock (eds.), *Historical Climate Variability and Impacts in North America*, DOI 10.1007/978-90-481-2828-0_8,

frost that plagued the northeast US and adjacent areas of Canada for so much of the summer of 1816. Wilson (1985) first extended the range of synoptic weather mapping from the limited US-centered observations used by previous researchers by using weather records from the Hudson's Bay Company and other weather diaries. This chapter extends her work with a search of American and European newspapers for additional information. I extracted weather data from the logbooks of British Royal Navy ships that were on station throughout the Atlantic region, including the Great Lakes of Canada and the US. The logbooks allow for a further expansion of the area of available data and for daily weather maps to be produced on a twice-daily basis. It is now possible to display the synoptic weather features for each day of the summer of 1816. Here, the period of 5–11 June 1816 is closely described, as this coincides with the arrival of cold air into the US and ends with the final morning of frost in New England before more seasonable weather returned.

2 Study Area

The area of study covers the eastern half of North America (eastward from about 95°W longitude) and adjacent waters, eastward into the Atlantic Ocean. It extends from the Caribbean Sea in the south to Hudson's Bay, Hudson Strait and south Greenland in the north (Fig. 1). Although the entire Atlantic basin was covered in the data search, the maps presented in this chapter extend out only to about 55°W longitude. The weather map analysis is based on the larger area (not depicted).

Current US weather symbols associated with official weather maps were used for plotting individual station data and the depiction of high and low pressure centers and frontal boundaries. Although limited pressure data were available, no attempt was made to draw isobars as the area of reliable coverage was limited to the New England region of the US and some individual station data were either not reliable and/or without sufficient metadata to properly adjust the data. The reader can use the displayed data to make his/her own interpretation of the pressure distribution.

3 The Synoptic Setting

3.1 Pressure Patterns

A period of extreme weather began to establish itself in the North Atlantic as early as 28 May when high pressure built over the waters west of Ireland and remained essentially stationary until 5 June. The weather pattern over North America featured a sharp cold wave on 29–31 May in Québec and New England. This was followed by two seasonably warm days, with temperature maxima exceeding 80°F (27°C) throughout New York and most of New England on 1 June. A portion of a quasi-stationary area of cold high pressure northwest of Hudson Bay moved southeast on 2–3 June and temperatures gradually fell as a cold front pushed south

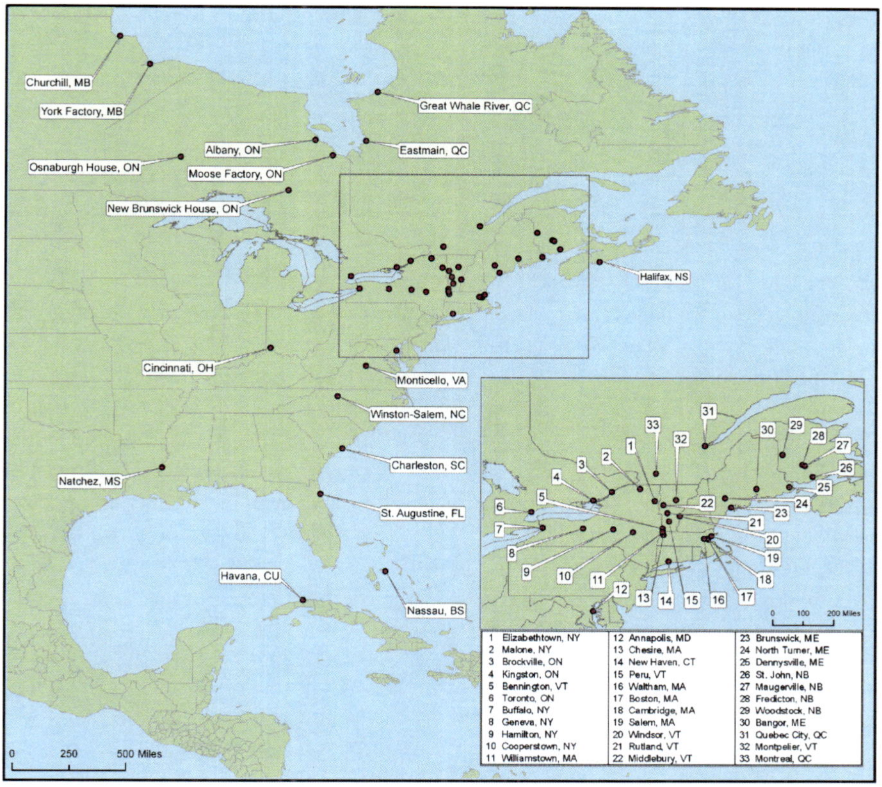

Fig. 1 Place-names of weather stations and other locations of weather reports cited in text

into the mid-Atlantic region of the United States. The night of 3–4 June brought widespread frosts to northern New England, Québec and interior New Brunswick. At Dennysville, Maine the minimum temperature on 4 June was 31°F (–1°C) (Lincoln MS) and at Woodstock, New Brunswick, the ground was white with frost (Dibblee MS). In a normal season, such a morning would mark the coldest morning of most Junes and be the last widespread frost in interior regions (U.S. Department of Commerce, 1968).

The narrow ridge of high pressure was already giving way to more seasonable weather. Ice pellets fell at Malone, New York (*National Standard* (Middlebury, VT), 3 July 1816) and rain showers at Montréal (McCord MS) on the morning of the 4th as warmer air aloft arrived. By afternoon, the surface warm front was advancing eastward across central Pennsylvania. Severe thunderstorms broke out over central Pennsylvania where over 1,000 acres of oats and rye were destroyed in Snyder County about 40 miles north of Harrisburg (*Farmer's Repository* (Charleston, WV), 3 July 1816). A thunderstorm with heavy rain passed over the USS *Washington* anchored at Annapolis, Maryland at 2:30 p.m. (*USS Washington* Logbook, US

National Archives). Thomas Jefferson observed no precipitation on this date at Monticello, Virginia (Jefferson MS).

3.2 Fronts

The northward moving rain shield near to and just south of the surface warm front covered most of New Brunswick, northern Maine, the St. Lawrence River valley and the northernmost counties of New York and Vermont by the morning of 5 June (Fig. 2). Rain at Cooperstown, New York at 9:00 a.m. (Cooper MS) was probably related to remnants of the previous day's thunderstorms over Pennsylvania. By early afternoon, the leading edge of the modified Pacific air mass was passing over eastern New York and Pennsylvania. Thunderstorms broke out ahead of and along the front from southern New Hampshire to at least western North Carolina. Some hail was reported in the Moravian communities near present-day Winston-Salem, North Carolina (Fries, 1947) but most areas along the front experienced only brief showers. In eastern Massachusetts, temperatures of 90°F (32°C) and higher were observed at the Boston area stations, with temperatures from 80–90°F (27–32°C) elsewhere with 77°F (25°C) readings along the southern New England coasts under the moderating influence of cool ocean waters.

3.3 Weather on 5 June

At 2:00 P.M. of 5 June, the leading edge of the polar air extended southwestward from the low pressure center (estimated central pressure 990 millibars) to just east of York (modern Toronto, Ontario) to northwest Ohio (this map is not shown, Fig. 2

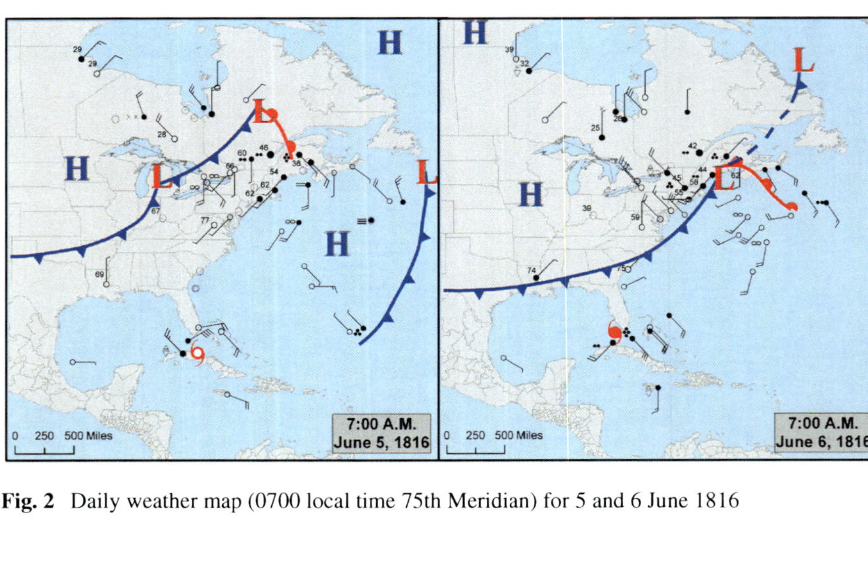

Fig. 2 Daily weather map (0700 local time 75th Meridian) for 5 and 6 June 1816

depicts the 7:00 a.m. conditions). Fresh gales were observed in Georgian Bay[1] (ADM 51 2355) and strong gales in western parts of Lake Erie (ADM 52 3933) as the front passed through the area. The mid-day temperature in central Ontario at New Brunswick House was 30°F (–1°C) with a few remnant snow flurries (New Brunswick House MS). All of New England, except the northeastern two-thirds of Maine, was southwest of the warm front.

Afternoon thunderstorms on 5 June which had formed ahead of the cold front had degenerated by that night to an area of showers over parts of Massachusetts, Connecticut and New Jersey. A few thunderstorms were still present over the Philadelphia (Colin MS) and southern New Jersey areas. Rain continued along and north of the warm front and in the immediate vicinity of the low pressure west of Montreal. For the first time, rain began to break out over parts of Nova Scotia (ADM 53 396). Fresh to strong breezes blew across eastern Lake Ontario (ADM 52 3933) and fresh to strong gales over Georgian Bay (ADM 51 2355) and central Lake Erie (ADM 52 4189).

3.4 Weather on 6 June

By 7:00 a.m. of 6 June, the center of low pressure was located in Maine [Fig. 2]. This would be about the southernmost position that the surface low would reach before moving to the east-northeast. The warm front passed north of Dennysville, ME where the morning temperature of 62°F (17°C) (Lincoln MS) was 16°F (9°C) above the average minimum. The original leading edge of the Pacific air mass was offshore of New England and re-emerged on land in the Virginia/North Carolina area. The southwest winds at Brunswick, Maine and Cambridge, Massachusetts suggests that the polar air mass lay to their west at 7:00 a.m. on 6 June and would soon catch up with the original cold front. [Although this could argue for two separate low pressure centers over western and eastern Maine, and the weakening of the western one, the lack of absolutely simultaneous observing times does not allow this placement with certainty.] In southeastern New England temperatures were slightly above average 55–60°F (~13–16°C) but had tumbled to 31°F (–1°C) at Malone, New York (*National Standard*, 3 July 1816). Cold air was entrenched in the Ohio valley and Great Lakes. Cincinnati, Ohio had a sunrise temperature of 39°F (4°C) (Jackson MS) and temperatures in the mid-twenties were observed in northeast Ontario and most of western Québec. The anomalies, however, were most extreme in the northern USA and adjacent regions of Canada.

Snow began to break out in the cold air mass from northwest to southeast. Snow and rain were first observed near Brockville, Ontario at 4:00 a.m. aboard HMS *Star*. At 8:00 a.m., the ship reported both snow and sleet, rain having ended, with a WNW fresh breeze (ADM 52 4617). No snow is mentioned at Kingston, Ontario at 8:00 a.m., although rain had ended by this time (ADM 52 4189).[2] HMS *Netley* at Kingston, Ontario reported a NW strong breeze (ADM 51 2605). HMS *Champlain*, a few miles north of the Vermont border at Isle aux Neuf, reported rain at 8:00 a.m.

(ADM 51 2398). Rain was falling at 7:00 a.m. at Middlebury, Vermont (*National Standard*, 19 June 1816) and snow at 5:00 a.m. at Malone, New York that was over by afternoon (*National Standard*, 3 July 1816). Elizabethtown, New York, reported a continuous fall of snow from 7:30 to almost 10:30 a.m. (*Albany Advertiser*, 22 June 1816). Snow was reported falling "in considerable quantities" at Geneva, New York (*New York Spectator*, 19 June 1816) and "falling rapidly" for the better part of the day at Hamilton, New York. (*American Watchman* (Wilmington, Delaware), 22 June 1816). Cooperstown, New York reported cloudy weather at 9:00 a.m. and snow at noon and 3:00 p.m. (Cooper MS).

At Bennington, Vermont, snow began to fall at 8:00 a.m. and continued "more or less" until shortly after 2:00 p.m. (Ludlum, 1966) and accumulated to 1.5 in. (3.8 cm) (*Albany Argus*, 11 June 1816). At nearby Williamstown, Massachusetts, snow fell several times during the day and "...about noon, lay for a minute on the ground – some hail, round snow, and flake snow" (Dewey MS). The Williamstown observer also recorded "Ground white in Peru, Windsor, Cheshire and the mountains west of us" (Dewey MS). Snow also covered the ground in Rutland, Vermont on the 6th (*Paulson's American Daily Advertiser* (Charleston, SC), 17 June 1816). Snow melted as it fell in the Montpelier area of central Vermont (*American Watchman*, 29 June 1816). Snow fell from 2:30 p.m. to about 4:00 p.m. at Bangor, Maine (Ludlum, 1966) and that evening accumulating snow was reported at Hallowell, Maine (*Dartmouth Gazette* (Hanover, NH), 19 June 1816).

In Québec, Montréal reported snow from 11:00 a.m. to 1:00 p.m. (*Québec Mercury*, 11 June 1816) and some snow fell at Québec City where the mountains to the north of the city were observed to be covered with snow in the afternoon once the cloud cover had lifted (*Québec Gazette*, 13 June 1816). At Isle aux Neuf, Québec snow was first reported at 2:00 p.m. (ADM 51 2398). Several inches (5 cm or more) were reported to have accumulated in unspecified areas of western New York (*Niles' Weekly Register*, 10 Aug. 1816). Snow accumulated one inch (2.5 cm) deep in the Catskills that afternoon and snow fell within ten miles of tidewater on the Hudson (Ludlum, 1966). Two to three inches (5–8 cm) of snow were reported in the mountains around Middlebury, Vermont on 6 June (*National Standard*, 12 June 1816). Relatively little precipitation of any type fell in southeastern New Hampshire, eastern Massachusetts, Rhode Island and all but northwest Connecticut. Skies were mostly clear to partly cloudy in these areas for the most part.

A close examination of the weather reports (including maps not depicted here) and reported timing of the snows suggests that there were two bands of snow. The first ran from upstate New York south into Pennsylvania and spread from west to east in the early morning hours.[3] Initially, it was a band of rain that that quickly turned over to a mix of precipitation and then all snow as colder air arrived. Accumulating snows fell quickly in heavy snow bursts with lighter flurries otherwise prevailing. In western and central New York and the mountains of western Pennsylvania, the snow began early in the day and continued at least until midday in the more western areas and into the afternoon elsewhere.

The second episode of snow developed north of the St. Lawrence River and spread south and east in the cyclonic flow behind the east-moving surface low. The

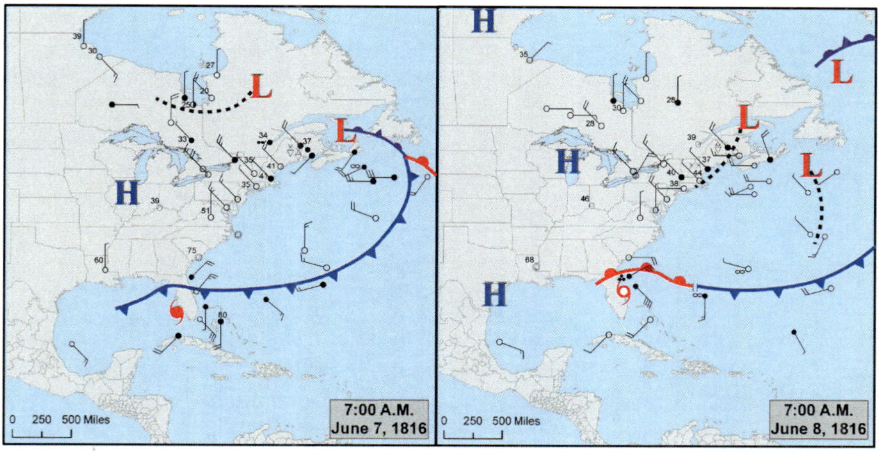

Fig. 3 Daily weather map (0700 local time 75th Meridian) for 7 and 8 June 1816

snow had ended by 1:00 p.m. and probably around the same time in Québec City, Québec. Snow reached Isle aux Neuf, Québec at 2:00 p.m. (ADM 51 2398) and Bangor, Maine by 2:30 p.m.. By evening, some of this snow area had penetrated into southwest Maine where it accumulated at Hallowell. Some snow also was reported, mixed with rain, in central New Brunswick. By the evening of the 6th, the snowfall had ended in all areas except in extreme northern Vermont, New Hampshire, adjacent areas of Québec and into northern Maine. During the night of 6–7 June, snow fell at Isle aux Neuf, Québec and at Woodstock, New Brunswick, where the ground was covered with snow on the morning of the 7th.[4] At St. John, New Brunswick, on the Bay of Fundy, snow and sleet was falling (ADM 51 2299) and snow showers were reported at Woodstock, Québec City, and North Turner, Maine (Dibblee MS; Sparks MS; Ludlum, 1966). Brunswick, Maine recorded 1.10 in. (28 mm) of rain as of the morning of 7 June (Cleveland MS). By the morning of 7 June, the occluded surface low was east of New Brunswick in the southern Gulf of St. Lawrence and moving slowly to the northeast (Fig. 3). Ice formed to a thickness of three-tenths of an inch at Malone, New York and the sunrise temperature was 29°F (–2°C) (*National Standard*, 3 July 1816).

3.5 Weather on 7 June

During the day on 7 June, snow was reported at Malone, New York (*National Standard*, 3 July 1816), Isle aux Neuf, Québec (ADM 51 2398) and Québec City, Québec (Sparks MS), Woodstock (Dibblee MS) and St. John, New Brunswick (ADM 51 2299) while a brief snow shower fell at Halifax, Nova Scotia at midday (ADM 53 396). Areas of northern New Hampshire and much of Maine also experienced snow showers. The snow fell the entire day at Québec City and "at 10 at night the ground was completely covered (Sparks MS)." In northern New Brunswick, snow accumulated from 2 to 4 in. in depth (5–10 cm) (Fisher, 1825). During the night of 7–8 June, this second snow of the month continued to affect northern New England. This event was less widespread but produced heavier amounts which also fell in the valley locations that had escaped any or only brief middling accumulations on 6 June.

3.6 Weather on 8 June

The snow had briefly ended at Isle aux Neuf, late in the afternoon on 7 June. By sunset, snow began again and continued until late in the evening on 8 June (ADM 51 2398). HMS *Star*, near Brockville, Ontario again reported some snow at 8:00 a.m. (ADM 52 4617) while Montréal, Québec reported some snow early in the morning (McCord MS) (Fig. 3). The snow had ended at Québec City, Québec by 8:00 a.m. (Sparks MS). Snow which had overspread upstate New York and northern New England after dark the evening before continued up until about noon or a little earlier in most of the area. Snow was falling as far south as Bennington, Vermont (Ludlum, 1966) and small snow showers were observed at Waltham and Salem, Massachusetts around 10:00 a.m. (Ludlum, 1966). A snowstorm was reported in the Old Forge, New York area on this date[5] (Brown and Walton, 1988). Snow fell for several hours in Portland (Ludlum, 1966) and Brunswick, Maine (Cleveland MS). In New Brunswick, snow overnight whitened the hills around Woodstock (Dibblee MS) while Maugerville reported one inch (2.5 cm) of snow on the ground (Miles MS).[6] In northern Vermont and New Hampshire, significant snows of at least five inches (13 cm) fell and were drifted to depths as great as eighteen inches (46 cm) (Ludlum, 1966). The snow lay the entire day in some valley locations. Ship reports on the inland waters of the eastern Great Lakes reported fresh northwesterly breezes. With early morning temperatures from 30–35°F (–1° to +2°C), wind chill values were about 10°F (–12°C). On the higher hills, wind speeds were probably even higher. For good reason this season was considered remarkable.

By the evening of the 8th, snow had ended over New England. Warmer daytime temperatures had turned snow over to rain in Maine and New Brunswick. During the night of 8–9 June, rain again changed to snow in northern Maine and interior New Brunswick. Snow again covered the hills around Woodstock on the morning of the 9th (Dibblee MS) (Fig. 4). Corn was killed by frost near Buffalo, New York on this morning (*Niles' Weekly Register*, 10 August 1816). The center of low pressure had now drifted over the southeast corner of Labrador.

3.7 Weather on 9–11 June

With the center of low pressure now moving out to sea, high pressure finally began to move southeast over Hudson Bay and toward New England. By the evening of the 9th, high pressure was centered over north central Ontario with a ridge axis

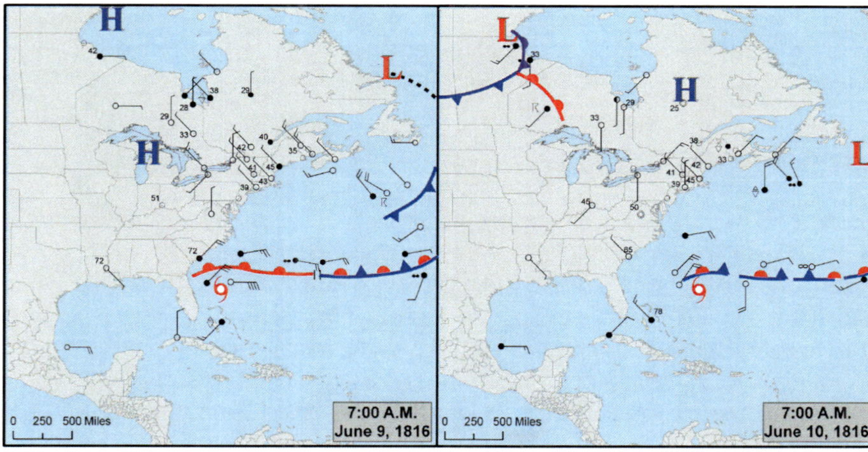

Fig. 4 Daily weather map (0700 local time 75th Meridian) for 9 and 10 June 1816

extending directly through New England. By the morning of the 10th, the high was over southwest Québec (Fig. 4). The temperature at Malone, New York dropped to 24°F (–4°C) (*National Standard*, 3 July 1816),[7] the coldest observed temperature in the USA in June 1816. During the day, the high centered itself over New England and by the morning of the 11th an elongated area of high pressure was draped over the Canadian Maritimes, New England and the mid Atlantic states (Fig. 5). This was the coldest morning of the cold wave in Maine, New Brunswick and southern New

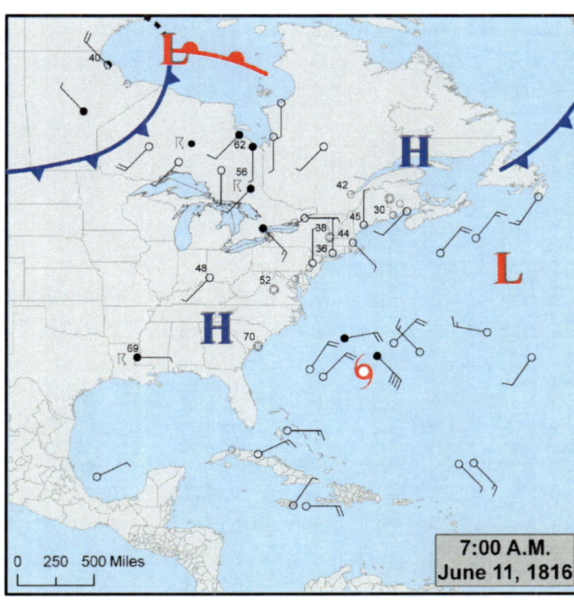

Fig. 5 Daily weather map (0700 local time 75th Meridian) for 11 June 1816

England and hard frosts again affected all of northern New England and New York again this morning. During the day high pressure drifted east and a southerly flow returned seasonable temperatures to the entire area.

4 Hurricane Weather

The origins of the 1816 season's first hurricane were in the southwest Caribbean in the general area of 15°N, 80°W–82°W. Evidence for disturbed weather occurs as early as 28 May, with a tropical depression possibly forming as early as 1 June and almost certainly by 2 June. Sustained rains set in at Port Royal (two miles west of the capital Kingston), Jamaica on 29 May and continued with little intermission until 7 June (ADM 52 4600). Winds from ships anchored in Jamaica were generally east to southeast and rarely more than a fresh breeze (ADM 52 4600; ADM 52 4308; ADM 52 4431; ADM 52 4242). The storm center appears to have passed near Little Cayman Island during the day on 4 June. The storm was probably only a tropical storm at this point as the wind field at Havana and Port Royal show virtually no difference from climatological values although Havana had a strong breeze (ADM 52 4648). Near Matanzas, east of Havana, *the Chilham Castle* reported gale force winds on the third of June (*Charleston Courier*, 15 June 1816).[8]

4.1 Weather on 4 June

The storm appears to have made landfall on the south coast of Cuba about sunset on the 4th.[9] Reports of ships wrecked and pushed ashore at Trinidad, Cuba suggest that the storm deepened before coming ashore and was a minimal hurricane at this point. At 8 p.m. on the 4th, Havana had a strong breeze from the NNW (ADM 52 4648) while ENE winds were blowing at Nassau (*Royal Gazette and Bahama Advertiser* [RGBA], 3 July 1816) and on the ship *Francis*, located at about 24°N 81°W (Francis Logbook).

4.2 Weather on 5 June

By 7 a.m. on the 5th, the weather was deteriorating rapidly in the Florida Straits as the storm reemerged over water at about 22° 30′ N, 78° 30′ W. Havana had a strong NW breeze with rain (ADM 52 4648) coming down in torrents and winds violent in squalls (*American Beacon and Commercial Diary*, 17 July 1816) while the *Francis*, located at about 24°N, 81°W had heavy gales from the NE with heavy rain (Francis Logbook). Both Nassau and Port Royal had fresh SE winds (RGBA, 3 July 1816; ADM 52 4600). Another ship, the *Gallatin*, at about 24° 30′ N, 80°W had a fresh E gale with heavy rain at daybreak. The *Gallatin* took measures to secure sails and deploy sturdier sails, but to no avail, as the winds increased and even the main sail gave way at noon as winds reached storm force (*Charleston Courier*, 17 June 1816).

By 9 p.m. the hurricane was moving northwestward and located north of Havana, Cuba at about 82°W. Havana was experiencing heavy gales from the WSW (ADM

52 4648) and the *Francis*, struggling not to run aground after losing her main topsail, foresail and top foremast staysail, was off the Florida Keys at about 24° 30' N fighting heavy SE gales and high seas (Francis Logbook). Early on the 5th, the *Gallatin* experienced hard gales and rain with a high cross sea and was shipping several seas. One wave at 3 p.m. "made a fair breach over her" as winds rose to "a perfect tempest" (Charleston Courier, 17 July 1816).

4.3 Weather on 6 June

By the morning of the 6th, winds at Havana were from the SW and only blowing fresh (ADM 52 4648). The rain continued heavier than before and 8 or 10 ships drifted into the Havana harbor while four or five ships were lost on the coast (*American Beacon and Commercial Diary*, 17 July 1816). By the evening of the 6th, the *Francis* was moving north near Bimini in a fresh south wind (*Francis* Logbook) while it was still raining with a SW moderate breeze in Havana (ADM 52 4648). In the meantime, a cold high pressure was pushing south into the US bringing unusually cold air south deep into the southern USA. The front was near Charleston, South Carolina and stretched to the west. The front in the western Gulf of Mexico would continue to move south while the portion nearest Florida would remain nearly stationary as the hurricane began to curve to the northeast. During the evening of the 6th, winds off the Georgia coast were fresh and from the northeast (*Supplement to the Royal Gazette*, 20–27 July 1816).

4.4 Weather on 7 June

During the early morning of the 7th, the hurricane appears to have made landfall in southwest Florida and crossed the peninsula during the day, moving to the east-northeast. Gales were now blowing from South Carolina to Cape Canaveral in central Florida. The ship *Huron*, 12 miles southeast of St. Simon's Lighthouse, Georgia, had stiff NE gales and "thick weather" (*Charleston Courier*, 28 June 1816). Off Cape Canaveral, the *Francis* encountered the hurricane again during the evening of the 7th. Very severe NE gales blew and the ship lost its waistboards, fore topsail, and the second foretopmast staysail and drifted on soundings (*Essex Register*, (Salem, Massachusetts), 26 June 1816) as the storm emerged off the coast. At Nassau, Bahamas a fresh SSE wind was blowing (*RGBA*, 3 July 1816) and a fresh WSW breeze blew at Havana, Cuba (ADM 52 4648).

The hurricane appears to have deepened in intensity during the evening of the 7th and early hours of the 8th. By 7 a.m. on the 8th, the *Francis* was encountering hurricane force SE winds at about 28° 30′ N, 78° 45′ W (Francis Logbook). The *Huron*, at about 30° 15′ N 81° 45′ W was facing "tremendous seas" and a heavy ENE gale. Even at Charleston, South Carolina there was a fresh east gale and the sea was running as high as in the gale [i.e., hurricane] of 1804 (*Charleston Courier*, 13 June 1816). At Nassau, winds were SW and strong (RGBA, 3 July 1816) but at Havana the storm was over with only a light breeze from W by S (ADM 52 4648). The storm continued eastward but subsequently weakened after it approached Bermuda.

The slow-moving hurricane was a consequence of a lack of "steering winds" aloft to move the hurricane at a faster rate. In turn, the extratropical low and associated cold front were likewise moving very slowly due to the enhanced meridional flow that is assumed to have prevailed at this time. Until this "blocking" weather pattern slowly relaxed to a more normal zonal "westerly to easterly" flow, the extreme weather in both Florida and New England persisted.

The unusual nature of the hurricane was commented on in both the Bahamas and in Florida. The Nassau *Royal Gazette and Bahama Advertiser* of 12 June reported

> We do not recollect such a continuance of boisterous weather at this season of the year as we have experienced for the last fortnight. It has been little less than a continued Gale, blowing successively from each point of the compass. We have already heard of several instances of its disastrous effects, and as far from the accounts given by those who have suffered from the weather, that we shall hear of many more....

From St. Augustine, Florida an account from a passenger on the ship-wrecked *Huron* reported that on 8 June they were taken to the Governor of Spanish Florida in St. Augustine from their landing point on Anastasia Island, 20 miles to the south. "The crops suffered severely in the gale and it is supposed that at least 1/3 of the cotton crop is destroyed – The oldest inhabitants say they never before experienced such a gale at this season of the year, to last 4 days. The inhabitants appeared much alarmed" (*Charleston Courier*, 28 June 1816).

The hurricane and snowstorm/cold air outbreak were an unprecedented simultaneous event. While hurricanes in the autumn months have interacted at time with cold air masses to produce snow in New England there is no precedence for one in the early part of the hurricane season. The slow movement of this hurricane, like the cold air masses to its north, is unlike any of the autumn hurricane/snow events which all involved much faster forward motion on the part of the hurricanes.

5 Conclusion

The largest data set of daily weather observations yet available for the summer of 1816 provide the most detailed look at the daily weather during the famous cold wave and snow event in New England from 5 to 11 June 1816. For the first time, a hurricane affecting Florida is shown to have been simultaneous with the snows in New England, an unprecedented event in American weather history. Both weather events were destructive to local crops and vegetation and such an event today would receive unprecedented press coverage.

The availability of ships' logbooks and abundant North American newspapers reveals that there is a tremendous amount of new information still unused in archives for documenting early American weather and climate. The study of extreme weather events, even one as well known as 1816, will benefit from utilizing these and other sources of data.

Acknowledgments The maps were produced from materials provided by the author to Richard Murphy, Department of Geography, Columbia, South Carolina and his assistance in drafting the maps is gratefully acknowledged. Some technical support was provided by NSF Grant ATM-0502105.

Notes

1. Admiralty (ADM) series documents are located in the UK National Archives, London (formerly the Public Record Office).
2. However, the local newspaper reported "some snow" fell on that day (Kevin Hamilton).
3. In upstate New York there were several accounts of the damage done to vegetation by the combination of cold and snow. "The outer leaves of some forest trees and shrubs, [such] as the sugar and soft maple, elm and basswood, the Indian-willow, moose-wood & white maple; of some fruit trees and shrubs, [such] as the damson and currant, the blades of Indian corn, beans and many garden plants, were killed by this storm of snow, west wind and the frost of the night succeeding. That the snow, and more especially the severe and chilling west wind had much agency in this business, is inferred from the great injury suffered by the west side of the tops of the trees, whilst their eastern sides escaped but little hurt." *National Standard*, 3 July 1816.
4. Dibblee's diary entry for 7 June: "Cloudy and cold as winter. Snow squalls all day. The snow fell last night so as to cover the ground – terrible indeed – Never knew snow in summer before.... River rising fast – No salmon – Never was there such weather – People ploughing and harrowing with great coats on – Wind high Norwest."
5. Brown and Walton, 1988, pp. 222–223.
6. Parts of northern New Brunswick received three to four inches of snow on 7 June and snowfall was reported to be general across the province on this day (see Peter Fisher, *History of New Brunswick*. Reprint of the 1825 original edition, The New Brunswick Historical Society, St. John, 1921). Snow and cold on 7 June is also mentioned in *Notitia of New Brunswick for 1836, and extending into 1837* by an unidentified author, printed in St. John in 1838.
7. Waterhouse noted on this date: "Although this morning as indicated by the thermometer was the coldest we have experienced since the 17th of April, water was but slightly skimmed with ice. So severe, however, was the frost that the tops (leaves and bows) of apple trees, seem to have suffered more than in any preceding night – Strawberry plants and pea vines are wilted, and have assumed a dark appearance." This mirrors the comment at Middlebury, Vermont: "The heaviest frost we have had for the month past, occurred the night of the 10th inst [i.e., 10–11 June]. My corn, potatoes, beans, vines & peas, which were just ready to blossom, were frozen quite hard – Most of my cucumbers, squashes, and other tender vegetables, were saved by sprinkling cold water on them before the sun was up." (*National Standard*, 19 June 1816).
8. Most likely these were peak wind gusts in squalls.
9. The *Adrianna* arrived in Charleston on 3 July after a 15-day journey from Trinidad, Cuba. The ship experienced a very heavy gale from the NE to SW, which lasted from the 3rd to the 6th as reported in the *Baltimore Federal Republican and Baltimore Telegraph*, 10 July 1816. A less complete report in the *Charleston Courier*, 3 July 1816 gives the date as the 14th of June but makes no mention of wind direction. The wind directions likely refer to the four-day period and not just the immediate winds near the storm center. Other reports from Trinidad, Cuba appear in the *Jamaica Mercury and Kings-Town Weekly Advertiser, Postscript to the Royal Gazette*, 29 June 1816.

References

Brown, H.A.L. and R.J. Walton (eds.), 1988. *John Brown's Tract: Lost Adirondack Empire.* Phoenix Publishers, Canaan, New Hampshire, 366pp.

Chenoweth, M., 1996. Ships' logbooks and 'The year without a summer'. *Bull. Amer. Meteor. Soc.*, **77**, 2077–2093.

Chenoweth, M., 2001. Two major volcanic cooling episodes derived from global marine air temperature, AD 1807–1827. *Geophys. Res. Lett.*, **28**, 2963–2966.

Fisher, P. *History of New Brunswick*, 1825, (1921 Re-print edition), 133 pp. [Available at the Nova Scotia Legislative Library, Halifax]

Fries, A.I. (ed.), 1947. *Records of the Moravians in North Carolina.* Vol. VII, 1809–1822. Raleigh, State Department of Archives and History.

Hamilton, K., 1986. Early Canadian weather observers and the 'year without a summer'. *Bulletin of the American Meteorological Society*, 67(5), 524–532.

Harrington, C.R. (Ed.), 1992. *The Year Without a Summer? World Climate in 1816.* Canadian Museum of Nature, Ottawa.

Hoyt, J.B., 1958. The cold summer of 1816. *Ann. Assoc. Amer. Geograph.*, 48, 118–131.

Ludlum, D., 1966. *Early American Winters, 1605–1820.* American Meteorological Society, Boston, 285 pp.

Perley, S., 1891. *Historic Storms of New England.* The Salem Press, Salem, Mass., 341 pp.

Post, J.D., 1977. *The last great subsistence crisis in the Western world.* Johns Hopkins University Press, Baltimore, 240 pp.

Stommel, H., and E. Stommel, 1983. *Volcano Weather, the Story of the Year Without a Summer.* Seven Seas Press, Newport, R.I., 177 pp.

Stothers, R.B., 1984. The Great Tambora Eruption in 1815 and Its Aftermath. *Science*, **224**, 1191–1198.

U.S. Department of Commerce, 1968. Climatic Atlas of the United States. ESSA, EDS, Washington DC, 80 pp.

Wilson, C.V., 1985. Daily weather maps for Canada, summers 1816–1818 – a pilot study. In: *Climatic Change in Canada 5*, C.R. Harington (ed.), Syllogeus Series, 55, Canadian Museum of Nature, Ottawa, 191–218.

Manuscript Sources Cited in Text

Cleveland, Parker, *MS Meteorological Observations,* Brunswick, Maine, 1807–1860. Parker Cleveland Papers, Bowdoin College Library, Special Collections, Brunswick, Maine.

Colin, Nicolas, *Journal of the weather from 1st of Jan. 1812 till the 27th of December 1819,* MS Collin, N. Nr 2, Kungliga Vetenskapsakademien, Centrum för Vetenskapshistoria, (Royal Swedish Academy of Science, Center for the History of Science), Stockholm.

Cooper, Isaac, *MS Meteorological Observations*, Cooperstown, New York, 1812–1818. New York State Historical Association, Cooperstown, NY.

Dewey, Chester, *MS Meteorological Observations*, Williamstown, Massachusetts, 1816–1838. Williams College Library, Archives and Special Collections, Williamstown.

Dibblee, Frederick, *Diary*, Microfilm No. F-363, Provincial Archives of New Brunswick, Fredericton.

Francis Logbook, M-656-1815F, James Duncan Phillips Library, Peabody Essex Museum, Salem, Massachusetts.

Jackson, Isaac H., *Weather Journal 1814–1848*, Isaac H. Jackson Papers, MSS qJ12RM, Cincinnati Museum Center, Cincinnati, Ohio.

Jefferson, Thomas, *MS Meteorological Observations*, Monticello, Virginia, 1810–1816. Massachusetts Historical Society.

Lincoln, Theodore, *MS Meteorological Observations,* Dennysville, Maine, 1816–1854. MS 2877, Misc. Box 40/3. Maine Historical Society, Portland, ME.

McCord Family, *MS Meteorological Observations*, Montreal, Canada, 1813–1826, McGill University Archives, Montreal.

Miles, Thomas Odbur, *Diary.* Graves Papers, Vol. 8, MC1156; Reel MC451. Harriet Irving Library, University of New Brunswick, Fredericton.

New Brunswick House Post Journal, 1815–1816. B.145/a/31-32. Hudson's Bay Company Archives, Winnipeg, Manitoba.

Sparks, Alexander, *MS Meteorological Observations*, Quebec City, Quebec, 1799–1819. McGill University Archives, Montreal.

March 1843: The Most Abnormal Month Ever?

John W. Nielsen-Gammon and Brent McRoberts

Abstract Weather observations taken four times daily across the present-day central and eastern United States permit a detailed reconstruction of monthly climate and daily weather during March 1843. Based on normalized departures of temperatures from the historic monthly mean, March 1843 may be regarded as the most anomalous month in recorded history for the central and eastern United States. In places, the average temperature for the month was more than 25°F (14°C) below normal, and the expected return frequency for such an anomaly is thousands of years. Using the detailed observations and recent analogs, the daily weather maps for March 1843 can be reconstructed. They show a storm track displaced well to the south of its normal position, with occasional snowstorms bringing winter weather to the Deep South, the Mid-Atlantic States, and New England. The extended severe winter weather caused hardship throughout the area.

Keywords March 1843 · Statistical climatology · Storm tracks · Deep South · Observer networks · Site exposure

1 Introduction

The geographical coverage of United States weather records during the 19th Century depends on both the geographical extent of the United States itself as well as the existence of organized professional or volunteer efforts to observe and record the weather. Volunteer weather records were supplemented in the early 1800s by a program run by the Surgeon General of the Army for army surgeons to make regular weather observations at US forts (Miller, 1931). Hopkins and Moran (2009,

J.W. Nielsen-Gammon (✉)
Department of Atmospheric Sciences, Texas A&M University, College Station, TX 77843-3150, USA
e-mail: n-g@tamu.edu

B. McRoberts (✉)
Department of Atmospheric Sciences, Texas A&M University, College Station, TX 77843-3150, USA

L.-A. Dupigny-Giroux, C.J. Mock (eds.), *Historical Climate Variability and Impacts in North America*, DOI 10.1007/978-90-481-2828-0_9,

Monitoring the climate of the Old Northwest: 1820–1895 of this volume) describe the growth of this network across the north-central United States, and Mock et al. (2007) describe its status in the late 1820s. As the United States' influence expanded westward in the 1820s and 1830s, it became possible to determine weather patterns across much of eastern North America for the first time.

Elias Loomis (1841) took advantage of the volunteer and organized weather observations to compile a record of a major storm system that traversed the United States in December 1836. His analysis of this storm appears to constitute the first comprehensive set of weather maps for a particular event on the synoptic scale in the United States. He subsequently conducted a separate analysis of two additional events in 1842 (Loomis, 1845), for the most part drawing conclusions from these limited analyses that hold up well in light of present-day meteorological knowledge.

By 1843 the fort network in the United States included various permanent or semi-permanent coastal and inland locations as well as more temporary forts on what then represented the frontier: Florida, the Canadian border, the Texas border, and the High Plains. Meteorological observations of temperature, wind, cloud conditions, cloud motion, precipitation, and (to a limited extent) barometric pressure were generally taken four times a day, at sunrise, 9 AM, 3 PM, and 9 PM solar time. With these observations, it is possible to create synoptic maps showing the location and evolution of weather systems.

Data records from these forts and various other individual data records have been scanned and placed online as part of the National Oceanic and Atmospheric Administration's Climate Data Modernization Program (CDMP), and digitization of this data record is in progress at the time of this writing. The available stations span the present-day United States east of about 96°W (Table 1). The non-fort records consist of a variety of volunteer observations, many of which utilized the Surgeon General forms while others recorded weather data on other forms or tables of their own devising, at different times of day. All such stations with data from March 1843 are listed in Table 1, and the station identifiers from Table 1 are plotted on the synoptic maps later in this chapter.

During March 1843, the weather records from these observations indicate a month of remarkable cold, with average temperatures possibly more unusual in the observed area (relative to normal conditions and the normal range of temperature variability) than in any recorded month before or since. Here, we document the unique aspects of the month and attempt to create weather maps for selected days during the month to gain a sense of what the weather patterns were like.

2 The Historical Data

The analysis of individual weather events during March 1843 uses the online scanned station records from the CDMP for the stations listed in Table 1. Most of these records consist of four-times-daily observations recorded on a monthly form bearing the label "*Form No. 3. Meteorological Register.*" The scanned form from

Table 1 Names and locations of forts and other stations for which scanned weather observations are available for March 1843 in the CDMP database

Station name	Alternative or present-day name	ID[1]	State	Lat(N)	Lon(W)
Mount Vernon Barracks	Mount Vernon Arsenal	MVA	AL	31.09	88.01
Fort Morgan		MOR	AL	30.23	88.02
Fort Smith		SMI	AR	35.39	94.44
Fort Trumbull		TRU	CT	41.34	72.09
Fort Leavenworth		LEA	KS	39.35	94.92
Fort Jesup		JES	LA	31.61	93.40
Fort Pike		PIK	LA	30.17	90.74
Fort Wood	Fort Macomb	WOO	LA	30.06	89.80
New Orleans (Barracks)	Jackson Barracks	NOB	LA	29.96	90.08
Baton Rouge		BAR	LA	30.44	91.19
Fort Fairfield		FAI	ME	46.77	67.83
Fort Kent		KEN	ME	47.25	68.59
Fort Preble		PRE	ME	43.65	70.23
Dearbornville Arsenal	Dearborn	DEA	MI	42.30	83.26
Detroit		DET	MI	42.33	83.05
Fort Severn	US Naval Academy	SEV	MD	38.98	76.49
Fort Adams		ADA	RI	41.48	71.34
Fort Sullivan	Eastport	SUL	ME	44.90	66.99
Hancock Barracks	Houlton	HAN	ME	46.13	67.84
Fort McHenry		MCH	MD	39.26	76.58
Fort Brady		BRA	MI	46.50	84.34
Fort Constitution		CON	NH	43.07	70.71
Fort Snelling		SNE	MN	44.89	93.18
St. Louis Arsenal		SLA	MO	38.59	90.21
Fort Gratiot	Port Huron	GRA	MI	43.00	82.42
Fort Mackinac		MAC	MI	45.85	84.62
Fort Columbus	Fort Jay	COL	NY	40.69	74.02
Buffalo Barracks		BUF	NY	42.90	78.87
Halifax		HAL	NS	44.63	63.59
Fort Moultrie		MOU	SC	32.76	79.85
Fort Towson		TOW	OK	34.03	95.26
Fort Washita		WAS	OK	34.10	96.55
Fort Gibson		GIB	OK	35.81	95.25
Fort Pickens		PIC	FL	30.33	87.29
Fort Shannon	Palatka	SHA	FL	29.65	81.63
St. Augustine	Ft. Marion; Castillo de San Marcos	MAR	FL	29.90	81.31
Fort Brooke	Tampa Convention Center	BRO	FL	27.94	82.46
Fort Atkinson		ATK	IA	43.15	91.95
Fort Croghan		CRO	IA	41.21	95.88
Fort Scott		SCO	KS	37.84	94.70
Fort Niagara		NIA	NY	43.26	79.06
Fort Hamilton		HAM	NY	40.61	74.03
Jefferson Barracks		JEF	MO	38.49	90.28

Table 1 (continued)

Station name	Alternative or present-day name	ID[1]	State	Lat(N)	Lon(W)
Fort Ontario	Oswego	ONT	NY	43.47	76.51
Fort Crawford		CRA	WI	43.04	91.14
Fort Winnebago		WIN	WI	43.56	89.43
Plattsburg Barracks	Plattsburgh Barracks	PLA	NY	44.68	73.44
Madison Barracks	Sackets Harbor	MAD	NY	43.95	76.11
Watervliet Arsenal		WVL	NY	42.72	73.71
West Point	West Point	WPT	NY	41.40	73.97
Allegheny Arsenal		ALL	PA	40.47	79.96
Carlisle Barracks		CAR	PA	40.21	77.18
Fort Mifflin		MIF	PA	39.88	75.21
Watertown Arsenal		WAT	MA	42.36	71.16
Fort Johnson		JOH	NC	33.92	78.04
Fort Macon		MCN	NC	34.70	76.68
Athens	Ohio Univ., Athens	OUA	OH	39.33	82.10
Newburyport		NEW	MA	42.81	70.87
Fort Monroe		MON	VA	37.00	76.31
Washington City	Hydrographical Office	WDC	DC	38.90	77.04
Gouverneur		GOV	NY	44.34	75.47
Flatbush		FLA	NY	40.65	73.96
Washington		WAR	AR	33.77	93.68
Steubenville		STE	OH	40.36	80.61
United States Naval Observatory	USNO	USN	DC	38.92	77.07
Cincinnati		CIN	OH	39.11	84.51

[1] Three-letter IDs are created here for the purpose of identifying stations on weather maps.

Fort Leavenworth (KS), the station that exhibited the largest normalized anomaly in March 1843 (see below), is shown in Fig. 1. Fort Leavenworth was established as a cantonment in 1827 to protect the Santa Fe Trail trade route with Mexico in the face of conflicts between traders and Native Americans (and, later, Texans). Soon the fort also served as a stabilizing influence among the Native American tribes that had been allocated land in the area after having been removed from the eastern United States. It rose to prominence as a major troop base during the Mexican-American War of 1846–1847 (Hunt, 1937).

The form includes columns for recording observations of the barometer, "thermometer attached" (to the barometer), "thermometer detached" (outdoor thermometer), clearness of the sky, wind (direction and speed), clouds (direction and speed), wet bulb, rain (timing and amount), and remarks. The form was used both by army surgeons and by a few civilian volunteer weather observers.

Metadata at the top of the form included the station name, the latitude and longitude of the station, and the altitude of the barometer above sea level or some fixed landmark such as a river. Longitude, when provided on the form, was most often given as the number of degrees west or east of Washington, DC, since the Greenwich Meridian was not yet in standard international use as a global reference

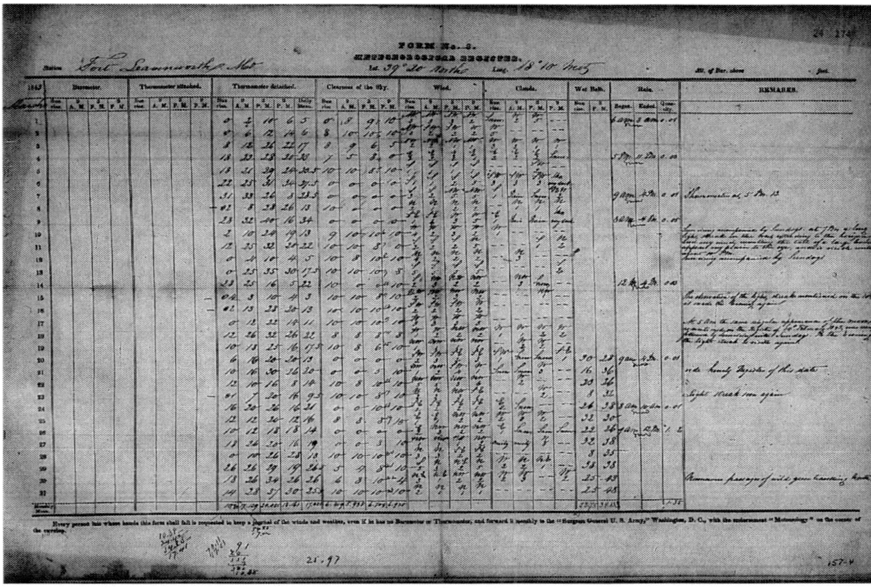

Fig. 1 The March 1843 weather records from Fort Leavenworth, scanned and posted in the CDMP database

point for longitude. The latitude and longitude values on the forms generally appear to be somewhat inaccurate, so the more precise locations in Table 1 were obtained through consultation of online historical records and topographic maps.

While detailed instructions are not present on the form shown in Fig. 1, a revised version of the form with instructions printed on the form was employed beginning April 1843 at some new stations. The detailed instructions note that the outdoor thermometer was to be located "in the shade and open air". Based on examination of 9 AM and 3 PM temperature observations at nearby stations, and comparison with normal expected temperature changes between 9 AM and 3 PM, it appears that the thermometers at Fort Fairfield (ME), Fort Gratiot (MI), Fort Mackinac (MI), Fort Scott (KS), and Washington (AR) may have had morning sun exposure, while the thermometers at Baton Rouge (LA), Fort Kent (ME), and Washington (AR) (Arkansas) may have had afternoon sun exposure. All temperatures will be presented in Fahrenheit in keeping with the meteorological practice at the time.

Besides sun exposure, numerous differences exist between temperature observation practices in 1843 and the present day. These differences include instrument siting, observation times, and the instruments themselves (Chenoweth, 1992; Chenoweth, 1993; Conner, 2009, Weather station history and introduced variability in climate data of this volume). Taken as a whole, such differences can easily lead to discrepancies of a degree C or more between historical observations and what would have been obtained by modern practice. This complicates the inference of long-term trends but is not a significant issue for the present work, considering the size of the temperature anomalies that were observed.

The wind direction was defined as "course from", consistent with modern meteorological practice. For example, a NW wind was blowing from the northwest toward the southeast. Quoting the April 1843 form, "The force of the wind is estimated in numbers 0 being a calm, 1 a very gentle breeze, 2 a gentle breeze, 3 a fresh breeze, 4 a strong wind, 5 a very strong wind, 6 a violent storm &c." These wind speed descriptions may be calibrated against the corresponding descriptions in the well-known Beaufort wind scale, but the resulting speed estimates appear to be too strong compared to modern-day climatological conditions at most stations. Thus, the wind force estimates are utilized here in their original numerical form, and wind speeds are plotted on weather maps using the following convention: direction indicator alone for force 0, a short barb for force 1, a long barb for force 2, one long barb and one short barb for force 3, and so forth, except that a pennant corresponds to missing wind force information. The direction indicator extends from the station in the upwind direction. The cloud motion (speed and direction) was reported in a similar fashion.

Sky cover was reported in what appear to be tenths of coverage, but in a manner opposite modern practice. As instructed by the form, "The clearness of the sky will also be marked in numbers, 0 representing entire cloudiness, 1 a slight degree of clearness and so on till 10, entire clearness." These reported numbers are assumed here to be tenths for the purpose of plotting the weather observations.

While no instructions were provided for "Remarks", observers noted the type of precipitation and other significant weather. This information was used to infer present weather at each station for the map times shown below. Precipitation amounts were given in inches, and the same units are reported here.

3 Methods

The analysis of the climatology of March 1843 compared to other months draws upon United States stations in the National Climatic Data Center's Global Historical Climatology Network (GHCN) (Peterson and Vose, 1997; Peterson et al., 1998). Average daily temperatures in March 1843 are available from 37 stations. Of these, most stations are located in the Northeast US. The GHCN stations used in estimating the magnitude of the regional temperature anomalies are listed in Table 2. To obtain an average temperature anomaly representative of the geographical domain in which data were available, data from only 4 stations in the Northeast US (those with especially complete weather records) are included. These retained stations, along with those elsewhere in the country, are listed in Table 2.

Monthly average daily-mean temperatures are computed for each of the stations in Table 2 over the full available period of record: 1835–2005. For each station and month, in addition to the overall monthly average, the standard deviation of monthly mean temperatures about that average is also computed. The departure of average temperature in a given month at a given station from that station's long-term mean monthly temperature is then normalized by dividing by the standard deviation. For example, a normalized departure of –1.0 means that the average temperature in

Table 2 GHCN stations utilized in monthly climatology

Station name	State	Station number	Lat(N)	Lon(W)
Key West/Int.	FL	42572201000	29.55	81.75
Savannah/Muni.	GA	42572207000	32.13	81.20
Baton Rouge	LA	42572231009	30.53	91.13
Princesse Anne	MD	42572402003	38.22	75.68
Portsmouth-Sciotoville	OH	42572425005	38.75	82.88
Hillsboro	OH	42572426002	39.20	83.62
Leavenworth	KS	42572446007	39.32	94.93
New York Central Park	NY	42572503001	40.78	73.97
Hanover	NH	42572604002	43.78	72.28
Minneapolis/St. Paul	MN	42572658000	44.88	93.22
Amherst	MA	42574491001	42.38	72.53
Blue Hill Observatory	MA	42574492000	42.22	71.12
Fort Scott	KS	42574542006	37.85	94.70
Wilmington	NC	42574699001	34.30	77.90
Natchez	MS	42574854004	31.55	91.38

that particular month is one standard deviation below the long-term climatology for that month. This normalization allows for direct comparison of the unusualness of weather at coastal vs. inland sites and in winter vs. summer.

The 1843 forts data are more geographically detailed than the 1843 GHCN data, so the forts data are used for determining the spatial pattern of temperature anomalies. Since true maximum and minimum temperatures are not available for each station, the 3 PM and sunrise temperature observations are used in their place.

A quantification of the difference between sunrise/3 PM and minimum/maximum monthly averages may be made with respect to Fort Snelling (MN), located near the core of the largest temperature anomalies discussed below. Fisk (1984) developed detailed regression models, taking into account all available temperature observations as well as length of day, wind direction, and cloud cover, to estimate maximum and minimum temperatures with respect to a midnight-to-midnight reporting period. In the case of March 1843, the estimated average maximum temperature was 1.1°F (0.6°C) warmer than the average 3 PM temperature, the estimated average minimum temperature was 3.6°F (2.0°C) colder than the average sunrise temperature, and the overall estimated average temperature was 1.1°F (0.6°C) colder than the average of the sunrise and 3 PM temperatures.

To determine temperature anomalies, these observations are compared to the PRISM 1971–2000 normals (Daly et al., 2004), which are of sufficiently high spatial resolution to resolve climatological temperature variations at coastal fort locations. Aside from any changes in what constitutes "normal" temperatures over the past 1–2 centuries, the biases described in the previous paragraph must be taken into account when the resulting maps are interpreted.

The construction of weather map analyses was aided by the identification of weather analogs from the period 1957–2006. Geographically-representative daily temperature observations from overlapping four-day periods were compared with the more recent analyses within the months of November through March, and the root-mean-square (RMS) difference computed. The use of a four-day window ensured that the analogs would be forced to agree both with the instantaneous temperature pattern and the evolution of that pattern. The fidelity of the analogs is measured objectively by the RMS differences and subjectively by the agreement of the analog isobar pattern with the March 1843 wind observations. On the weather maps shown below, isobars are schematic and based on the wind patterns and analogs; the few available barometric pressure observations were insufficiently calibrated to be useful in the direct analysis of pressure.

4 An Outlier Month

4.1 Mean Temperature Anomalies

The largest normalized standard deviations for temperature for all months, 1835–2005, among all fifteen GHCN stations included in this study, are shown in Table 3. The most impressive anomaly, 4.52 standard deviations below the respective mean, occurred at Leavenworth (KS) in March 1843. The other two stations in the northwestern portion of the area of data coverage, Minneapolis/St. Paul (MN; formerly Fort Snelling) and Fort Scott (KS), also had anomalies more than 4 standard deviations below the mean. Only one other station-month, Savannah/Muni. (GA) in November 1845, departed more than 4 standard deviations from the mean.

Table 3 Greatest normalized anomalies from monthly mean, selected GHCN stations, 1835–2005

Rank	Station	State	Month	Year	Anomaly
1	Leavenworth	KS	March	1843	–4.52*
2	Savannah/Muni.	GA	November	1845	–4.07*
3	Minneapolis/St. Paul	MN	March	1843	–4.05
4	Fort Scott	KS	March	1843	–4.05
5	Key West/Int.	FL	July	1836	–3.94*
6	Minneapolis/St. Paul	MN	June	1842	–3.79*
7	Baton Rouge	LA	April	1840	+3.64*
8	Key West/Int.	FL	October	1835	–3.63*
9	Portsmouth-Sciotoville	OH	April	1844	+3.60*
10	Baton Rouge	LA	April	1844	+3.57
11	Key West/Int.	FL	August	1870	+3.56*
12	Wilmington	NC	May	1844	+3.56*
13	New York Central Park	NY	October	1836	–3.52*
14	Hanover	NH	February	1981	+3.51*
15	Savannah/Muni.	GA	March	1841	+3.51*
16	New York Central Park	NY	June	1836	–3.51*
17	Hanover	NH	July	1868	+3.50*
18	Savannah/Muni.	GA	October	1919	+3.50*

*Greatest normalized anomaly in that particular month and year.

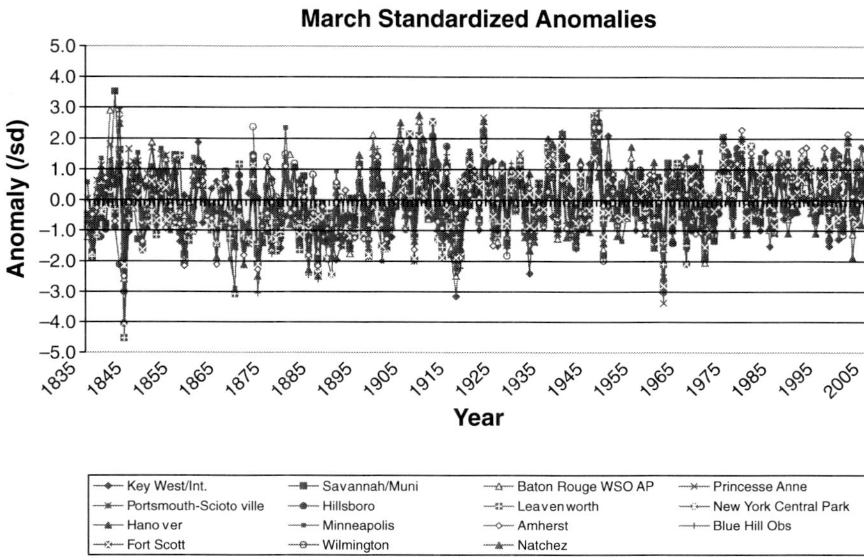

Fig. 2 March standardized anomalies (anomalies from the 1835–2005 mean divided by the 1835–2005 standard deviation) for each of 15 selected GHCN stations in the present-day central and eastern United States

To provide a visual sense of how unusual March 1843 was, the normalized departures from the mean of all 15 stations are plotted for the month of March from 1835 through 2005 in Fig. 2. The 1843 anomaly of –4.52 stands out, as does the 1841 anomaly of +3.51. Anomalies beyond –3.0 are exceeded in four other Marches (1867, 1872, 1915, and 1960), while no other years have March anomalies above +3.0.

With 27,127 individual monthly observations, a normal distribution would yield an expected value of 2 departures exceeding 4 standard deviations and 0.2 departures exceeding 4.5 standard deviations. Based on these statistics, and assuming a stable climate, the expected return period for a 4.5 standard deviation monthly anomaly at any given location is 10,000–15,000 years. In that sense, if the data from (Fort) Leavenworth are correct and can be described by a normal distribution (but see below), the average temperature in March 1843 may have been the most unusual at that location of any month there since the last ice age!

Of the 18 monthly anomalies exceeding 3.5 standard deviations, 14 occur in the decade 1836–1845. This could be due to any of the following three causes: (1) that particular decade may have had remarkably extreme weather; (2) changes in "normal" over the past 170 years may have caused what would have been normal in the mid 19th century to be unusual compared to a longer record; or (3) early weather data collection may have been subject to greater biases or inconsistencies, leading to unusual weather records. Rosendal (1970) proposed heretofore undocumented volcanic eruptions as an additional possible cause, but such eruptions, if they took place, remain undocumented.

Explanation #2 is easy to dismiss. While the climate has certainly undergone changes over the past two centuries, the magnitude of the early temperature excursions are much larger than any changes in average conditions. For example, in March, the average normalized anomaly among all 15 stations during the period 1836–1845 was –0.05. This finding also indicates that systematic biases among weather records were relatively small.

The remaining viable explanations are real variability or inconsistent/erroneous records. If the early weather records really were inconsistent or erroneous, one would expect little agreement among neighboring stations in the event of extremely anomalous recorded conditions. Fig. 2 highlights the consistency in the sign and magnitude of anomalies among the 15 stations during most of the period. Viewed in that regard, the March 1841 +3.51 anomaly at Savannah is suspicious, because the next largest normalized anomaly is less than 1.0.

Table 4 Months with the largest normalized temperature anomalies (NA) averaged across all stations

Month	Year	# Stns[1]	NA	Largest[2]	station	smallest[3]	station
Dec	1989	15	–2.42	–3.14	Hanover (NH)	–1.47	Key West (FL)
Mar	1843	15	–2.31	–4.52	Leavenworth (KS)	–0.16	Key West (FL)
Apr	1857	10	–2.25	–3.34	Hillsboro (OH)	–1.17	Hanover (NH)
Dec	1876	13	–2.22	–2.49	New York CP (NY)	–1.85	Hanover (NH)
Mar	1945	14	+2.14	+2.77	Blue Hill Obs (MA)	+1.39	Fort Scott (KS)
Oct	1869	9	–2.13	–3.35	Natchez (MS)	–1.19	Hanover (NH)
Mar	1960	15	–2.11	–3.37	Princess Anne (MD)	–1.05	Hanover (NH)
Feb	1895	13	–2.11	–2.96	Wilmington (NC)	–0.90	Minneapolis (MN)
May	1917	15	–2.10	–2.95	Amherst (MA)	–0.88	Minneapolis (MN)
Jan	1977	15	–2.06	–3.01	Hillsboro (OH)	–1.10	Hanover (NH)
Jan	1857	10	–2.04	–2.55	New York CP (NY)	–0.93	Baton Rouge (LA)
Dec	1917	15	–2.04	–2.64	Hanover (NH)	–1.37	Natchez (MS)
Oct	1947	14	+2.02	+3.03	Blue Hill Obs (MA)	–0.18	Key West (FL)
Oct	1836	11	–2.02	–3.52	New York CP (NY)	–0.99	Portsmouth (OH)[4]
Jan	1918	15	–2.01	–2.78	Portsmouth (OH)	–1.08	Key West (FL)

[1]Number of stations available in a given month out of the 15 listed in Table 2.
[2]The largest departure from average among the available stations for that month.
[3]The smallest departure from average (or largest departure of the opposite sign) among the available stations for that month.
[4]The smallest recorded on Oct 1836, –0.89 at Amherst, is inconsistent with surrounding stations so is regarded as suspicious.

To further explore the issue of consistency, it is useful to consider the normalized anomalies averaged over all 15 stations. Table 4 shows the most extreme monthly average conditions across the entire network. The average anomalies are much more homogeneously distributed through time than the individual station anomalies, with only one year from the 1830s listed in Table 4, one from the 1840s, and no decade with more than two such years. This even distribution implies that many (though not all; see Dupigny-Giroux, 2009, Backward seasons, droughts and other bioclimatic indicators of variability of this volume) of the extreme individual anomalies recorded in the early part of the data record and shown in Table 3 are at least partly erroneous.

On the other hand, March 1843 appears prominently in Table 4, second only to December 1989 in the unusual nature of its cold weather. Cold weather was experienced in March 1843 throughout almost the entire set of stations. Perhaps the only thing keeping March 1843 from being the most anomalous in Table 4 was the fact that the unusually cold weather did not reach Key West (FL). Unlike many other exceptional records in the 1830s and 1840s that were limited to a small number of stations, the March 1843 anomaly is confirmed by regionally-representative stations throughout the eastern United States.

While the unusual events in Table 4 are distributed evenly in time, they are not evenly distributed throughout the year. No warm-season events (June–September) were unusual enough over the entire network to make the list. The events are also not evenly distributed with respect to sign: 13 out of 15 feature negative temperature anomalies. This indicates that the anomaly distribution is skewed, so that extremely cold temperatures are more common than extremely hot ones. The return period estimate for Fort Leavenworth (KS) earlier in this section likely exaggerates the unusual nature of March 1843.

4.2 Daytime and Nighttime Temperature Anomalies

The 3 PM temperatures at locations north and west of the Ohio River averaged below freezing for the month of March 1843 (Fig. 3a). Departures from normal were greatest at the edge of the High Plains, at Fort Leavenworth (KS) and Fort Croghan (IA), where 3 PM temperatures were at least 28°F (16°C) below the normal maximum temperature. If the Fisk (1984) correction at Fort Snelling applies, the true anomaly was around 27°F. Temperatures were at least 10°F below normal maxima everywhere except in parts of Florida, New York, and New England. The extension of anomalously cold temperatures southwestward along the Appalachian mountains in Fig. 3a (and elsewhere) is based on observations in Washington DC and Pennsylvania and expectations of the dominance of cold-air damming (e.g., Bailey et al., 2003) with the March 1843 storm track (see below).

The sunrise temperatures are below normal minimum temperatures everywhere except northern Maine (Fig. 3b). At Fort Snelling (MN), the sunrise temperatures averaged –5°F (–21°C), compared to the modern normal minimum temperature of 21°F (–6°C). This is all the more remarkable when one utilizes the Fisk

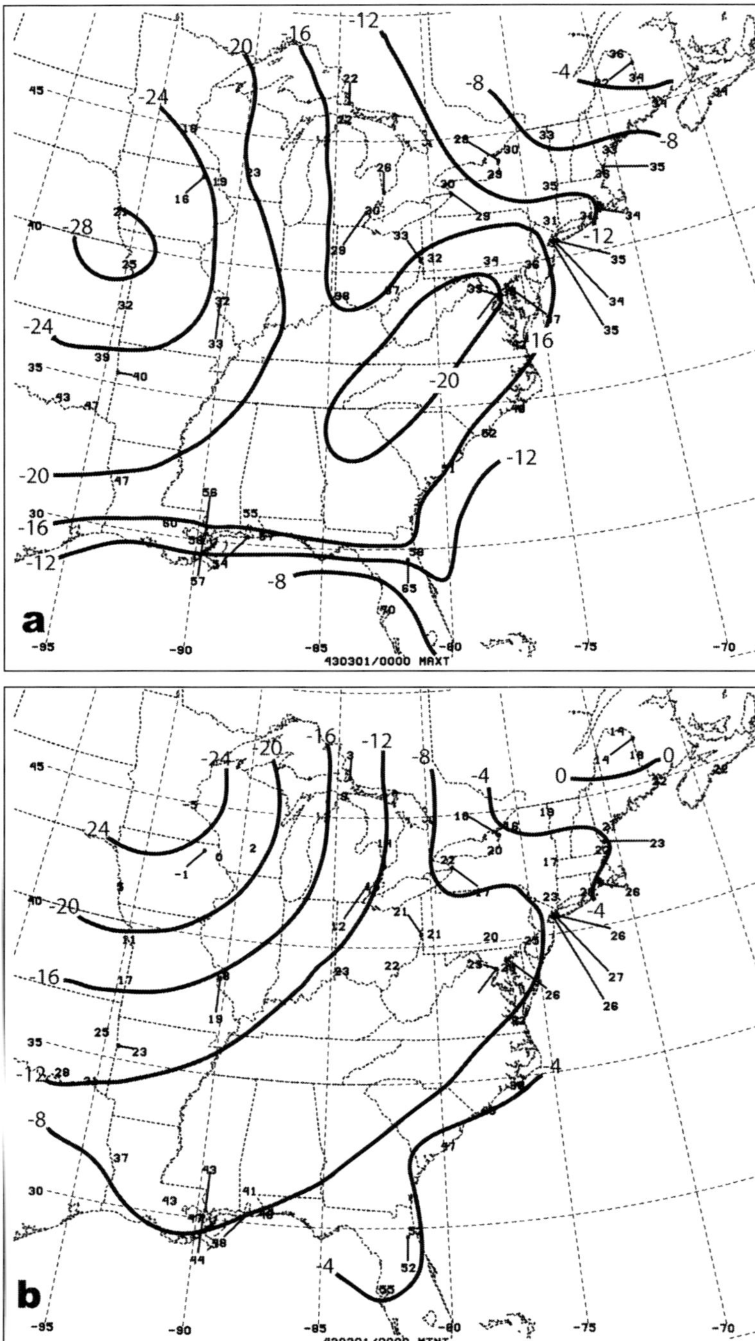

Fig. 3 (continued)

(1984) estimate for midnight-to-midnight average minimum temperature for March 1843: minimum temperatures in March 1843 were close to 30°F (17°C) below normal.

Taking the average of the anomalies shown in Fig. 3 as an estimate of daily mean temperature anomalies gives mean temperature anomalies of –24°F (–13°C) across southern and western Minnesota, western Iowa, extreme northwestern Missouri, northern Kansas, and at least the eastern portions of Nebraska and South Dakota. The Fisk (1984) adjustments for Fort Snelling (MN) imply that the true average temperature anomaly was at least 1°F (0.6°C) colder. Rosendal (1970) extrapolated the pattern of his anomaly analysis to infer departures of –30°F (–17°C) across North Dakota, but the analysis in Fig. 3 is consistent with a maximum departure of –27°F (15°C) (adjusted to –28°F) in southeastern South Dakota.

The month of March 1843 combined truly exceptional temperatures in the north-central United States with network-wide normalized anomalies that were the second-largest on record. It seems fair to say that, in the sense considered here, March 1843 had some of the most consistently unusual temperatures of any month in the historical data record in the eastern half of the United States.

One familiar period of cold temperatures prior to the comprehensive historical record is the so-called "Year Without a Summer", 1816. However, as discussed by Stommel and Stommel (1983), Chenoweth (2009, Case studies of the summer weather of 1816 in North America of this volume), and Dupigny-Giroux (2009, Backward seasons, droughts and other bioclimatic indicators of variability of this volume), the summer of 1816 was marked by the occasional occurrence of exceptionally cold temperatures in early June, early July, and mid-August, but it did not feature sustained cold temperatures. In that respect, it was no match for March 1843.

One final measure of the unusualness of the weather at Fort Leavenworth (KS): according to the Leavenworth GHCN data, the average temperature in March 1843 was colder than 139 Decembers out of 140, 141 Januarys out of 152, and all 152 Februarys. At Fort Snelling (MN), according to Fisk (2007), March 1843 was colder than all but 1% of Decembers, 10% of Januarys, and 3% of Februarys. So March 1843 would have been remembered as unusually cold even if it had occurred in the middle of the winter!

Fig. 3 (**a**) Mean March 1843 3 PM temperatures (°F, plotted) and their departures from 1971 to 2000 mean maximum temperatures (°F, contoured). Data from some stations have been displaced to prevent overlap; straight lines connect these station data to the true location of the station. (**b**) Mean March 1843 sunrise temperature (°F, plotted) and their departures from 1971 to 2000 mean minimum temperatures (°F, contoured)

5 The Weather During March 1843

5.1 Temperature Thresholds

The sustained nature of the March 1843 cold temperatures was impressive by any standard. At Fort Leavenworth (KS), for example, the 3 PM temperature was always at least 12°F (7°C) below the normal March maximum temperature.

One common mark of frigid weather is when the temperature drops below 0°F (–18°C). In March 1843, this occurred as far south as Fort Smith (AR) (Fig. 4). Such frigid events occurred occasionally in inland portions of the Northeast as well, but they were remarkably common in the upper Mississippi River region. Fort Snelling's (MN) thermometer registered below zero (°F) on 22 out of 31 March days, and five other stations in the area recorded temperatures below zero on at least 10 separate days during the month.

Another important temperature threshold, this one relevant for the melting of snow cover, is the number of days the temperature remained at or below freezing (32°F, or 0°C) (Fig. 5a). At Fort Snelling (MN), the temperature never reached

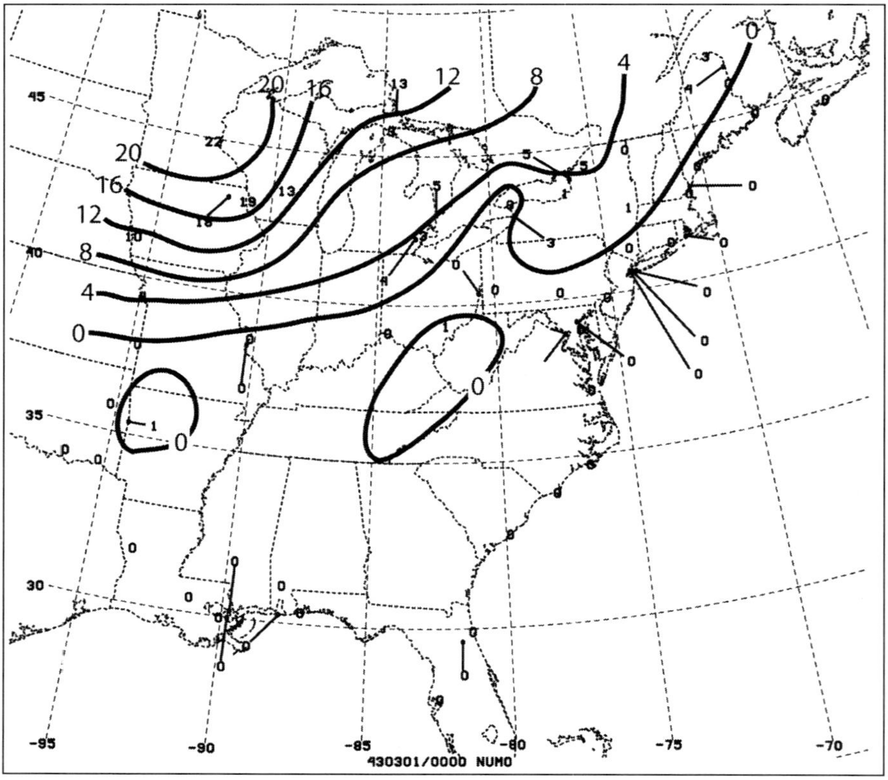

Fig. 4 Total number of days in March 1843 in which the temperature dropped below 0°F (–18°C). In this and the subsequent two figures, no attempt has been made to account for likely local variations caused by the Great Lakes

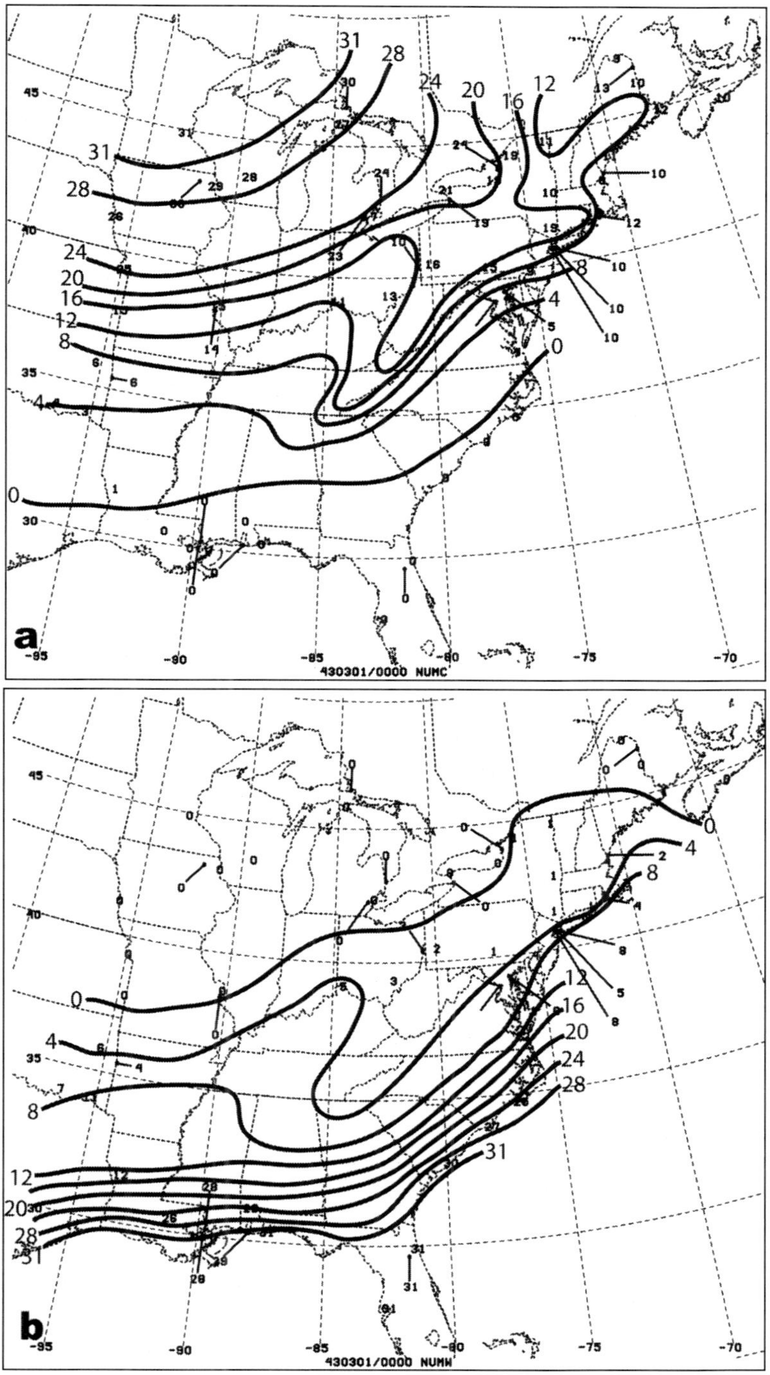

Fig. 5 (**a**) Total number of days in March 1843 in which the temperature was never above 32°F (0°C). (**b**) Total number of days in March 1843 in which the temperature was never below 32°F (0°C)

freezing, part of a string of 60 consecutive days below freezing that also included all of February (Ludlum, 1968, p. 154). In the Great Lakes region, temperatures remained below freezing on at least half the days, and a full day below freezing was registered as far south as Fort Jesup (LA). The winter snow did not leave the ground in southern Michigan and Ohio until early April (Ludlum, 1968, p. 154). In the Northeast, even at coastal locations, the temperature typically remained at or below freezing for 10 or more days out of the month. Snowstorms in March and early April contributed to a 3–4 foot (1 m) deep snowpack in mid-April stretching from upstate New York to Maine (Ludlum, 1968, p. 45).

A third temperature threshold, critical for tender vegetation and agricultural planting, is the number of days in which the temperature remains at or above freezing (Fig. 5b). In March 1843, a freeze-free month was found only in Florida and along the immediate coast from Georgia to Texas. Days in which the temperature dropped below freezing were common throughout the month even within most parts of southern coastal states (except Florida). In the rest of the country days that did not freeze were few and far between.

5.2 Synoptic Maps

The persistent cold weather and the enhancement of the usual north-south temperature gradient across the southern United States suggest that a semi-permanent frontal zone and storm track was established from Texas across the Carolinas and into New England. A few synoptic maps have been chosen to illustrate the weather patterns that were present during most of the month. These maps include the two most notable snowstorms during the month and a particularly cold late March day across the eastern United States.

Ludlum (1968, pp. 43–44, 104) reports that the storm of 16–17 March 1843 brought 15″ (0.4 m) of snow to Shelbyville, TN, 8″–13″ (0.2–0.3 m) in Little Rock, AK, Memphis and Nashville, TN and generally 12″ (0.3 m) across Virginia, Washington, Baltimore, MD, Philadelphia, PA and New York City, NY. The surface map for 3 PM on 15 March 1843 (Fig. 6) reveals that the northeastern United States was still under the influence of a previous storm system that had moved into Eastern Canada and was causing strong cold advection. Temperatures were generally below freezing even at this hour upstream of the Appalachians, but the coldest temperature was +4°F (–16°C) at Fort Atkinson (IA), under sunny skies! (The wind directions at Fort Atkinson are generally inconsistent with surrounding stations and are attributed to observer error.)

The preceding cold front is presumed to have pushed well out into the Atlantic, given the cold temperatures throughout the Northeast, and a secondary trough is analyzed through the Mid-Atlantic States. The original cold front probably did not pass through Florida, because the strong temperature gradient there and in the Deep South implies a synoptic-scale frontal zone. The wind direction at Fort Brooke (FL) is presumed to be incorrect by 180°.

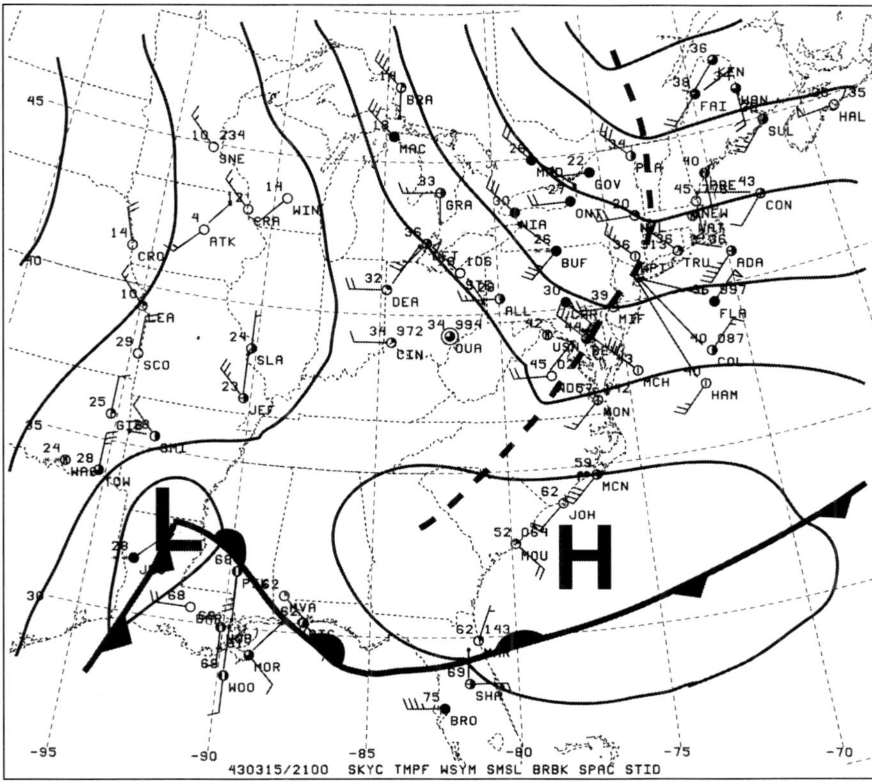

Fig. 6 Reconstructed surface weather map, 3 PM local time March 15, 1843. Wind barbs correspond to subjective wind scale described in Section 2, while other aspects of station plots are conventional. Isobar analysis is semi-schematic and is intended to have a roughly 4 hPa isobar spacing

The temperature gradient is at its most impressive in the vicinity of the low pressure system in Louisiana. In the warm sector, temperatures are in the upper 60s (°F) (~20°C), while at Fort Jesup (LA) the temperature is 28°F (–2°C). The diary of Adolphus Sterne, the postmaster of Nacogdoches, Texas, about 140 km west of Fort Jesup, records the remarkable weather on that day (some punctuation modified for clarity):

> 'What not a day may bring forth!' may well be said of this day – it was cloudy but warm in the morning like the Climate of Italy – at 9 A. M. the wind suddenly changed to the north, a severe Storm Thunder and lightning, with a tremendous rain, growing colder and colder every second – rain continued till noon when it commenced Snowing, and at 4 P. M. what was Italy this morning is now changed to Seberia snowing and freezing – the western mail was made up and I had determined to take it to Judge Terrells myself rather then there should be a failure, but, [with] the sudden change of weather, tremendous rain, cold, and snow, I determined not alone not to go myself, but would not [even] have turned a common curr dog out of Doors... (Smither, 1931).

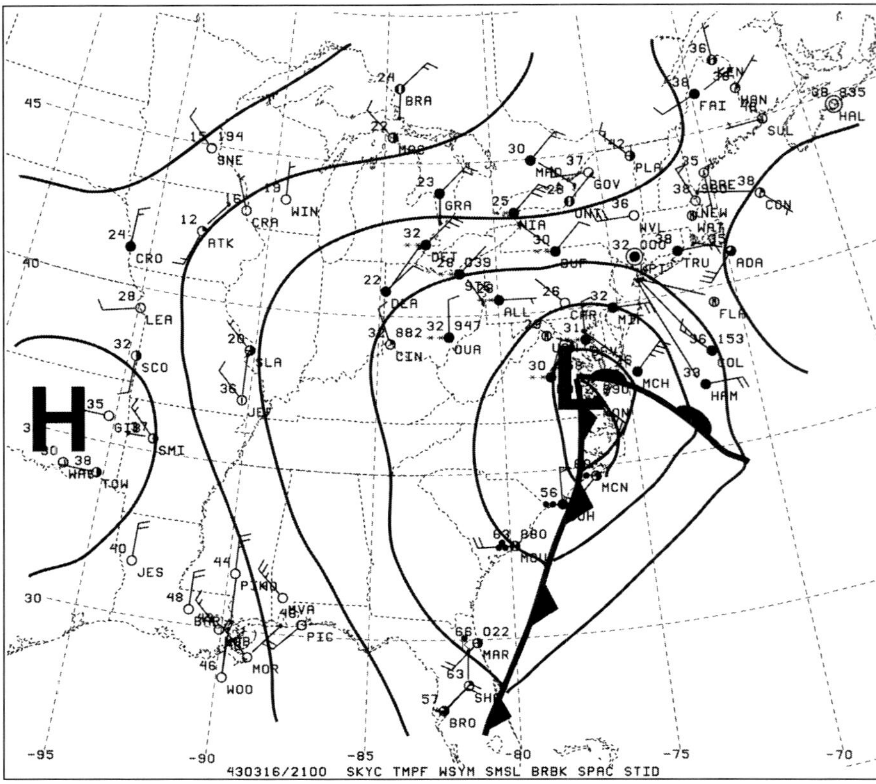

Fig. 7 Reconstructed surface weather map, 3 PM local time March 16, 1843

Similar reports of thunder or heavy rain accompanied the cold front's advance across Louisiana and Alabama later that evening. By the following afternoon the low pressure system had moved across the southeastern states and had just reached Chesapeake Bay (Fig. 7). The lack of a prominent high pressure system to its north caused the circulation ahead of the low to be confined to a relatively small distance from the low. This factor, along with its fairly rapid movement, allowed the snowfall to be limited to approximately a foot (0.3 m) despite the near-ideal path followed by the storm and the cold temperatures over land.

This period of time, 15–16 March 1843, was by one measure the most unusual period of weather during this most unusual month. The analog identification for this period yielded potential analogs that were in much poorer agreement with the corresponding observed temperatures than were analogs for other periods of time during March 1843. Likewise, the pressure patterns for the analogs generally did not include the snowstorm itself. Both the remarkable intensity of the temperature gradient in the southern United States and the rapid motion of the storm system on

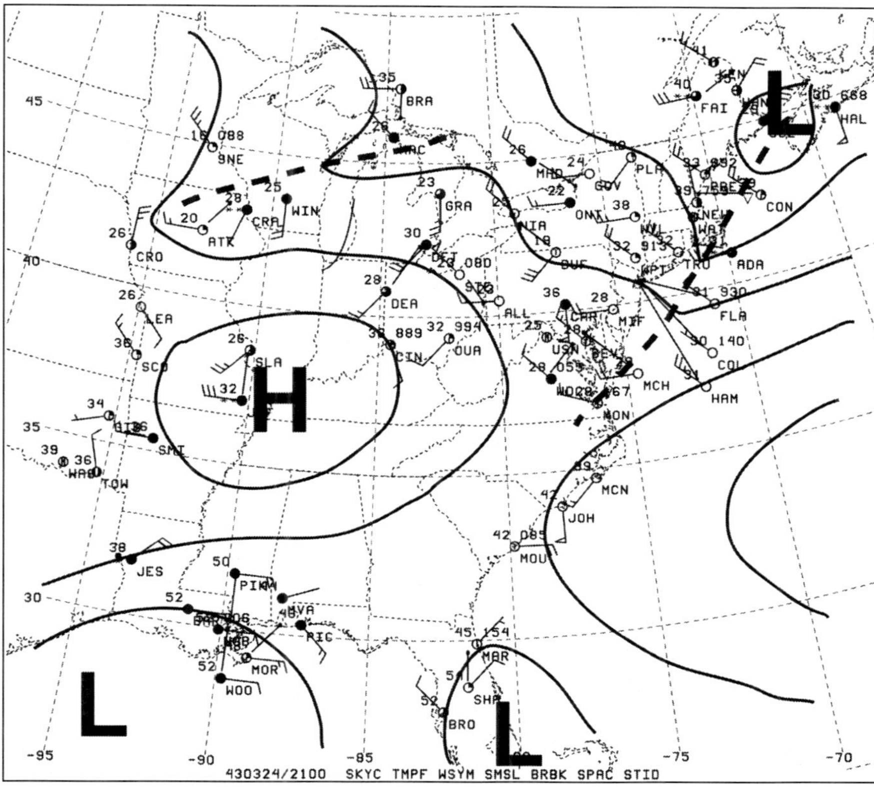

Fig. 8 Reconstructed surface weather map, 3 PM local time March 24, 1843

the heels of a larger-scale cyclonic flow are unusual aspects of the weather situation, a combination that does not seem to have been observed in the era of modern weather analysis.

The weather map for 3 PM 24 March 1843 (Fig. 8) features several unusual aspects, yet close analogs were found. First, the low pressure center in the Gulf of Mexico illustrates that the storm track was at times very far south for the month of March. This particular low would produce the final major snowfall of the season for the Southern US. Adolphus Sterne by this time appears to have grown weary of the extended winter weather, describing the weather on March 24 as "...very cold to day, rain Sleet, Snow, and most rascally weather..." (Smither, 1931). In Natchez, Mississippi, however, at least some people maintained a positive attitude, as recorded by William Johnson, a barber and landholder: "...a short time after dinner we had the snow commence falling and in the course of three hours the town was perfectly white. Then the citizens commenced to throw snowballs." (Hogan and Davis, 1951)

In the Northeast, the blustery day was one of the coldest of the month. Temperatures at 3 PM were in the mid to upper 20s (°F) (a few degrees below

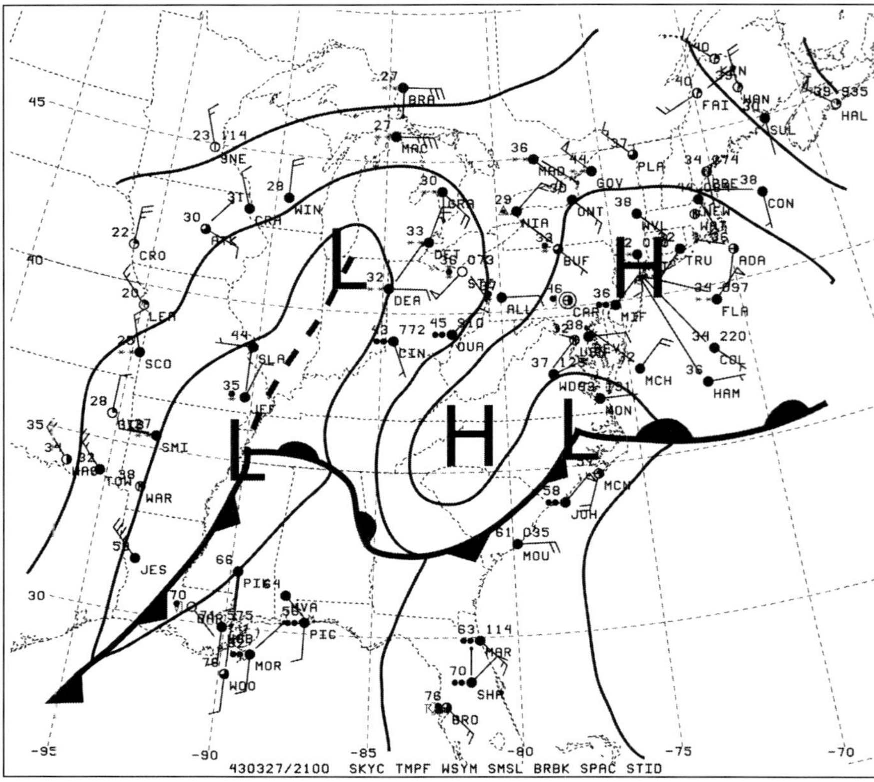

Fig. 9 Reconstructed surface weather map, 3 PM local time March 27, 1843

0°C) from Washington, DC to Philadelphia (PA). This is far from typical weather for the fourth day of spring. Conversely, Fort Brady (MI) observed its only temperature above freezing for the entire month. Since temperatures nearby are colder, it appears that the westerly wind must have taken the air across a large expanse of still-open water on Lake Superior prior to reaching the Fort.

The final map is for 3 PM on 27 March (Fig. 9). The complex weather pattern includes a primary low pressure center near Memphis, TN, an inverted trough extending to the north and producing some snow in Michigan, and a secondary low pressure center along the coastal front in North Carolina. All of these features appear in various temperature analogs for this day and are consistent with the wind and weather observations. The upper-level maps corresponding to these analogs indicate a large, mobile upper-level trough moving across the southern Plains at this time.

This storm represents the end of the extreme southward displacement of the storm track. Rather than following a more southern course, the storm tracked east-northeastward toward the central Appalachians. Ludlum (1968, p. 44) concludes from the Fort Johnson (NC) observations that the center of the storm passed east of Cape Hatteras, but observations from Fort Macon (NC) and Fort Monroe (VA) belie that assessment. Instead, Fort Johnson was probably influenced by a persistent coastal front, while the primary low and the Carolina coastal disturbance tracked toward the northeast and merged on March 28 over southeastern Pennsylvania.

The large geographic extent of the overall cyclonic circulation brought a great deal of moisture to the Northeast, while the track of the storm (essentially close to present-day Interstate 95 from Pennsylvania to Maine) meant that in the more populated areas most of the precipitation fell as rain. The highest total reported by Ludlum (1968, p. 44) was 4.45″ in Brunswick, Maine. In Gardiner, Maine, H. Gardiner reported:

> The storm of wind, snow and rain of this day, (March 28th) I think somewhat exceeds every storm of the season. With the wind E.S.E. to S.E., in five or six hours we had eight or nine inches of very solid snow. In 13 or 14 hours more we had a rain, which if it had valen in snow, as the temperature almost permitted, there would have been 18 inches or 2 feet more. As it is – we had in our yards and gardens about four feet of snow this morning. (Ludlum, 1968, p. 45)

6 Summary of the Month

March 1843 was but the final phase in the extended period of winter weather that had begun as early as November 1842. Ludlum (1968, pp. 154–155) describes widespread livestock and animal losses in the Midwest and notes that the winter of 1842–1843 showed the Midwestern farmers that winters could be as brutal there as in New England. In the Northeast, at least one writer found the winter without precedent:

> To sum up the whole: the writer of this article after more than 50 years observation, and who has on the subject a pretty clear recollection, has never known a period of five months so distinguished for cold, snow, & tempest in the aggregate, and so much snow remaining at the end of March as in March 1843. (H. Gardiner, quoted by Ludlum, 1968, p. 45).

The observers at military forts might have been expected to provide some remarks on the unusual weather, but most of their remarks were instead devoted to an equally unusual phenomenon, the Great Comet of 1843. This sungrazer comet was remarkable for the length of its tail, and it was visible shortly after sunset from early March through the rest of the month. The presence of the comet provided some small additional incentive to make the regular 9 PM weather observations during this cold month.

The observer at Fort Jesup (LA) did provide a summary of the month, which serves as a suitable description of the impact of March 1843 across most of the southern United States:

> This month as will be seen by the record has been one of extraordinary severity, the cold having been greater and more constant than that of (many?) of the winter months recorded or recollected by the oldest inhabitants. Vegetation which had previously appeared was retarded or destroyed and fruit trees which had bloomed in February stripped of their blossoms & most of the gardens have to be replanted and corn fields resown.

Can something like March 1843 happen again? Changes in global temperatures may in the future prove to make such an event less likely, but at this point we can only note that based on the limited historical record such an event can be expected less than once a century, and the core of such an anomaly will strike any particular location much less than once a millennium. Nonetheless, an observer on the brink of 1843 blessed with similar knowledge of statistical climatology as today would have had little expectation that such an event was imminent.

Acknowledgments About a decade ago, many of the weather records for this month were brought to the attention of John Griffiths, who served as Texas State Climatologist for nearly thirty years. Dr. Griffiths then passed the records along to the lead author. This paper is dedicated to John Griffiths' memory. Roberto Gasparini assisted with an earlier version of this study, and his help is gratefully acknowledged, and Karen Andsager provided valuable assistance.

References

Bailey CM, Hartfield G, Lackmann GM, Keeter K, Sharp S (2003) An objective climatology, classification scheme, and assessment of sensible weather impacts for Appalachian cold-air damming. Weather Forecast 18:641–661

Chenoweth M (1992) A possible discontinuity in the US historical temperature record. J Clim 5:1172–1179

Chenoweth M (1993) Nonstandard thermometer exposures at US cooperative weather stations during the late nineteenth century. J Clim 6:1787–1797

Daly C, Gibson WP, Doggett M, Smith J, Taylor G (2004) Up-to-date monthly climate maps for the conterminous United States. Proc., 14th AMS Conf. on Applied Climatology, 84th AMS Annual Meeting Combined Preprints, Amer. Meteorological Soc., Seattle, WA, January 13–16, 2004, Paper P5.1, CD-ROM.

Fisk CJ (1984) Reconstruction of daily 1820–1872 Minneapolis-St. Paul, Minnesota temperature observations. M.S. Thesis, University of Wisconsin-Madison

Fisk CJ (2007) The First Fifty Years of Recorded Weather History in Minnesota (1820–1869) – A Year-by-Year Narrative Account. http://home.att.net/~station_climo/purpose.htm. Cited 16 June 2007. Alternate source: Fisk CG (1994) The first 50 years of continuous recorded weather history in minnesota (1820–1869) – a narrative chronology. Minnesota Historical Society

Hogan WR, Davis EA (1951) William Johnson's Natchez: the ante-bellum diary of a free Negro. Source Studies in Southern History, Vol. 1. Louisiana State University Press, Baton Rouge. Quoted in Ludlum (1968, p. 105)

Hunt E (1937) History of Fort Leavenworth 1827–1937 (2nd Edition, with updates by W. E. Lorence), reprinted by The Fort Leavenworth Historical Society (1981)

Loomis E (1841) On the storm which was experienced throughout the United States about the 20th of December, 1836. Trans Am Philos Soc New Ser 7:125–163

Loomis E (1845) On two storms which were experienced throughout the United States, in the month of February, 1842. Trans Am Philos Soc New Ser 9(2):161–184

Ludlum DM (1968) Early American Winters II: 1821–1870. American Meteorological Society, Boston

Miller ER (1931) The evolution of meteorological institutions in the United States. Mon Weather Rev 59:1–6

Mock CJ, Mojzisek J, McWaters M, Chenoweth M, Stahle DW (2007) The winter of 1827–1828 over eastern North America: a season of extraordinary climatic anomalies, societal impacts, and false spring. Clim Chang 83:87–115.

Peterson TC, Vose RS (1997) An overview of the Global Historical Climatology Network temperature database. Bull Am Meteorol Soc 78:2837–2849

Peterson TC, Vose R, Schmoyer R, Razuvaev V (1998) Global Historical Climatology Network (GHCN) quality control of monthly temperature data. Int J Clim 18:1169–1179.

Rosendal HE (1970) The unusual general circulation pattern of early 1843. Mon Weather Rev 98:266–270

Smither H (1931) Diary of Adolphus Sterne. Southwest Hist Q Online 35(2). http://www.tsha.utexas.edu/publications/journals/shq/online/v035/n2/contrib_DIVL1800_print.html. Cited 8 May 2007 University site

Stommel HM, Stommel E (1983) Volcano weather: the story of 1816, the Year Without a Summer. Seven Seas Newport RI

Part III
The Role of Station History in Understanding the Instrumental Record

Weather Station History and Introduced Variability in Climate Data

Glen Conner

Abstract This chapter presents station histories as an important reference from which to identify or eliminate causes of variability and to explain or confirm the behavior of climatic data. Some apparent variability in climate data was introduced by changes in the weather station that recorded them. Significant changes in the observation rules occurred as the different networks evolved: the Surgeon General's, the Smithsonian's, the Signal Service's, the Weather Bureau's and the National Weather Service's. Each of them established rules for what to observe, what time to observe, how often to observe, and how to record. Location of stations, both initial and subsequent, changed in response to those rules. The resulting moves from frontier to populated areas, rural to urban, surface to roof top, and manual to automated produced changes in the data. Instrumentation changes and the exposures of those instruments caused differences in the measurements they made. In all of the networks, there were frequent changes in the observers. Their qualifications varied and in later years the observations became a corporate effort instead of an individual one. Each observer change presented an opportunity for the data to be unintentionally but systematically altered.

Keywords Station history · Exposure · US Army Surgeon General · Smithsonian Institution · Signal Service · Climate Database Modernization Program

1 Introduction

Climate varies in both time and space, including across intraannual, interannual, and decadal timescales within the instrumental record.. However, some prominent variations in climate data can be introduced into the record by changes at the weather station itself. The non-climatic causes should be ruled out before the

G. Conner (✉)
Department of Geography and Geology, Western Kentucky University, Bowling Green, KY 42101, USA
e-mail: glen.conner@wku.edu

L.-A. Dupigny-Giroux, C.J. Mock (eds.), *Historical Climate Variability and Impacts in North America*, DOI 10.1007/978-90-481-2828-0_10,

research analysis proceeds to examine other possibilities more specific in relation to climate.

Significant changes in the observation rules occurred in the United States as the major different networks evolved within the last two hundred years: the Surgeon General, the Smithsonian Institution, the Signal Service, the Weather Bureau, and the National Weather Service. Each of them established rules for what to observe, what time to observe, how often to observe, and how to document the observations. Some station moves were dramatic, such as moves from rural to urban sites, from surface to rooftop exposures, and more recently from manual to automated affected trends in the data. In all of the networks, there were frequent changes in the observers which also could affect data quality from possible personal biases in reporting.

Climatologists seek to discover data variations that are persistent, becoming more or less frequent, or show spatial changes. The purpose of identifying the causes of such differences is not to correct the data as in a quality assurance effort but to understand the real reason that the data changed. This chapter presents the most common sources of introduced variability and change in climate data. These sources are changes in station location, site selection criteria, instrumentation, instrument exposure, weather observers, observation time, and methods of calculation means. Station histories are the key to identifying these and other sources of introduced variance.

1.1 Metadata and History

The World Meteorological Organization has provided guidelines for metadata (Aguilar, 2003). The goal was to produce metadata that would allow adjustments to data to more accurately describe climate. Their definition was "Data about data, necessary to correctly understand and use meteorological data." Note that the emphasis is on data not narrative. In digital datasets, metadata are entered during the quality control or assurance phase of digitization, perhaps appearing as flags in the digital record. History is a chronological record that describes, explains, and comments on past events or, in our case, climate observations. The definition assumes a narrative form of presenting those descriptions, explanations, commentary, and data. The term history includes the term metadata but not vice versa.

2 The Main Observational Networks in the United States

A station's history is inseparable from the network that the station served. The networks have influenced some of the nonclimatic changes in the data of a given location, and sometimes different national networks have had overlapping time periods. For further details on the creation and rationale of the observing networks, see Hopkins and Moran in this volume.

2.1 Surgeon General Network

The US Army Surgeon General developed the first national climate network in 1814 (Lawson, 1855) and used the surgeons at Army Posts as their observers (Smart, 1894). The reasons for developing such a network were clearly stated in Surgeon General Lovell's report in 1826. The report stated that the network was formed to determine whether a change in the climate occurred, and if so, how far it depended upon cultivation of the soil and the density of population. Surgeon General Lawson wrote in 1851 that the network had assisted in determining the influences that the progress of civilization had wrought on climate, temperature, and atmospheric phenomena. The Army network followed the expansion of the settlement frontier with observations at the new Army posts and thus provided the first weather information recorded in many states. By 1853, data were being collected at 97 Army posts (ESSA, 1970).

2.2 Smithsonian Institution Network

The Smithsonian Institution began receiving observations from their new network in 1849 (Miller, 1931). Their first participants were observers with experience in recording weather. Later efforts considered geographic distribution of the observation sites but almost all were near the observers' homes. The network consisted of 616 Smithsonian observers just prior to the Civil War. The Smithsonian network was designed to collect information on climate, with aspirations of developing forecasts using data collected by way of telegraph.

2.3 Signal Service Network

The US Congress created the Signal Service within the Army Signal Corps in 1870 with the primary objective of weather forecasting (Miller, 1931) The Signal Service was expected to make observations and issue warnings of impending storms, making particular use of the telegraph network to acquire near real time reports of weather observations sent to a central office in Washington. The first such reports were telegraphed on 1 November 1870 from 28 locations, and the first national daily surface weather map was distributed.

The Signal Service supplanted the Smithsonian network and the primary role changed from the Smithsonian's observations of climate to the Signal Service's observations of weather. Later, the Smithsonian observers were invited to submit their data to the Signal Service as voluntary observers. By 1884, the Signal Service had 458 stations reporting and most were volunteers (Signal Service, 1887). The Surgeon General network continued during the Signal Service years and provided some observations to them.

2.4 Weather Bureau Network

A major change in observations occurred when the weather functions of the Signal Service were disestablished in 1891 (Signal Office, 1891). Its observational network morphed into the new Weather Bureau formed within the Department of Agriculture, one that supported agriculture's needs (U.S. Senate, 1890). Although the emphasis in forecasting shifted to an agricultural interest, the initial locations and observers, mostly in downtown areas, were the same ones used by the Signal Service. Although their mission was focused on agriculture, the Weather Bureau offices did not move to rural areas away from their Signal Service locations. Instead, an extensive Cooperative Observer Network was developed that eventually exceeded 12,000 observers nationwide. The Forecast Offices remained downtown until the early 1940s when the rapid development of aviation began to take center stage. The offices then moved to the cities' airports and most downtown sites were abandoned during the next few years.

2.5 National Weather Service Network

The Weather Bureau was moved into the US Department of Commerce in 1940 and in 1970 was renamed the National Weather Service. Its observational network evolved toward automated observations. Over 800 Automated Surface Observation System (ASOS) sites and over 600 Automated Weather Observing System (AWOS) sites are operational. In the 1980's, over one thousand observing stations were selected for designation as the Historical Climatology Network.

The moves of the Forecast Offices away from the airports in the late 1990's was a radical change. In many cases, the moves were to locations that had no observational equipment. Some of those locations were rural, others suburban, and some collocated with the radar. Eventually, observations were resumed at most of those stations.

3 Documentation of Station History

Station histories are vital in the search for causes of observed changes in climate data. The interest and appreciation of the significance of station history evolved slowly from the nineteenth century. From the mid-1850s, most observers would occasionally comment about the instruments, the exposure, or the environment. The Smithsonian Institution brought an increased interest in the instrumentation involved and the Signal Service added the Army's penchant for record keeping. The Weather Bureau, in particular its development of the Cooperative Network, kept records most specifically for historical purposes.

3.1 Pre-1891 Station Histories

The observation forms used by the US Army Surgeon General's network listed the location, elevation, and the observer's name. There was no formal history section but comments were entered in the remarks. Neither the observer nor the equipment was described in the detail that we now desire.

The Smithsonian Institution (Smithsonian Institution, 1848) reserved a full page on the monthly observation form for comments by its observers, remarks on their instrumentation and exposure on that page, and descriptions of their observation site.

The Signal Service produced an early history form titled "Index of Meteorological Observations" for each state. Each form listed the latitude and longitude, the elevation, the length of the climate record in years, the elements that were recorded, and the agency that received the records. The ending date for the historic data was 1 January 1890. The Signal Service Observer Sergeants kept detailed records of virtually everything they did. Their correspondence, reports to the Chief Signal Officer, and interviews with the press provide a wealth of information from which to construct station histories.

3.2 Weather Bureau Station Histories

After the Weather Bureau was established in 1890, the emphasis on commentary was diminished and was relegated to entries in blocks on a form. Forms were created for the specific purpose of recording historical information. The Description of Cooperative Observer's Station and Instruments (Form 4029) was put into use in the mid 1920s. This form was a detailed description of the equipment in use, the location of the shelter and rain gage, and a description of the environment. It was the earliest attempt to expand the types of information recorded for historical purposes alone.

The digital data record began in 1948, and the Weather Bureau in the early 1950s prepared a summary of each cooperative station's history on its Substation History (Form 530) (Fig. 1). These metadata forms were initiated just as the State Climatology Programs were being implemented in each state. These forms included the station name, county, state, latitude, longitude, elevation, and added a description of exposure, a list of instruments used, where the data were published, and the names and periods of observations of each of the observers. The most important addition was the Index Number of the station. That number would allow for retrieval of the digital data.

For the first order station histories, the Station History (Form 500) was used. Its content was similar to the Form 530 used for the cooperative observers.

3.3 National Weather Service Histories

Station history management changed dramatically in the digital age. Thus, metadata were included in digital datasets as was information on how the digital data were formatted. However, the existence of detailed metadata did not obviate the need for station histories. The National Weather Service filled that need by using tabular forms instead of narrative histories.

The Report on Substation (B-44 Form) was a remarkable departure from previous station history forms. The front side of the form was much like previous

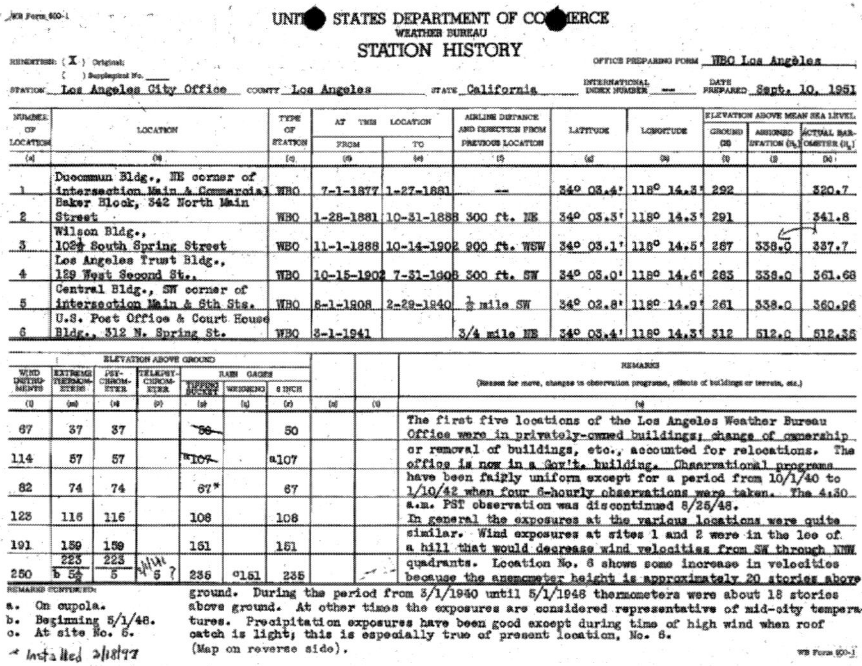

WB Form 530-1

UNITED STATES DEPARTMENT OF COMMERCE
WEATHER BUREAU
STATION HISTORY

REVISION: (X) Original; ( ) Supplement No. ____ OFFICE PREPARING FORM WBO Los Angeles

STATION Los Angeles City Office COUNTY Los Angeles STATE California INTERNATIONAL INDEX NUMBER --- DATE PREPARED Sept. 10, 1951

NUMBER OF LOCATION (a)	LOCATION (b)	TYPE OF STATION (c)	AT THIS LOCATION FROM (d)	TO (e)	AIRLINE DISTANCE AND DIRECTION FROM PREVIOUS LOCATION (f)	LATITUDE (g)	LONGITUDE (h)	ELEVATION ABOVE MEAN SEA LEVEL: GROUND (i)	ASSIGNED STATION (H_p) (j)	ACTUAL BAROMETER (H_b) (k)
1	Ducommun Bldg., NE corner of intersection Main & Commercial	WBO	7-1-1877	1-27-1881	---	34° 03.4'	118° 14.3'	292		320.7
2	Baker Block, 342 North Main Street	WBO	1-28-1881	10-31-1888	300 ft. NE	34° 03.5'	118° 14.3'	291		341.8
3	Wilson Bldg., 102½ South Spring Street	WBO	11-1-1888	10-14-1902	900 ft. WSW	34° 03.1'	118° 14.5'	267	338.0	337.7
4	Los Angeles Trust Bldg., 129 West Second St.	WBO	10-15-1902	7-31-1908	300 ft. SW	34° 03.0'	118° 14.6'	263	338.0	361.68
5	Central Bldg., SW corner of intersection Main & 6th Sts.	WBO	8-1-1908	2-29-1940	½ mile SW	34° 02.8'	118° 14.9'	261	338.0	360.96
6	U.S. Post Office & Court House Bldg., 312 N. Spring St.	WBO	3-1-1941		3/4 mile NE	34° 03.4'	118° 14.3'	312	512.0	512.36

ELEVATION ABOVE GROUND: WIND INSTRUMENTS (l)	EXTREME THERMOMETERS (m)	PSYCHROMETER (n)	TELEPSYCHROMETER (p)	RAIN GAGES: TIPPING BUCKET (q)	WEIGHING (r)	8 INCH (s)	(t)	(u)	REMARKS (Reason for move, changes in observation programs, effects of buildings or terrain, etc.) (v)
67	37	37		~~50~~		50			The first five locations of the Los Angeles Weather Bureau Office were in privately-owned buildings; change of ownership
114	57	57		~~a107~~		a107			or removal of buildings, etc., accounted for relocations. The office is now in a Gov't. building. Observational programs
82	74	74		67*		67			have been fairly uniform except for a period from 10/1/40 to 1/10/42 when four 6-hourly observations were taken. The 4:30
123	116	116		108		108			a.m. PST observation was discontinued 8/25/48. In general the exposures at the various locations were quite
191	159	159		151		151			similar. Wind exposures at sites 1 and 2 were in the lee of a hill that would decrease wind velocities from SW through NNW
250	223 / b 5½	223 / 5	5 ?	235	c151	235			quadrants. Location No. 6 shows some increase in velocities because the anemometer height is approximately 20 stories above

REMARKS CONTINUED:

a. On cupola.
b. Beginning 5/1/48.
c. At site No. 6.
* Installed 2/18/77

ground. During the period from 3/1/1940 until 5/1/1948 thermometers were about 18 stories above ground. At other times the exposures are considered representative of mid-city temperatures. Precipitation exposures have been good except during time of high wind when roof catch is light; this is especially true of present location, No. 6. (Map on reverse side).

WB Form 530-1

Fig. 1 Weather Bureau Form 530, station history for Los Angeles, California, 1954

forms recording information about the station. The back of the form required the drawing of a map of the observation site. It also required driving directions for reaching the station. Those maps make the B-44 forms the most valuable for station history.

Table 1 is a partial list of the forms used by the Weather Bureau and the National Weather Service that are directly beneficial to the stations' histories. In addition to the forms listed, there is the potential for historical information considered important by observers to be included in the remarks section of all observation forms from prior eras.

3.4 Content of Station Histories

Recently, the National Climatic Data Center's Climate Database Modernization Program prepared selected station histories to aid in climatic analysis (Dupigny-Giroux et al., 2007). Those historical narratives present biographical sketches of observers, documentation of locations, identification of instrumentation, discussion of observations, and photographs and maps that support the researcher's need to determine data quality assessment. Most city and state libraries have collections of photographs of prominent buildings in a city. Weather instruments on rooftops are

Table 1 Station history forms

Form number	Title	Form number	Title
WB 1058	Report of elevation and position of instruments	WB 4005	Inspection of substation
WB 1130	Surface weather observations	WB 4029	Description of cooperative observer's station and instruments
WB 1144	Station record	WB 4302	Report on climatological and/or crop substation
WB 4064	Inspection of airport and airways stations instrumental equipment	WB 4203	Report on substation
WB 4065	Description of topography and exposure of instruments	WB 4304	Report on hydroclimatic substation
WB 450	Description of topography and exposure of instruments	WB 530	Substation history
WB 500	Station history	WB 531	Report on substation
WB 54.3.	Barometer correction card	WB 6055	Inspection of substation
WS A-1	Station description and instrumentation	WS 23	Substation inspection
WS A-4	Station history	WS B-44	Cooperative station report
WS B-33	Station inspection report	WB 4005	Inspection of sub station
WS B-40	Barometer correction card	WB 4029	Description of cooperative observer's station and instruments
WBAN10	Surface weather observations	WB 4302	Report on climatological and/or crop substation
WS A-4	Station history	WB 4203	Report on substation

sometimes visible in those photographs of prominent city buildings (Fig. 2) that show location and exposure.

3.4.1 Site Maps and Diagrams

Few maps of the observation sites are available in the records, except for the B-44 forms. Historic maps are available but seldom coincide with the observation site in either temporal or spatial context. Even though the exposures of instruments may be known, the surrounding environment of the site often is usually not. Therefore, maps of the observation site and its environment should be included in station histories if possible. The Global Positioning System is used now to identify the location and elevation of a station.

4 Observation Site Selection Criteria Changes

The early interest in developing observational networks focused on the understanding of climate rather than on weather. A French writer, C. F. Volney (1804), toured Kentucky, Ohio, and New York and wrote that everywhere he visited, he heard

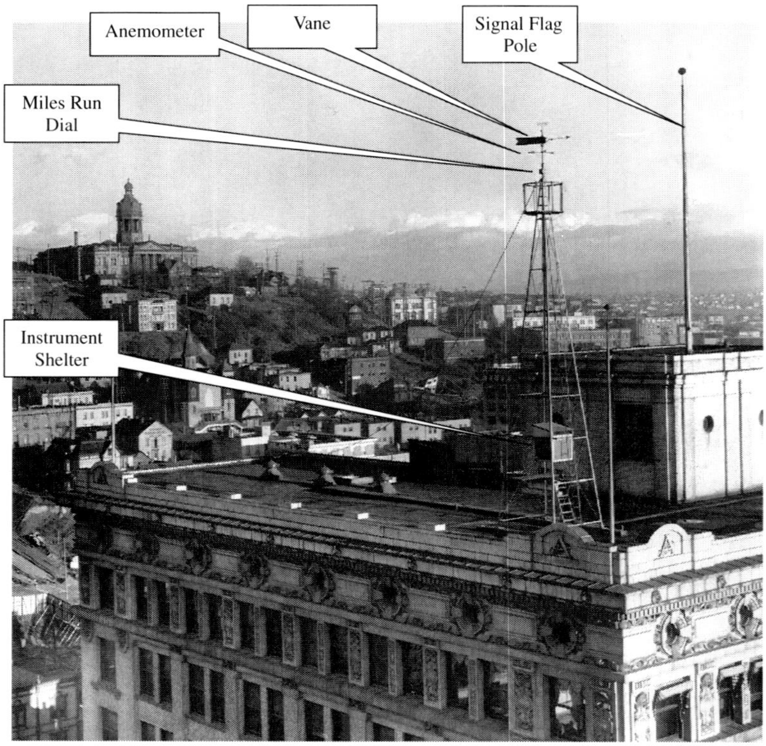

Fig. 2 Rooftop instruments on the Alaska building, Seattle, Washington between 1905 and 1911

reports of climate change. The change was a trend to longer summers, later autumns, shorter winters, and less snow. Those changes were represented as rapid and sudden in proportion to the amount of land that was cleared. These reports prompted concern of the early climate networks about whether an observation site would provide data representative of the climate change that was occurring. Subsequent networks developed different criteria for site location as new theories on climate change and meteorology emerged (Conner, 2007).

4.1 Surgeon General Network Sites

Army Post Surgeons were directed to make the weather observations at US Army Posts. Most observation sites were near the surgeons' quarters or near the hospital, if there was one. All of the observers were physicians, all trained as scientists, and motivated by an interest to discover the relationship between climate and disease. All the observations were taken from closely similar sites: on an Army post, located near the Surgeon's office or hospital with similar exposures, and observed at the same times of day.

4.2 Smithsonian Network Sites

The initial Smithsonian network in 1847 incorporated observers who already provided information to other collectors such as the networks in New York and Pennsylvania. The initial network's voluntary nature obviated the need for site selection criteria. Although observation sites were not selected for climatological reasons, they were very uniform in spatial attributes. Most instruments were in the backyards of the observer's home, in rural or in small towns, and observations were made at prescribed fixed times. The Smithsonian concluded that country locations were preferable over city ones because of the heat radiated by the city, what is now known as the urban heat island effect (Hazen, 1885).

4.3 Signal Service Network Sites

The Signal Service network was the first to have published site selection criteria. It required access to telegraph lines that followed the railroads and the usual dissemination of forecasts visually by forecast flags or postings on bulletin boards. Those requirements limited the site location to downtown city areas. Directions for exposures exacerbated the effects of downtown locations by stating that wind measurements were best when mounted on the highest buildings.

All of the initial Signal Service data came from sites that were new, with prescribed exposures and frequent inspections to assure conformity; essentially uniform sites. Typically, the site was in the middle of a city, on the roof of one of the highest buildings with instruments located as far as possible from live chimneys.

Voluntary observers, including some who were former Smithsonian observers, were added to the Signal Service network. These volunteers' locations were typically rural or suburban.

4.4 Weather Bureau Network Sites

The Signal Service sites and their site selection criteria were absorbed into the Weather Bureau network sites and continued without significant change from 1890 to 1905. Subsequently, the Weather Bureau's Station Regulations stated that the office building should be higher than the surrounding structures, assuring that instruments would remain on top of downtown buildings. In the case of the Burlington, VT building, the site was leased from the University of Vermont (UVM) and the building constructed under the terms that it would be used for Weather Bureau operations only. The land is the highest elevation in Burlington. The building and land reverted to UVM when the Bureau relocated to the Burlington Municipal airport. The Cooperative Observer program was expanded and most of those sites were rural or suburban. This program provided excellent data that were relatively free of the effects of urbanization.

The Weather Bureau moved its observations to the city airports in the early 1940s, usually to the roof of administration buildings. Most of the downtown locations were abandoned within a few years. The Weather Bureau observations then became a corporate endeavor when hourly observations began. The observers worked in shifts and the observer changed with the shift change. Some of the cooperative stations (e.g., Municipal Water Companies, State Police Stations, etc.) used multiple observers as well. Increasingly, the emphasis was on collecting data useful in daily weather forecasting as opposed to climatological applications.

4.5 National Weather Service Network Sites

The Weather Bureau was renamed the National Weather Service (NWS) in 1967. In the 1980s, over one thousand NWS stations (mostly cooperative observers) were selected for designation as the Historical Climatology Network (Karl et al., 1990). The selection criteria included the period of record, percent of missing data, number of station moves, and other station changes that could affect data homogeneity and spatial coverage. Most of those selected were rural, cooperative stations.

The current Climate Reference Network's site selection criteria required that its sites must remain largely stable for 50 years or more, be located in fairly pristine environments, have clearance from obstructions in clear terrain, have good exposure for instruments, and be separated from micro-climate inducing influences ranging from small ponds to urbanization. The inclusion of the stability timeframe in the site criteria and the restriction of those criteria to only climate factors were a marked departure from all the previous networks.

5 Station Location Changes

Relocation of a station plays the most prominent role in introducing observational problems in data. A few moves were made to achieve exposure of wind instruments but more often were not related to observation problems, such as a change in observers, a reduction in the rental cost, or for some other non-climate reason. Figure 3 depicts the moves in the downtown area of Seattle, Washington.

The information on all of the observation forms included the name of the station, its post office name, its latitude and longitude, and elevation. The geographical locations on the forms should be verified but, in general are accurate. Changes in the station location can induce changes in the data as illustrated in Fig. 4. The 90th percentile of daily cooling values is depicted for the spring season March through May. Daily cooling is measured by subtracting today's low temperature from yesterday's high temperature, adjusted for time of observation. The cooling value is recognized as a measure of microclimate effects on the temperature record. Each station location is represented by a unique symbol on the graph. The horizontal lines

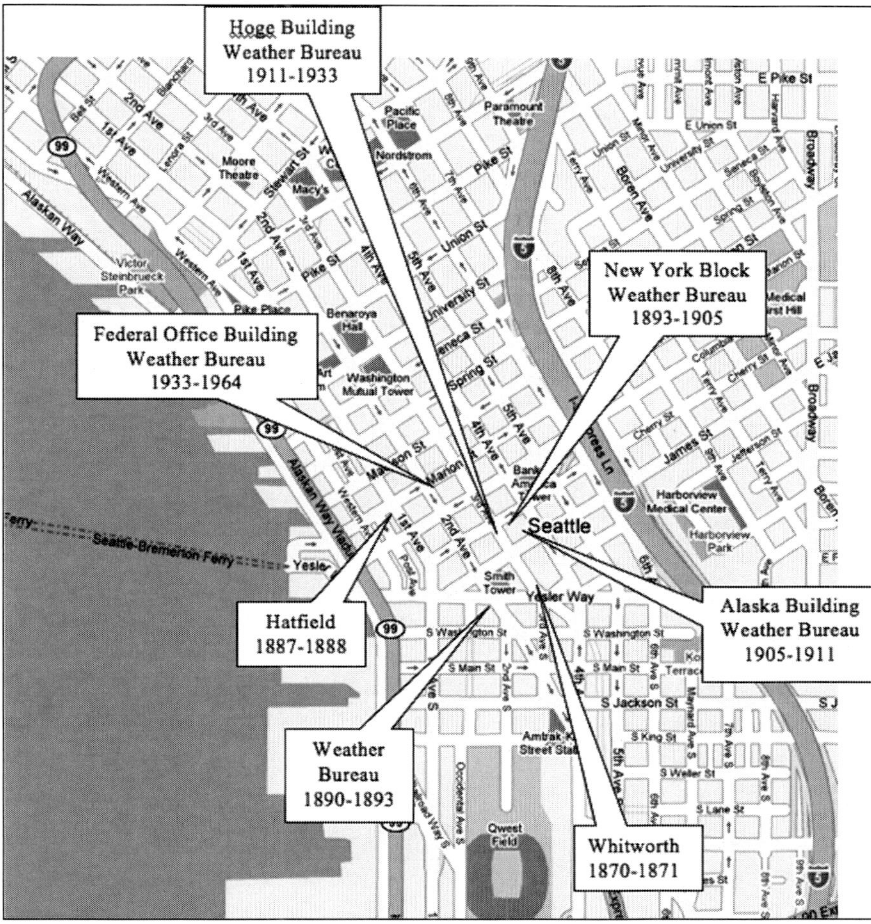

Fig. 3 Downtown locations of observations in Seattle, Washington 1887–1964

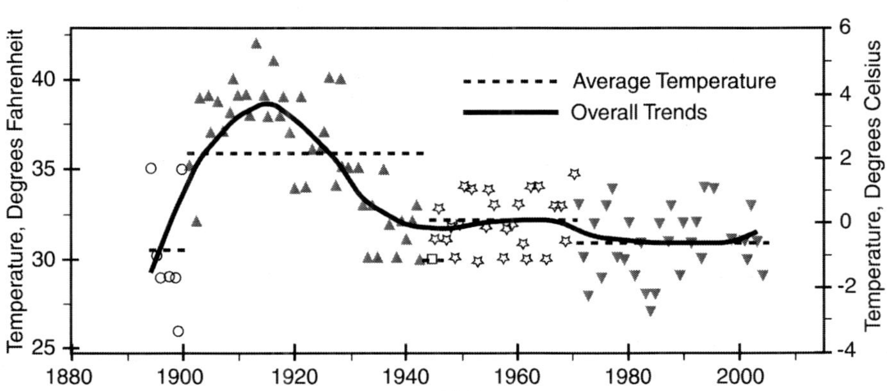

Fig. 4 Induced changes in the 90th percentile of daily cooling values for the spring season, March through May, at Bowling Green, Kentucky

represent the mean of the data for each location and clearly show the effect of relocation. There were significant increases from the first location to the second during the 1900–1940 period. The 1940–1970 period was a reduction toward the original location level. The last location further reduced the values to near the original level.

5.1 Station Number Assignment

The Weather Bureau's Merrill Bernard, Chief of the Climatological and Hydrologic Services Division, began what he called their "mechanization program" in 1947. The effort was the result of the punch card technology that was at the cutting edge at that time. Data were keyed causing holes to be punched in a card. A punch card reader would then decipher the holes in the cards, use the deciphered data for calculations, and print summaries on special typewriters.

Station names presented a problem in the punch card process because of length and duplications among states. The Weather Bureau Tabulation Unit at New Orleans assigned "punched card numbers" for all. Its rules for cities that had more than one station were either to use a locally accepted name for each, add a postscript number to the name, or add distance and direction from the post office.

Primary names were to be those used in the Rand-McNally atlas. Postscripts were added, for example "WB city" or "WB Airport," after Weather Bureau names. A change in station number was authorized when the station moved five miles or more, or its elevation changed by 100 ft or more, or the post office or community name changed.

5.1.1 The Basic Pattern of Station Numbers

Status Report No. 12, dated 10 May 1948 reported that the assigned numbers for the United States had six digits, the first two of which identified the state (01 for Alabama, 48 for Wyoming, and 49 for the District of Columbia). In September 1948, two digit numbers were assigned to "extra continental" [sic] stations (50 for Alaska and 51 for Hawaii). At the same time, two digit identifiers were assigned for North American and Caribbean countries.

The last four numbers of the station number identified the station. The original numbers were assigned according to the station's relative position on the "Index of Cities and Towns" published in the 65th edition of the Rand-McNally Atlas. A station number 1734 was about 1734/9900 of the distance between the first and last names in the state's index. A minimum of five numbers separated stations within the same city and a minimum of eight numbers separated stations adjacent in the index. If a station moved significantly, a new station number would be used and the old number would not be reused for any other location. The old number was reused if the old site was reestablished.

This interesting and precise procedure allowed the subsequent assignment of new station numbers over the past 60 years without exhausting the numbering system. The Station Index in each state's Annual Summary 1948 of the Climatological Data contained a list of stations under the still relatively new station numbers and station names. It was the only issue that contained an additional column headed "Former Station Names."

5.2 Station Heritage

A station's history includes changes in its station number. The replaced number becomes part of the station's heritage. Some of those changes, such as a change in the post office name, have no impact on climate analysis. In some climatological studies, the change in location or elevation may have an acceptable impact. In other cases, the need to extend the record back to an earlier time may require compensatory adjustments as to which heritage sites are acceptable for a study.

6 Instrumentation Changes

The instruments used in observations have changed as the networks evolved and as technology improved. The provision of instruments differed between stations depending upon the purpose of the station. The earliest daily observations by the Surgeon General Network in 1819 used only a thermometer and a weather vane. Army posts that made weather observations were provided with a rain gauge in 1836 for daily precipitation measurements (Lawson, 1855).

The receipt of maximum and minimum thermometers, barometers with attached thermometers, psychrometers, anemometers, and other instruments, as well as replacements were recorded. An example from the Milwaukee Climate Record Book used by the Weather Bureau documents additions and replacements of observational equipment, in this case maximum thermometers (Table 2). Note that the

Table 2 Maximum thermometer replacement at Milwaukee

	In use	
Number	From	To
17595	17 Jan 1917	10 Feb 1917
*17870	10 Feb 1917	5 May 1922
*15847	5 May 1922	1 Aug 1922
*24442	1 Aug 1922	24 Aug 1925
**24709	24 Aug 1925	4 Feb 1927
*28125	4 Feb 1927	25 Aug 1927
***19993	25 Aug 1927	27 Aug 1927
*24209	27 Aug 1927	15 Jun 1928

* Broken
** Probably defective
*** Rapid Retreater

replacement of a defective thermometer or one that was a “retreater” was an opportunity to introduce errors in the data.

7 Instrument Exposure Changes

There were rules for exposure that date back to the US Army Surgeon General’s network. The surgeons at the forts measured temperature from a thermometer in their “thermometer box” mounted on an exterior north-facing wall. The Smithsonian Institution developed shelters for thermometers, hygrometers, and self-registering thermometers. Those louvered double roofed shelters were to be mounted in windows or placed on supports over sod. The Surgeon General network adopted them and the Signal Service and Weather Bureau continued their use. A seminal paper on thermometer exposure (Hazen, 1885) examined all aspects of the problems related to instrument exposure.

Concern for proper measurement of precipitation led to rules for the standardization and exposure of rain gauges. Like all rules, they changed. An early Surgeon General rule was to affix the rain gauge to a post at a height of 8 ft above the ground and located away from elevated surfaces at a distance equal to or greater than its height above ground level (Lawson, 1855). The Smithsonian Institution for a time buried a gauge in the ground with only 4 in. of gauge extending above ground. The Signal Service used a conical gauge for a period before adopting the standard gauge. Each network provided different rules for snow collection and measurement.

7.1 Site Diagrams

The Signal Service began the practice of inspecting its observation stations in 1870, requiring drawings by the Signal Service inspector and including information on the exposure of instruments. An example of the Weather Bureau’s arrangement of instruments on the roof in Indianapolis, IN shows the value of drawings (Fig. 5). The shelter containing the thermometers was 194 ft above the ground level overlooking twelve attic vents that released heat during the summer. The anemometer was 36 ft higher on a tower (Fig. 6). Another example from San Diego, CA depicts the Signal Service instrument shelter mounted in a second story window on a north-facing wall. It contained the thermometers, hygrometer, and barometers. The wind direction dial, connected to the vane on the roof, was mounted on the ceiling.

The Weather Bureau made few such diagrams but continued with inspections that contained verbal descriptions of the sites. In recent years, the National Weather Service required site diagrams but did not routinely enforce the rules. They occasionally included photographs and provided written descriptions of the exposures.

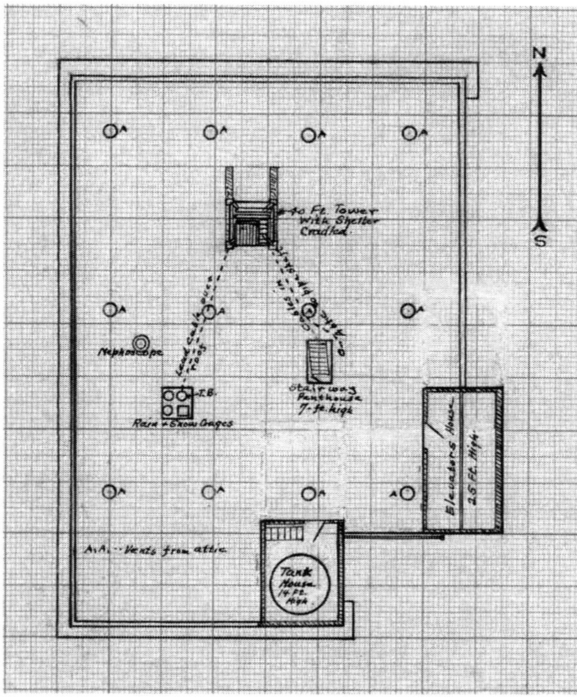

Fig. 5 Rooftop layout on the consolidated building in Indianapolis, Indiana in 1932

8 Observer Changes

A potential for altering climate data exists each time a change in observers occurs at a weather site. These potentials differ somewhat as related to the following aspects.

8.1 Observer Qualifications

Many of the early observers were professionals in other fields. For example, the US Army used the surgeons at Army Posts as observers (Smart, 1894). The Smithsonian Institution began its climate network in 1847 using many of the observers who had already been reporting climate observations to Professor James Coffin of Lafayette College in Pennsylvania (Rives, 1997). The Smithsonian developed its climate network and solicited observers who were both experienced and equipped. In later years, the Army's Signal Service required trained observers and opened a meteorology school at what is now Fort Myer, Virginia (Signal Service, 1887). The Weather Bureau used post-secondary graduates with meteorology training as they became more available. Variations in climate data attributable to changes in observers appear as a scaled increase or decrease. The most common observer errors (e.g., entering the max in the min block) are occasional and have little effect on climatological

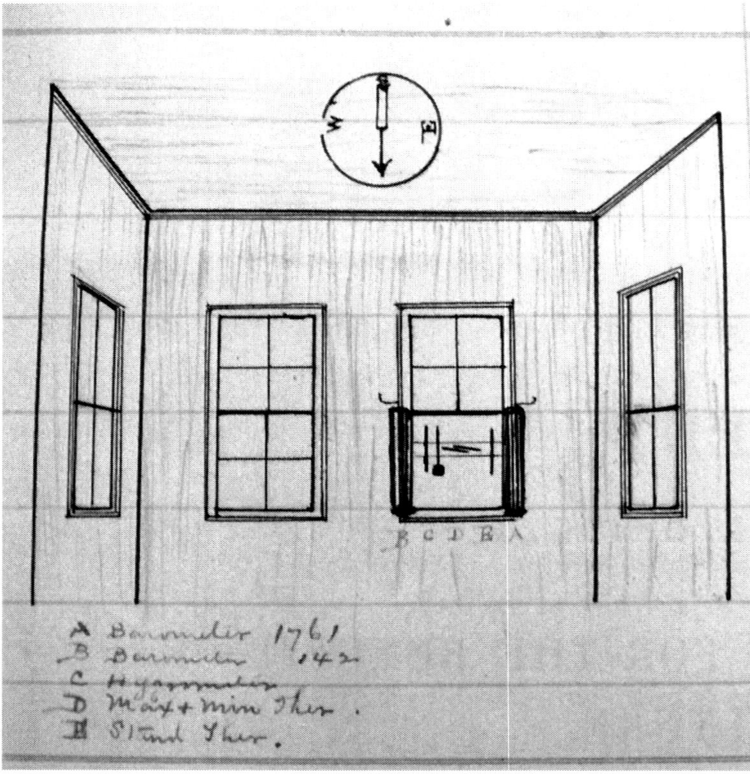

Fig. 6 Office layout in the western union building in San Diego, California, 1878

calculations. Systemic errors are uncommon but detectable (e.g., reading the min from the wrong end of the index marker).

8.1.1 Observers by Occupation

Observation forms contain the observer's name, Post Office address, and his or her title if there was one. Changes in observers are thus identifiable from those forms. In 1859, a survey of the occupations of 49 of the 54 observers in California, Kentucky, and Pennsylvania reveal that 22% were doctors, 22% were professors or teachers, and 10% were other professionals (lawyers, engineers or druggists). The remaining 46% of observers included other trusted members of the communities such as notaries public, farmers, ministers, merchants, and tradesmen. The observers were neither uneducated, unreliable, nor undependable. Among them was Cleveland Abbe from Lansing, Michigan a young tutor at Michigan University who would become one of the most famous meteorologists in US history. The backgrounds of the observers are strong reasons to have confidence in the quality of the data they produced. (Conner, 2004).

9 Observation Time Changes

Changes in observation time can alter the recorded daily temperature and thus the monthly mean by as much as $\pm 0.72°C$ presenting a possible misleading change in temporal climatic trends. Therefore, climate analysis requires that the observation time be corrected. Different fixed observations times as well as different types of "time", however, can complicate the best practice for creating and applying these corrections (Schaal and Dale, 1977).

9.1 Time of Beginning the Observational Day

The time of the beginning of an observational day may introduce changes in the data. The National Weather Service currently uses midnight to midnight as its observation period and is consistent with the date. Only places with automated observations or corporate observers find this period acceptable in terms of convenience. An alternate observational period is the 24 h period beginning at some prescribed local time, a practice which varied among the different observational networks. Sometimes, it also changed within a given network. Most observation periods began at 7 a.m., while other stations (mostly agricultural), using 5 p.m. to begin the twenty-four period were less common.

9.2 Local Time Determination Prior to 1883

Local time was used for all observations before 1883. Local time was determined by observing solar noon. Sun time (sometimes called meridian time) varied by longitude. In a unique example, San Francisco, CA hired the local weather observer to be its official timekeeper in 1865 (Tennent, 1890) The observer made transit observations of the sun three times each week and updated his clock. He rang a bell at noon each day, with the first bell sound marking the exact time of noon, so that others could reset their clocks.

9.2.1 Surgeon General's Observations Times Before 1883

Local meridian time was used in the Surgeon General's network for its first 66 years. For example, the surgeons at Newport Barracks, Kentucky in July 1825 observed and recorded the weather at sunrise, 2 p.m., and sunset. The first two of those readings had the advantage of usually being near the lowest and highest temperature of the day, making them closely compatible with today's observations. The 2 p.m. time could be approximated by sundial or clock. However, sunrise time varies with latitude and season. Using Sun time, observations in Kentucky at sunrise varied from current Standard Time by about two and a half hours from an early 5:20 a.m. (CST) in summer to a late 7:55 a.m. (CST) in winter. Sunset observations varied similarly.

9.2.2 Smithsonian Observation Times

All Smithsonian Institution data were collected using local meridian time. The prescribed observation times were 7 a.m., 2 p.m., and 9 p.m. local time. This convention avoided the variance of sunrise and sunset times but introduced new temperature variance because 7 a.m., in relation to sunrise and noon, is seasonally much earlier or later.

9.2.3 Signal Service Observation Times

Beginning in 1870, the Signal Service used meridian time to take advantage of data transmission by telegraph. Observers were required to use the 75th meridian time to assure that the maps represented instantaneous conditions across the country. The Signal Service also retained local times in order to maintain the long periods of climate record created by the Smithsonian Institution. The result was that poor Sergeant Watkins at Sacramento CA, for example, had to take readings seven times per day beginning at 4:37 a.m. and ending at 9:00 p.m. local time. The three times daily requirement for using only local meridian times was soon reinstated.

9.3 Standard Time

Most states and the Signal Service soon adopted the use of Standard Time after it was first agreed to by the United States in 1883. The time zone boundaries were altered to accommodate local interests. The general trend was to move the western boundaries of time zones farther west. For example, the Eastern Standard Time Zone's eastern boundary has moved from eastern Ohio to the west side of Indiana. The far northwestern boundary of Michigan and its Eastern Standard Time Zone is almost at the center longitude of the Central Standard Time Zone. This positioning introduced a variance from solar noon of about 1.5 h within that zone.

9.3.1 Weather Bureau and Time

All Weather Bureau observations used Standard Times. The creation of "war time" and "daylight saving time" did not affect weather observations because neither was ever used.

9.4 Time Considerations

For the nineteenth century observations based on Sun Time before 1883, time was a continuous variable on a spatial scale. In other words, 24 h of time was centered on solar noon just as the peak of solar radiation was. Therefore, data from the pre-1883 period offer a period without the temperature changes due to the introduction of Standard Times. Climatologists who study the data from the post-1883 period must be sensitive to the variation in observation time relative to solar noon caused by Standard Time. Researchers should look for it to show up as a longitudinal variation in temperature not otherwise explained.

10 Subjective Observations

The Surgeon General, Smithsonian Institution, and the Cooperative Observer Program encouraged observers to routinely comment on weather phenomena in a space on the observer form labeled Remarks or Casual Phenomena. Those remarks may be the most important tool in the verification of data that appear to be incorrect. For example, the observation report of five inches of both snowfall and snow depth in Springfield, Kentucky on 20 May 1894 likely would be rejected during quality assurance if only data were reported. Even though the snow came at a time that seemed too late in the spring, the observer's remarks of severe tree damage from snow accumulation on the leaves provides additional reliable information. Climate researchers should be aware that observer remarks are sometimes available to explain suspicious data.

10.1 The Remarks Section of Observer Reports

The Army Post Surgeon was directed to keep a diary of the weather and to note the climate and diseases prevalent in the vicinity (Smart, 1894). The emphasis was on subjective observations and, in effect, data supplemented the remarks.

Remarks often provide vital historical information. Dr. Samuel D. Martin from Pine Grove, Kentucky wrote on his February 1865 form,

> In March 1865, he remarked, "You will observe I have altered the time of the morning observation in this paper to 6 o'clock which is only an approximation to the time. The observations were made as soon as it was light enough to see how to make them, which was generally about six o'clock in the morning."

Few observers today have the difficulty experienced by the observer in Chloride, Arizona, who wrote on his September 1889 form, "Owing to the threatened outbreak of the Wallapais [Indians] the rain gauge was abandoned for several days & was only visited the 2nd day after the rains so that the returns for 17th, 18th and 19th are not accurate." His remarks explain what was an anomaly in the dataset.

By 1941, the remarks section of the monthly Cooperative Observers' Meteorological Record had been relegated to a small block in the corner of the form. The subsequent absence of remarks leaves the researcher with only the metadata and history.

11 Changes in the Calculation of Mean Temperature

Some data on the observation forms were calculated rather than observed. The daily mean temperature was one such calculation. Changes in the method of calculation introduced intentional changes in the result. It had long been understood that the sum of hourly temperature observations divided by 24 produced an acceptable daily mean. However, hourly observations were not possible at stations with only one observer and no recording instruments. The early climatologists knew that the mean daily temperature typically occurs twice during each day's temperature oscillation. They knew that the times of those occurrences varied significantly both spatially and temporally. Therefore, it was not possible to schedule observations at those times in advance, thus requiring an acceptable surrogate method.

Several methods for calculating the mean were used (McAdie, 1891) before the widespread use of the maximum and minimum thermometers (called "self registering" in the early days). The maximum and the minimum temperatures were added and the sum divided by two to produce the daily mean. The method for calculating the mean may not be readily apparent by comparing a daily mean with the observed temperatures. For example, the Smithsonian Institution's calculation used observations made at 7 a.m., 2 p.m., and 9 p.m. The 9 p.m. value was doubled and added to those at 7 a.m. and 2 p.m. That sum divided by four produced the daily mean.

Station histories and the original observer forms are the primary sources for determining how the means were calculated. Before using the recorded or published means, the method for calculation should be known.

12 Summary

Perturbation, variability, or change in a station's climate data may have non-climatic causes that are imperative to understand when utilizing weather data for studying climatic change. Some changes are rather abrupt such as those caused by a change in location. Other changes, such as the increase in intensity of an urban heat island, are gradual. Both of these examples are datasets that correctly recorded the changes that actually occurred but may provide some incorrect climatic signals. Even perturbation, variability, or change detected in spatially disparate climate data may have non-climatic causes. For example, systemic decisions to move observations to rooftops or to airports may have that effect.

A running mean may be superimposed over a graph of the annual march of daily mean temperature to aid in analysis. However, the smoothed line may hide realistic characteristics. All observed climate datasets contain both temporally and spatially scaled variance. Station histories provide the best, and most often the only, source from which to judge what produced the variance. To avoid misinterpretation of the causes of the variances, researchers should resist the temptation to statistically remove the variances before they begin their analysis.

References

Aguilar E et al. (2003) Guidelines on climate metadata and homogenization. WMO TD no.1186, World Meteorological Organization, New York

Bernard M (1947) Status report No. 12, dated 11 September 1947, Climatological and Hydrologic Services Division, Weather Bureau, Department of Commerce, Washington

Conner G (2004) Nineteenth Century Weather Observers: A Whodunit. Paper 7.14, Preprints, 14th Conference on Applied Climatology, American Meteorological Society, Boston, http://ams.confex.com/ams/pdfpapers/70253.pdf

Conner G (2007) Criteria for Good Weather Observation Site from 1826–2006. Paper 1.1, Extended Abstracts, 16th Conference on Applied Climatology, American Meteorological Society, San Antonio, Texas, http://ams.confex.com/ams/pdfpapers/118210.pdf

Dupigny-Giroux L-A, TF Ross, JD Elms, R Truesdell, and SR Doty (2007) NOAA's Climate Database Modernization Program. Bulletin of the American Meteorological Society 88(7), 1015–1017

ESSA (1970) Environmental Science Services Administration Vol. 5, No. 1, January 1970, US Government Printing Office, Washington

Hazen H (1885) Thermometer Exposure. Professional papers of the Signal Service No, XVIII, War Department Signal Office, Washington City

Karl TR, CN Williams, Jr., FT Quinlan, and TA Boden (1990) United States Historical Climatology Network (HCN) Serial Temperature and Precipitation Data. Environmental Science Division, Publication No. 3404, Carbon Dioxide Information and Analysis Center, Oak Ridge National Laboratory, Oak Ridge, TN

Lawson T (1851) Army meteorological register for twelve years, 1831–1842. Surgeon General's Office. C. Alexander Printer, Washington

Lawson T (1855) Army meteorological register for twelve years, 1843–1854. Philadelphia, War Department, Washington.

Lovell J (1826) Meteorological register for the years 1822, 1823, 1824, & 1825 from observations made by the surgeons of the Army at military posts of the United States. US Surgeon General's Office. Edward de Krafft, Washington

McAdie A (1891) Mean temperatures and their corrections in the United States. Signal Office, Washington. In Studies of weather types and storms. Weather Bureau, US Department of Agriculture, Washington

Miller E (1931) The evolution of meteorological institutions in the United States. Mon Weather Rev 59(1), 1–6

Rives F (1997) Joseph Henry, father of weather service. The Joseph Henry Papers Project, History Division, Smithsonian Institution

Schaal LA and RF Dale (1977) Time of Observation Temperature Bias and "Climatic Change." J Appl Meteorol 16(3), 216–222

Signal Office (1891) General Order No. 25. Signal Office, War Department, Washington City, June 30, 1891.

Signal Service (1887) General instructions to observers of the Signal Service. Government Printing Office, Washington. 1887.

Smithsonian Institution Annual Report (1848) Misc. No. 48. Smithsonian Institution Archives, Washington, D.C.

Smart C (1894) Handbook for the hospital corps of the U.S. Army and State Military Forces. William Wood Publisher, New York

Tennent T (1890) Tennent's Nautical Almanac. Thomas Tennent, San Francisco.

U.S. Senate (1890) Report of the Senate Committee on Military Affairs. 51 Cong., 1 Sess, S. Rept 402 (serial 2704), Washington

Volney CF (1804) View of the climate and soil of the United States of America. C. Mercier and Company, 6 Northumberland Court, London, England.

Monitoring the Climate of the Old Northwest: 1820–1895

Edward J. Hopkins and Joseph M. Moran

Abstract Systematic gathering of climate data in the Old Northwest began in the second decade of the 1800s, prompted by curiosity and practical concerns. Climate information was needed for agriculture and commerce, to investigate the possible relationship between weather and human diseases, and to develop a scientific understanding of storms. In addition, the climate record fueled speculation that the climate was changing and human activities associated with settlement were contributing factors. The first widespread networks of weather/climate stations in the Old Northwest were operated by the US Army Medical Department, the Army Corps of Topographical Engineers, and the Smithsonian Institution. By the mid-1870s, the early climate stations and telegraph-linked weather stations became part of a new national weather service operated by the Army Signal Service (forerunner of today's National Weather Service) whose primary aim was short-term weather forecasting. Examination of these first weather/climate observation networks reveals how instrumentation and observation techniques evolved with important implications for the study of climate change.

Keywords Old Northwest · Observing networks · Cleveland Abbe · Increase A. Lapham · Forts data

1 Introduction: Historical Perspective

This chapter focuses on the efforts to gather weather and climate information in the portion of the United States originally known as the *Old Northwest* (and later the *Northwest Territory*) spanning the period 1820–1895. The Old Northwest

E.J. Hopkins (✉)
Department of Atmospheric and Oceanic Sciences, University of Wisconsin-Madison, Madison WI 53706, USA
e-mail: ejhopkin@facstaff.wisc.edu

J.M. Moran (✉)
American Meteorological Society, Suite 800, Washington, DC 20005-3826, USA
e-mail: moranj@uwgb.edu

L.-A. Dupigny-Giroux, C.J. Mock (eds.), *Historical Climate Variability and Impacts in North America*, DOI 10.1007/978-90-481-2828-0_11,

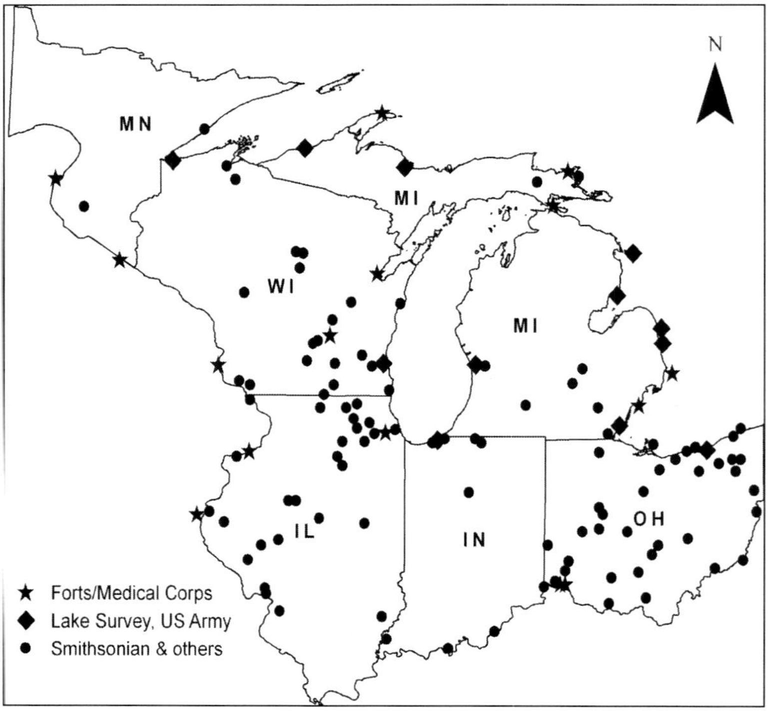

Fig. 1 Map of the Old Northwest encompassing lands west and south of the Great Lakes, northwest of the Ohio River, and east of the Mississippi River. Locations of early weather stations are marked

encompassed lands west and south of the Great Lakes, northwest of the Ohio River, east of the Mississippi River, and west of Pennsylvania (Fig. 1). Today, this area includes the states of Ohio, Michigan, Indiana, Illinois, Wisconsin, and northeastern Minnesota. During the 17th century, explorers from New France first entered the Old Northwest and later were followed by missionaries and settlers, using the region's numerous waterways as primary travel routes.

Jean Nicolet (1598–1642) was probably the first European to set foot in the Old Northwest, landing on the shore of Green Bay about 10 mi (16 km) northeast of the present City of Green Bay, WI. In 1673, the Jesuit missionary Jacques Marquette, S.J., (1637–1675), Louis Jolliet (1645–1700) a French native of Québec and their Native American guides crossed what is now Wisconsin by birch-bark canoe. They "discovered" the Mississippi River as well as the divide separating rivers and streams draining eastward into the Great Lakes and the St. Lawrence River from those flowing westward into the Mississippi River, eventually reaching the Gulf of Mexico.

Clashes between French and British interests over the fur-rich Old Northwest culminated in the French and Indian Wars. With the end of hostilities in 1763, New France was ceded to Great Britain and became New Québec. Later, as a provision of the 1783 Treaty of Paris which ended the American Revolution, Great Britain ceded the Old Northwest to the newly formed United States. The Northwest Ordinance, passed by the US Continental Congress in 1787, established the Northwest Territory and provided the blueprint for its subdivision into new territories and eventually states. In 1803, Ohio became the first state to be carved from the Northwest Territory, followed by Indiana (1816), Illinois (1818), Michigan (1837), Wisconsin (1848), and Minnesota (1858). The U.S. military built forts in the Northwest Territory (some along the Great Lakes as early as the 1790s) mainly to enforce a presence in the region. However, quarrels with British traders persisted and contributed to the War of 1812 which ended with the 1814 Treaty of Ghent firmly establishing US sovereignty over the former Old Northwest.

In the mid to late 19th century, many factors spurred people to leave their homeland for America including political instability, limited employment opportunities, religious persecution, and crop failure. Attracted to the Old Northwest by fertile soils and the potential for work in mining, the timber industry, shipbuilding and later, manufacturing, settlers arrived via the Great Lakes, Cumberland Gap, Ohio Valley, and Mississippi River. In many cases, adverse weather conditions contributed to crop failures that drove farmers from Western Europe to places like Illinois, Michigan and Wisconsin. Weather was a factor in the 1840s Irish potato famine that caused a mass exodus to North America. From the 1840s through 1880s, crop failure was one of the reasons for emigration by people from the German-speaking regions of Europe and the Scandinavian countries.

Newcomers to the Old Northwest found themselves in a place where the climate differed dramatically from that of their homeland. Winters featured bone-numbing cold waves and deadly blizzards while summers brought soil-parching heat waves and drought. Seasonal swings in temperature were greater than the immigrants were accustomed to. And nothing could have prepared them for severe storms – especially tornadoes – that must have terrified the new arrivals. But in time, they adapted to the climate of the Old Northwest much as the region's indigenous peoples had done before them.

Some of the earliest descriptions of the weather and climate of the Old Northwest appear in diaries and journals maintained by individuals. According to Thomas Jefferson (1743–1826), the average air temperature decreased from coastal Virginia westward to the Allegheny crest and then increased westward from there (Hill, 2005). This westward warming was refuted by Cincinnati physician Daniel Drake (1785–1852) who in 1815 published the first climatology of Ohio based on data from 1789 through 1813 (Alexander, 1924; Conner, 2004a). An early continuous weather record comes from College Hill near Cincinnati, OH. Isaac H. Jackson maintained a monthly mean temperature record from January 1814 through December 1848 (Conner, 2004b). Jackson's observations pre-date those by the Army Medical Department in the vicinity by 11 years (Newport Barracks, KY) although the military conducted 11 months of weather observations in 1790–1791 at Fort Washington (Cincinnati, OH). Although weather records compiled by individuals can provide valuable insight on local climate, records produced by organized weather/climate networks have the advantages of standard instruments and observational practices. Hence, networks generate climate data that are more readily summarized and interpreted for broad geographical areas.

2 Army Medical Department Weather/Climate Network

The possible link among human diseases, weather and its seasonal fluctuations was a serious concern for the military because even well into the 20th century, more soldiers lost their lives to illness than combat. For this reason (and to learn more about the climate of the continental interior), the Army Medical Department established the first national weather observing network. During the War of 1812, as part of a reorganization of the US Army Medical Corps, Surgeon General James Tilton (1745–1822) ordered Army surgeons to "keep a diary of the weather" at Army posts (Fleming, 1990). Tilton issued his general order on 2 May 1814, but ongoing hostilities delayed compliance. In 1818, Joseph Lovell (1788–1836) succeeded Tilton as Surgeon General and, with the approval of Secretary of War John C. Calhoun (1782–1850), issued the first formal instructions for taking weather observations (Smart, 1894).

The first weather data (for March–June 1816) were submitted by Benjamin Waterhouse, Army surgeon at Cambridge, MA, although some uncertainty exists as to who actually made the observations (Chenoweth, 1996). In 1819, weather reports from other posts began trickling into the Army Medical Department. By the late 1830s, sixteen army posts had compiled at least 10 complete, but often not consecutive, years of weather data. By the end of the American Civil War, weather records had been compiled for varying periods at 143 Army posts. More than 120 Army medical personnel were still sending in monthly weather reports in 1874, the year the network was transferred to the US Army Signal Service.

The Army post's chief medical officer or surgeon was responsible for weather observations but often the task was assigned to a hospital steward or orderly. Weather observations were entered in a standard journal and quarterly summaries sent to the Army Medical Department in Washington, DC. Lovell began summarizing weather data and in 1826 published the *Meteorological Register for the years 1822–1825, from observations made by the Surgeons of the Army at the military posts of the United States*. For this reason, Lovell rather than Tilton is usually credited with founding the national system of weather observation (Landsberg, 1964). Later, Thomas Lawson, Surgeon General from 1836 to 1861, directed publication of the *Meteorological Registers* covering the period 1826–1854 (Lawson, 1840, 1851, 1855). Much of the daily weather data are available via microfilm from the US National Archives at College Park, Maryland.

Initially, post surgeons had only a thermometer and perhaps a wind vane for taking readings three times daily, at 7 a.m., 2 p.m., and 9 p.m., local sun time. Surgeons also reported the day's prevailing wind and weather. In the "remarks" column of the journal, they commented on the health of the troops, any extreme weather, and phenological or other natural events. In 1836, most posts were supplied with a DeWitt conical rain gauge (Smart, 1894; DeWitt, 1832). About the same time, surgeons began reporting prevailing wind and weather for morning and afternoon. In 1842, the Army Medical Board issued higher quality instruments and revised instructions for their use. Beginning the following year, temperature, cloud cover (in tenths), and wind direction were recorded four times daily, at sunrise, 9 a.m., 3 p.m., and 9 p.m., local sun time. Some Army posts also provided barometer and hygrometer readings. In 1855, observation times shifted back to the original three per day, presumably for a better estimate of mean daily temperature (Smart, 1894).

Many frontier Army posts were located in or near the Old Northwest (Fig. 1). Most of the frontier forts in Ohio, Indiana, and some in Illinois predated Tilton's 1814 order for weather observations. For this reason, most of the Army Medical Department's climate data in the Old Northwest come from forts in Michigan, Wisconsin, and Minnesota. However, the records from these forts vary in length and have significant gaps (Table 1). The westward shift of the frontier and the need for soldiers elsewhere (coupled with the downsizing of the Army) were among the reasons for gaps in the record (Jung, 1995; Prucha, 1964).

The longest and most complete of the Army climate records in the Old Northwest comes from Fort Snelling situated on a bluff overlooking the confluence of the Minnesota (St. Peter) and Mississippi Rivers near present-day St. Paul, MN (Grice, 2005; Baker et al., 1985). Soldiers arrived in the area in August 1819. Until the new fort was ready for occupancy in 1824, troops first camped along the St. Peter River (about 1 mi southeast of the new fort) where temperature readings began in

Table 1 US Army Medical Department Weather/Climate Stations in the Old Northwest

Army post	Location	Period of record[a]
Fort Armstrong	Rock Island, IL	1820–1836
Fort Dearborn	Chicago, IL	1821–1823, 1832–1836
Fort Edwards	Warsaw, IL	1823–1824
Fort Brady	Sault Ste. Marie, MI	1823–1825, 1827–1828, 1830–1856, 1872–1892
Fort Wilkins	Copper Harbor, MI	1844–1846, 1867–1870
Fort Mackinac	Mackinac Island, MI	1826, 1831–1836, 1842–1892
Fort Shelby	Detroit, MI	1820–1826
Detroit Barracks	Detroit, MI	1839–1851
Dearbornville Arsenal	Wayne County, MI	1836–1848
Fort Wayne	Detroit, MI	1862–1892
Fort Gratiot	Port Huron, MI	1831–1836, 1840–1846, 1849–1852
Fort Howard	Green Bay, WI	1821–1841, 1849–1852
Fort Crawford	Prairie du Chien, WI	1820–1825, 1828–1845, 1848–1849
Fort Winnebago	Portage, WI	1829–1845
Fort Snelling	St. Paul, MN	1819–1858, 1867–1892
Fort Ripley	Morrison County, MN	1849–1877

[a] Not necessarily complete years of data.

October 1819. In May 1820, the garrison was relocated to Camp Coldwater (about 1.0 mi or 1.6 km northwest of the previous camp) and weather observations continued albeit with some interruptions. Fort Snelling's climate record was continuous from 1 April 1824 through April 1858. Following a nine-year hiatus when the fort was unoccupied, observations resumed on 1 April 1867 and continued until the final observations were made on 29 February 1892.

At Fort Snelling, rainfall records began in July 1836, air pressure readings on 5 July 1842, and humidity measurements on 3 April 1843. The thermometer was unsheltered at least through 1858. Self-registering thermometers for determining maximum and minimum temperatures were first used on 1 January 1870. As was typical at Army posts, anemometers were not in use prior to the 1860s so that until then the wind speed was estimated (e.g., from the wind's effect on leaves and trees).

Fort Ripley, Minnesota's second frontier fort, overlooked the western bank of the Mississippi River (in present-day Crow Wing County) at the western border of the Old Northwest, well to the northwest of Fort Snelling. The principal purpose of Fort Ripley's garrison was to maintain peace among the Ojibwa, Dakota Sioux, and Winnebago peoples. The fort was first occupied on 13 May 1849 and closed on 11 July 1877. Weather observations began in July 1849 and the record was continuous through June 1857 (Boulay, 2006). Following a several month abandonment of the fort, weather records resumed in November 1857. Although observations continued until the fort closed, 6 months of data are missing (during 1865–1866 and March–November 1869). Self-registering thermometers were in place in December 1869 and instruments were first sheltered the following year. Aneroid barometer readings began in September 1872.

Wisconsin's frontier forts consisted of Fort Howard (Green Bay), Fort Winnebago (Portage), and Fort Crawford (Prairie du Chien). All three began as isolated outposts along the Fox-Wisconsin waterway linking the Upper Great Lakes and the Mississippi River. Of the three, the Fort Howard climate record is the longest and most complete. Fort Howard was one of many army posts established to enforce US authority over the fur trade (Prucha, 1964). It was erected in 1816–1817 on the former site of French (1680–1760) and British (1761–1796) posts on the northwest bank of the Fox River just above its mouth at Green Bay. Weather observations began on 8 August 1821 and continued until 30 June 1841 when the garrison was withdrawn for duty in Florida (Seminole War) and later Texas (Mexican War). Fort Howard was reoccupied in 1849 and weather observations resumed on 1 October 1849 and continued through 31 May 1852. That year, the War Department ordered the fort abandoned.

In early 1821, some troops were removed from Fort Howard and temporarily garrisoned at a new site, later known as Camp Smith, about 3 mi (5 km) upstream along the Fox River. Camp Smith was on the southeast bank about 75 ft (23 m) above the river. Army officials in Washington, DC, however, saw no strategic or health advantages to the new site and citing high construction costs, ordered troops back to Fort Howard, abandoning Camp Smith. Troops occupied Camp Smith for about one year, and for a three-month period (November 1821–January 1822), simultaneous daily weather observations were taken at Fort Howard and Camp Smith.

Forts Crawford and Winnebago were established for the same purpose as Fort Howard. The first Fort Crawford was constructed in 1816 at the former site of British forts on what is now St. Feriole Island in the Mississippi River, in Prairie du Chien, WI. Flooding in 1828 forced the relocation of the fort to higher ground overlooking the Mississippi River, about 2 mi (3 km) above the mouth of the Wisconsin River. Weather records cover the periods: January 1820–June 1823, November 1823–March 1825, January–September 1828, July 1829–September 1845, and November 1848–April 1849. Fort Winnebago opened on 9 October 1828 at the portage between the Fox and Wisconsin Rivers, near present-day Portage, WI (Fig. 2). Weather records are continuous from January 1829 to September 1833 and January 1834 to August 1845. Fort Winnebago was abandoned on 10 September 1845 and Fort Crawford on 9 June 1856, although the latter was a recruiting station during the Civil War.

Frontier forts in Michigan included Fort Mackinac (on Mackinac Island in the Straits of Mackinac), Fort Brady (Sault Ste. Marie), Fort Wilkins (on Lake Superior near Copper Harbor), Fort Gratiot (at the exit of Lake Huron on the St. Claire River), and several forts near Detroit, including Fort Shelby, Detroit Barracks, Fort Wayne, and Dearbornville Arsenal. At Fort Mackinac, occupied by American forces from 1796 until 1895, weather observations were taken in 1826, 1831–1836, and 1842–1892. Three forts were called Fort Brady, dating from 1822, the late 1860s, and 1890s to 1944. The weather record spans the intervals: 1823–1825, 1827–1828,

Fig. 2 The Surgeon's Quarters, the only remaining building of Fort Winnebago, near Portage, WI. The post surgeon was responsible for taking weather observations and the record extends from January 1829 through August 1845 with a several month hiatus in late 1833

1830–1856, and 1872–1892. Fort Wilkins was garrisoned for relatively short periods: 1844–1846 and 1867–1870, and weather observations were taken from 1844 through 1846. Fort Gratiot was occupied from 1814 through 1821 and again from 1828 to 1879. In 1813, American forces captured a British fort in Detroit and named it Fort Shelby; this installation remained open until 1828. In 1838, Detroit Barracks opened, to be replaced by Fort Wayne in 1861.

Illinois frontier forts included Fort Dearborn (located at present-day Chicago), Fort Armstrong (at the foot of Rock Island in the Mississippi River), and Fort Edwards (on a bluff overlooking the Mississippi River near present-day Warsaw, about 40 mi (65 km) above Quincy, IL (Fig. 3)). The first Fort Dearborn, erected in 1803, burned to the ground in 1812. The second Fort Dearborn was built in 1816 and garrisoned until 1823 and again from 1828 until 1837. The weather record spanned the periods 1821–1823 and 1832–1836. Fort Armstrong was occupied from 1816 to 1836 with weather observations taken from 1820 through 1836. Fort Edwards operated from 1817 to 1824, but had only 2 years of weather records, 1823–1824.

Data from the Army Medical Department's weather observing network were the basis for the first authoritative description of the nation's climate. Samuel Forry (1811–1844), an assistant Army surgeon from 1836 to 1840, authored *The Climate of the United States and its Endemic Influences* (1842), drawing on weather data from 31 stations, most having less than 10 years of records. Forry's primary

Fig. 3 This monument marks the location of Fort Edwards overlooking the Mississippi River near Warsaw, IL. US Army medical personnel took weather observations here from 1823 though 1824

interest was the relationship between weather and human health. For him, climate "embraces not only the temperature of the atmosphere, but all those modifications of it which produce a sensible effect on our organs [and] constitutes the aggregate of all the external physical circumstances appertaining to each locality in its relation to organic nature" (Forry, 1842, p. 127).

3 Smithsonian Weather/Climate Network

Beginning in 1849, the Smithsonian Institution recruited civilian volunteers from all walks of life to monitor the weather. With the addition of state and private weather services, the number of volunteer observers grew from 150 at the end of 1849 to perhaps as many as 600 at times during the 25 years of operation. Smithsonian volunteers were located throughout the US (including the Old Northwest), as well as in Canada, Mexico and Latin America. Most Smithsonian volunteers gathered weather observations for the climate record and mailed monthly reports to the Smithsonian in Washington, DC; many of these records are also available from the US National Archives. Within a few years, another smaller group of volunteers were telegraphing daily weather observations to the Smithsonian for weather forecasting. Formation of this climate/weather observing network was the first scientific endeavor of the Smithsonian and its first secretary, Joseph Henry (1797–1878) (Millikan, 1997). The principal goals of the project were to describe the climate of North America and learn more about storms crossing the nation (Fleming, 1990, pp. 75–93).

The Smithsonian supplied its volunteer observers with instruments, standard reporting forms, and instructions for taking observations. Observers were organized into three classes depending on types of instruments: class one observers were issued a barometer, thermometer, wind vane, rain gauge, and in some cases a hygrometer; class two had the same instruments except for the barometer; class three observers had no instruments. All observers estimated wind speed and direction, type and amount of cloud cover, and time and duration of precipitation. Similar to the Army Medical Department's observers, the Smithsonian volunteers also noted natural phenomena such as phenological events. Numerous weather observers in the Old Northwest participated for varying periods in the Smithsonian weather network. The long-term College Hill, OH climate record was maintained by Smithsonian volunteer observers from January 1854 until it was taken over by the Signal Service in 1870 (Conner, 2004b). See Fleming (1990, pp. 175–184) for brief profiles of a sample of Smithsonian observers, listing their location, occupation, years of observation, and instruments.

Smithsonian thermometers were not self-registering so that mean daily temperature was computed from several regularly scheduled daily instrument readings. At first, instruments were read four times daily, at sunrise, 9 a.m., 3 p.m., and 9 p.m., local sun time. In 1853, observation times shifted to 7 a.m., 2 p.m., and 9 p.m. Reliance on local sun time meant that observations were not simultaneous even regionally so that the data were more useful for climatic purposes than for weather studies.

Joseph Henry's research on electromagnetism contributed to the invention of the electric telegraph (Millikan, 1997; Hughes, 1994). Henry realized that telegraphy enabled rapid transmission of weather observations and recognized the value of simultaneous observations for weather forecasting. In 1849, Henry persuaded the heads of several telegraph companies to authorize transmission of weather reports at no charge. In return, Henry supplied thermometers and barometers to telegraph operators in major cities. In the morning, telegraph operators wired the latest weather observations to the Smithsonian. Based on these near real-time observations, Henry prepared the first *current* national weather map in 1850 and later regularly displayed the daily weather map for public viewing in the Great Hall of the Smithsonian building (Langley, 1894, p. 217). On 1 May 1857, the *Washington Evening Star* published the nation's first weather forecast – likely prepared by Henry and James P. Espy, a pioneer storm researcher in the War Department. By 1860, 42 telegraph stations were participating in the Smithsonian network – all but three located east of the Mississippi River (Fleming, 1990, p. 145).

Lorin Blodget (1823–1901), a former Smithsonian observer at Chautauqua, NY, helped analyze the flood of weather data pouring into the Smithsonian. Henry was impressed by Blodget's statistical skills and in 1851 gave him responsibility for synthesizing and interpreting all available US climate records. Blodget also helped the Army Medical Department organize their data and in 1855 authored a report summarizing the nation's climate illustrated with maps of isotherms and isohyets. At the time, only 18 stations nationwide had at least 30 years of climate data. Positive response to his report encouraged Blodget to expand his findings into his classic text, *Climatology of the United States* (Blodget, 1857).

During the Civil War, the military had priority use of telegraph lines, disrupting the Smithsonian telegraph-based network for the duration especially in the South. Compounding this situation, in 1865 a fire at the Institution destroyed some 31 months of weather records between 1849 and 1863. Although weather observations for climatic purposes continued throughout the War, budget cuts and loss of observers to military service reduced the number of reporting stations to under 300. During Reconstruction, more farmers joined the network so that by the early 1870s the number of volunteer observers recovered to near pre-War highs. However, in 1872, budget problems forced Henry to arrange for the transfer of the Smithsonian volunteers to the new network operated by the Army Signal Service; the transfer was completed in 1874 (Miller, 1930).

4 Other Weather/Climate Networks

In 1817 (following the infamous "Year without a Summer"), Josiah Meigs (1757–1822), Commissioner of the General Land Office in Washington, DC, ordered the registers of the nation's twenty regional land offices (including those in Michigan, Ohio, Indiana, and Illinois) to take systematic observations of temperature, wind, and weather (Fleming, 1990; Landsberg, 1964). Unfortunately, these records have never been located and may be lost.

The US Army Corps of Topographical Engineers (created in 1838 and incorporated into the Corps of Engineers in 1863) undertook the *Survey of Northern and Northwestern Lakes* in 1841 initially gathering topographical and hydrographical data (Schubert, 1988). Captain George G. Meade (1815–1872), later Union Commander at the Battle of Gettysburg, directed the Lake Survey from 1857 to 1861 and added meteorological observations setting up as many as 25 reporting stations along the Great Lakes' shoreline. Lake Superior stations included Superior City, WI, Ontonagon, MI, and Marquette, MI; on Lake Michigan were Milwaukee, WI, Michigan City, IN, and Grand Haven, MI. Lake Huron stations consisted of Thunder Bay, Ottawa Point, Forestville, Sanilac, Fort Gratiot, and Detroit in Lower Michigan. Monroe, MI, Cleveland, OH, and Buffalo, NY were stations on Lake Erie. Observations were taken daily at 7 a.m., 2 p.m., and 9 p.m., local sun time, and included temperature, humidity, precipitation, evaporation, air pressure, cloud cover, and wind. In addition, lake levels were recorded. Observations were mailed to the Survey Office in Detroit, MI, and by 1861 shared with the Smithsonian Institution. Lake Survey data were published in the *Annual Reports of the War Department* and after 1868, in the *Annual Reports of the Chief of Engineers*.

Lake Survey observations revealed much about how weather influenced the Great Lakes. For example, comparison of weather data with lake-level variations demonstrated that winds blowing persistently from the same direction caused water to pile up at the downwind end of the lakes. Observations also confirmed that most storms that influence Great Lakes' weather generally arrived from the west or southwest. In 1874, the US Army Signal Service's storm-warning network absorbed the Lake Survey stations.

Astronomer Cleveland Abbe (1838–1916) set up a telegraphic weather network based at Cincinnati's Astronomical Observatory (Miller, 1931; Willis and Hooke, 2006). Abbe, Director of the Observatory from 1868 to 1871, argued that astronomical observations would benefit from a better understanding of the atmosphere (Abbe, 1916). With short-term funding from the Cincinnati Chamber of Commerce, the network began operating on 1 September 1869 with weather reports telegraphed from St. Louis, Chicago, Leavenworth, KS, and Cincinnati. Abbe's *Weather Bulletin of the Cincinnati Observatory* included trial 24-hour weather forecasts, first issued on 2 September 1869. Eventually, the network grew to more than 17 stations located mostly west and south of Cincinnati before merging with the Signal Service network in 1870.

During the 1800s, some individuals formed their own network of correspondents who shared weather diaries and weather observations. Notable among these was James P. Espy (1785–1860). Espy's network began in 1834 and by 1842 some 110 volunteer observers were participating (Jenne and McKee, 1985). His primary objective was to gather observational data in support of his theory of storms (*The Philosophy of Storms*,1841). His ideas, although incorrect, stimulated considerable public debate with fellow scientists from the mid-1830s to the mid-1840s (Fleming, 1990, pp. 23–54). Espy's network eventually merged with the Smithsonian network.

5 US Army Signal Service

Between 1870 and 1891, the US Army Signal Service conducted systematic weather and climate observations throughout much of the settled regions of the nation (including the Old Northwest). Data were gathered under more standardized conditions, gaps in the record were fewer and shorter, the number of simultaneous weather observations increased, and a better understanding of weather systems made possible the first regular weather forecast service. Increase A. Lapham (1811–1875) and Brig. General Albert J. Myer (1828–1880) were early key players in these advances in weather and climate observation.

Lapham was a self-educated natural scientist whose interest in weather, especially on the Great Lakes, dated at least to his arrival in Milwaukee in 1836. In 1840 he learned of Espy's theory of storms in a letter from his brother Darius (Miller, 1931). Lapham was Milwaukee's first volunteer Smithsonian observer, recording weather conditions at his Milwaukee home for varying periods from 1 March 1849 to 31 December 1871 (Conner, 2006). Overlapping his Smithsonian service, Lapham was an observer in Espy's network from 1849 to 1853 and the Lake Survey from 1859 to 1871. (While away from home, his wife Ann took observations.)

Lapham's experience as a weather observer and his life-long interest in the natural environment were the basis for his several publications on Wisconsin's physical geography and climate. In his 1844 book, Lapham surmised the moderating influence of the Great Lakes on the region's climate (Hayes, 1995). Lapham wrote: "The Great Lakes have a very sensible effect upon our climate, making the summers less hot and the winters less cold than they would otherwise be." In 1867, Lapham and colleagues speculated on the climatic implications of the rapid clearing of Wisconsin's forests (Lapham, 1867).

Lapham was troubled by the loss of life in shipwrecks caused by storms sweeping across the Great Lakes. By studying the work of Espy and others, he learned that storms generally approach the Great Lakes from the west or southwest and intensify with falling air pressure. Lapham argued for a network of telegraph-linked weather stations that would give advance warning of storms taking aim at the Great Lakes. He demonstrated the feasibility of such a warning system by analyzing Smithsonian data from 13 to 17 March 1859, tracking a storm system that crossed the Texas coast and traveled to Lake Michigan and then on to the Atlantic coast and Newfoundland. In 1861, Lapham collaborated with Asa Horr (1817–1896), a physician and Smithsonian observer in Dubuque, IA, on telegraphic techniques for forecasting weather, especially involving air pressure variations. Lapham argued that the lives and property saved by a telegraphic storm warning system would more than compensate for the cost of operating the system.

On 8 December 1869, Lapham presented his proposal (*memorial*) for a storm warning service to Milwaukee Congressman Halbert E. Paine (U.S. Congress House of Representatives, 1869), citing published reports of fatalities from shipwrecks on the Great Lakes. In 1868, storms damaged or sunk 1164 vessels with the loss of 321 sailors and passengers and $3.1 million in property damage. The following year,

1914 vessels were damaged or sunk, the death toll was 209, and property damage totaled $4.1 million.

Paine found Lapham's argument compelling, and on 16 December 1869, he introduced a bill (H.R. 602) into Congress calling for a storm-warning service. Letters of support were submitted by J.K. Barnes, Surgeon General; Joseph Henry; Brig. General Albert. J. Myer, Chief Signal Officer of the US Army; and Elias. Loomis, Yale College (Miller, 1930; U.S. Congress House of Representatives, 1870). Working with Senator Henry Wilson of Massachusetts, Paine re-introduced the bill as a Joint Resolution of Congress (H.J. Res. 143) on 2 February 1870. The Resolution called for the Secretary of War

> to provide for taking meteorological observations at the military stations in the interior of the continent, and at other points in the States and Territories of the United States . . . and for giving notice on the northern lakes and on the seacoast, by magnetic telegraph and marine signals, of the approach and force of storms.

On 4 February 1870, the Paine-Wilson Joint Resolution passed Congress without debate and was signed into law by President Ulysses S. Grant on 9 February 1870.

Brig. General Albert J. Myer saw the new storm-warning network as an opportunity to save the Signal Corps from the budget axe. In 1860, Myer replaced Army couriers with an innovative system of signals consisting of flags and torches. In so doing, he revolutionized military field communications and founded the US Army Signal Corps. Except for a brief hiatus from 1863 to 1867, Myer served as Chief Signal Officer until shortly before his death in 1880. Following the Civil War, Congress reduced the size of the Army to save money and some began questioning the need for the Signal Corps. Legislation passed in 1863 authorized the Corps only during the Civil War and the number of officers and enlisted men in the Signal Corps plunged by almost 90% between 1864 and 1865 (Raines, 1996). Myer petitioned Paine to make weather observation the new mission of the Signal Corps. (Myer was an army surgeon in Texas and familiar with the Army Medical Department's weather network.) Paine, a Civil War veteran, favored the War Department for weather observation believing that military discipline would ensure timely and reliable observations. On 15 March 1870, Secretary of War William W. Belknap assigned national weather observing duties to the Signal Corps (thereafter known as the Signal Service).

Myer named the new weather service, *The Division of Telegrams and Reports for the Benefit of Commerce*, and a few years later added the words *and Agriculture* to the title. Myer purchased weather instruments, began training personnel in weather observation and telegraphy, and arranged for telegraph service. On 1 November 1870, 24 Signal Service observer-sergeants began taking weather observations at localities stretching from the eastern seaboard westward to Cheyenne in Wyoming Territory. Eight of the original stations were located in the Old Northwest: Milwaukee, WI, St. Paul, MN, Duluth, MN, Chicago, IL, Cincinnati, OH, Toledo, OH, Cleveland, OH, and Detroit, MI. Each station was equipped with a barometer, thermometer, hygrometer, anemometer, wind vane, and rain gauge.

Many civilians worked for the Signal Service; among them was Increase Lapham. On 8 November 1870, Lapham was appointed Assistant to the Chief Signal Officer. Based in Chicago, Lapham supervised the Great Lakes storm warning service and on his first day of duty, issued a storm-warning bulletin, forecasting strong winds for Lake Michigan:

> Chicago, November 8, 1870, Noon. High wind all day yesterday at Cheyenne and Omaha; a very high wind this morning at Omaha; barometer falling, with high winds at Chicago and Milwaukee today; barometer falling and thermometer rising at Chicago, Detroit, Toledo, Cleveland, Buffalo, and Rochester; high winds probable along the Lakes.

Lapham (1871) noted that five lake ports reported high winds ranging from 24 to 38 mph the following day, verifying his forecast.

Lapham relinquished his Signal Service appointment in May 1872 and returned to Milwaukee. Cleveland Abbe was appointed the first chief meteorologist in the Signal Service's Washington, DC office on 3 January 1871 and his distinguished career spanned three decades. According to Willis and Hooke (2006), "Abbe, by example and precept, established the scientific standards for the [weather] service."

The new weather network expanded rapidly. Although the original enabling legislation applied only to the Great Lakes and coastal zone, the Appropriations Act of 1872 extended weather reports and storm warnings to the entire nation. In 1873, the Signal Service began stringing dedicated telegraph lines westward, eventually reaching the Southwest and Pacific Northwest. By 1881, Signal Service telegraph lines attained their maximum total length of about 5,000 mi (8,000 km). In 1874, the Signal Service absorbed weather/climate stations operated by the Army Medical Department, Smithsonian Institution, and US Army Corps of Engineers. By 1880 the number of stations reporting daily weather observations by telegraph totaled 110, including many in the Old Northwest. At the same time, the number of volunteer observers, medical officers at Army posts, and state weather service personnel who took weather observations for the climate record topped 500 nationwide. On 1 January 1872, the Signal Service began monitoring river levels and by spring of that year, started a river-forecast service.

Between November 1870 and the end of 1884, Signal Service weather observers were responsible for two sets of daily observations (Weber, 1922). One set was taken simultaneously at all stations and telegraphed to the Signal Service Office in Washington, DC where meteorologists plotted and analyzed daily weather maps. The second set of observations was for the climate record. Observers recorded temperature, relative humidity, air pressure, wind speed and direction, precipitation, cloud cover, and weather conditions. As of 1 January 1885, the special observation times for climatic purposes were discontinued because of widespread introduction of self-registering thermometers.

The Signal Service prepared its first weather map on 1 January 1871 and its first daily weather forecast (initially called *probabilities*) on 19 February 1871. Probabilities were issued three times daily from Washington, DC for 8 geographical districts and specified expected weather conditions, temperature, winds, and pressure. Beginning in October 1872, the regular prediction period was 24 h and probabilities were issued for 9 districts nationwide. The prediction period was extended to 32 h in July 1885, 36 h in July 1888, and 48 h on 1 August 1898. Beginning in May 1886, predictions were made for the individual states (instead of districts) and on 1 April 1889, the term "forecast" was used for the first time. By 1890, weather forecasts were made and issued at local Signal Service stations.

Myer's successor as Chief Signal Officer in 1880, Col. William B. Hazen (1830–1887), emphasized weather observation for storm warning and forecasting plus basic research on weather systems such as thunderstorms and tornadoes. During Hazen's administration, Sgt. John P. Finley authored the pioneering report *Character of Six Hundred Tornadoes*, a comprehensive work on tornadoes and their climatology for the period 1794–1881 (Finley, 1884). Responding to disastrous floods, 43 special rainfall stations, including many on major tributaries in the Old Northwest, were established to supplement river stations, using observers who filed special reports of excessive rain events and snowfall data.

Hazen and H.H.C. Dunwoody (1842–1933) revitalized climatological services beginning in 1881 (Miller, 1930). Hazen and his colleagues knew the importance of climate information for agriculture and developed a plan for state weather/climate services, calling for at least one observer per county. By 1892, all states had such a service and the number of volunteer observers topped 2000. Observers submitted reports to the chief of the state service who published monthly summaries. State summaries were sent to the Chief Signal Officer for inclusion in the *Monthly Weather Review* (then published by the Signal Service). By the 1890s, the State Weather Service Division became the Climate and Crop Service.

Federal budget cuts in 1883–1884 forced the closing of some Signal Service weather stations. By the mid 1880s, Congressional leaders were questioning whether weather observation was the proper purview of the military and whether the Signal Service's weather duties were detracting from its military functions (National Archives, 1942, p. 30–38). Although diminishing resources challenged the weather observing network, Brigadier General Adolphus W. Greely (1844–1935), who became Chief Signal Officer in 1886, ordered signal flags flown for expected wind direction, intensity of an approaching storm, and cold wave warnings. By the close of 1886, signal flags flew at 260 locations nationwide (Bradford, 1999).

6 US Weather Bureau

With a shrinking military in the late 1880s, civilians made up an increasing percentage of Signal Service employees and volunteer observers were increasingly relied upon for gathering climate data. Finally, on 3 December 1889 President Benjamin Harrison (1833–1901) called for transfer of the weather service out of the War Department. In 1890, Congress passed the Organic Act, which assigned weather observing duties to the US Department of Agriculture effective 1 July 1891. Thus the weather service shifted from military to civilian control and Signal Service enlisted personnel had the option of accepting an honorable discharge to join the new weather service, which most of them did. This agency, named the US Weather

Bureau (USWB), was mandated to provide weather and climate guidance for agricultural interests. In 1970, the US Weather Bureau became the National Weather Service.

The Organic Act not only established the US Weather Bureau, but also provided for the national Cooperative Observer Network (Littin, 1990; Thomas, 1979). The Cooperative Observer Network is rooted in the Smithsonian Institution volunteer network and weather/climate services sponsored by the individual states. Today's cooperative observers are citizen volunteers who record daily maximum and minimum temperatures and precipitation totals for climatic, agricultural, and hydrologic purposes. Some also report snowfall, depth of snow cover, or river levels. Volunteer observers are supplied with calibrated instruments and data management services. In addition to the National Weather Service, the US Army Corps of Engineers, and the US Departments of Agriculture, Transportation, and the Interior sponsor some cooperative observers. Individuals, corporations, colleges, and utilities also operate cooperative stations. The Organic Act of 1891 was responsible for a substantial increase in the number of Cooperative Observer stations in the Old Northwest, from 407 in 1890 to 567 by 1900. Moran and Hopkins (2002) describe the subsequent evolution of weather/climate monitoring in the Old Northwest.

7 Conclusions

The period 1820 to 1895 witnessed the westward shift of the nation's frontier and settlement of the Old Northwest. Weather/climate networks operated by the Army Medical Department, Army Corps of Topographical Engineers, Smithsonian Institution, and the Army Signal Service gathered data on the climate of the region. Besides being useful for agriculture and commerce, these data fueled speculation that settlement and the accompanying deforestation and cultivation of the land were responsible for climate change. In an interesting parallel with contemporary discussions regarding global climate change, 19th century scientists disagreed on the type and extent of the climatic impact of human activities (Fleming, 1998). In his analysis of *the Meteorological Register for 1822–1825* prepared by Lovell, Smart (1894) addressed the question of climate change due to settlement. He noted that opinions on the subject were contradictory with "some contending that as the population increased and civilization extended the climate became warmer, others that it became colder, and others that there was no change."

A lengthy instrument-based climate record – especially one that extends back to pre-settlement days – is valuable in the study of climate change and the potential role of human activity. Such a record provides a comprehensive view of the potential range of climate variability and a perspective on the present climate. But as is evident in this chapter, interpreting the 19th century climate record from the Old Northwest must take into account many factors that bear upon the integrity of the record. Through the period of record, the sophistication and reliability of instruments improved, the location and exposure of instruments changed, and the density of weather/climate stations increased. These factors caution against a simplistic comparison of early and more recent climate data in the search for signals of climate change.

References

Alexander WH (1924) Climatological History of Ohio. Ohio State University Engineering Experiment Station, Columbus, OH

Abbe C (1916) A short account of the circumstances attending the inception of weather forecast work by the United States. Mon Weather Rev 44:206–207

Baker DG, Watson BF, Skaggs RH (1985) The Minnesota long-term temperature record. Clim Change 7:225–236

Blodget L (1857) Climatology of the United States and the temperate latitudes of the North American continent. L. Lippincott, Philadelphia

Boulay P (2006) History of weather observations, Fort Ripley, Minnesota, 1849–1990. Midwestern Regional Climate Center, Climate Database Modernization Program, NOAA's Climatic Data Center, Asheville, NC

Bradford M (1999) Historical roots of modern tornado forecasts and warnings. Weather Forecasting 14:484–491

Chenoweth M (1996) The U.S. army medical department's involvement in the early history of weather observing: was Benjamin Waterhouse the first army weather observer? Bull Amer Meteor Soc 77:559–563

Conner G (2004a) History of weather observations, Cincinnati, Ohio, 1789–1947. Midwestern Regional Climate Center, Climate Database Modernization Program, NOAA's Climatic Data Center, Asheville, NC

Conner G (2004b) History of weather observations, College Hill, Ohio, 1814–1889. Midwestern Regional Climate Center, Climate Database Modernization Program, NOAA's Climatic Data Center, Asheville, NC

Conner G (2006) History of weather observations, Milwaukee, Wisconsin, 1837–1948. Midwestern Regional Climate Center, Climate Database Modernization Program, NOAA's Climatic Data Center, Asheville, NC

Darter LJ (1942) List of climatological records in the national archives. National Archives, Washington, DC

DeWitt S (1832) Description of the nine inch conical rain gage. Silliman's J (Amer J Sci) 22: 321–324

Espy, JP (1841) The philosophy of storms. Little and Brown, Boston.

Finley JP (1884) Report on the character of six hundred tornadoes. Professional Papers of the Signal Service 7, Office of Chief Signal Officer, War Department, Washington, DC

Fleming JR (1990) Meteorology in America, 1800–1870. Johns Hopkins University Press, Baltimore

Fleming JR (1998) Historical perspectives on climate change. Oxford University Press, New York

Forry S (1842) Climate of the United States and its endemic influences. J. and H.G. Langley, New York

Grice GK (2005) History of weather observations, Fort Snelling, Minnesota, 1819–1892. Midwestern Regional Climate Center, Climate Database Modernization Program, NOAA's Climatic Data Center, Asheville, NC

Hayes PG (1995) Increase A. Lapham: a useful and honored life. Wis Acad Rev 41:10–15

Hill J (2005) Kentucky weather. The University Press of Kentucky, Lexington

Hughes P (1994) The great leap forward. Weatherwise 47:22–27

Jenne RL, McKee TB (1985) Data. In: Houghton DD (ed.) Handbook of applied meteorology. Wiley, New York

Jung PJ (1995) Soldiering at Fort Howard, 1816–1841. Voyageur 12:26–35

Landsberg HE (1964) Early stages of climatology in the United States. Bull Amer Meteor Soc 45:268–275
Langley SP (1894) The meteorological work of the Smithsonian Institution. Bulletin 11:216–220, Department of Agriculture, Weather Bureau, Washington, DC
Lapham IA (1867) Report on the disastrous effects of the destruction of forest trees, now going on so rapidly in the state of Wisconsin. State Historical Society of Wisconsin, Madison
Lapham IA (1871) Meteorological observations 1870–1871. Papers of Increase A. Lapham, Vol. IX, Box 28, State Historical Society of Wisconsin, Madison
Lawson T (1840) Meteorological registers for the years 1826–1830 (1822–1825 appended). Haswell, Barrington and Haswell, Philadelphia
Lawson T (1851) Meteorological register for twelve years from 1831 to 1842 inclusive. Alexander, Washington, DC
Lawson T (1855) Army meteorological register for twelve years from 1843 to 1854 inclusive. Alexander, Washington, DC
Littin B (1990) Citizen weather observers. Weatherwise 43:254–259
Miller ER (1930) Tradition versus history in American meteorology. Mon Weather Rev 58:65–66
Miller ER (1931) New light on the beginnings of the weather bureau from the papers of Increase A. Lapham. Mon Weather Rev 59:65–70
Millikan F (1997) Joseph Henry's grand meteorological crusade. Weatherwise 50:14–18
Moran JM, Hopkins EJ (2002) Wisconsin's weather and climate. University of Wisconsin Press, Madison
Prucha FP (1964) A guide to the military posts of the United States, 1789–1895. State Historical Society of Wisconsin, Madison
Raines RR (1996) Getting the message through: a branch history of the U.S. army signal corps. Army Historical Series, United States Army Center of Military History, Washington, DC
Schubert FN (ed.) (1988) The nation builders, a sesquicentennial history of the corps of topographical engineers 1838–1863. Office of History, United States Army Corps of Engineers, Fort Belvoir, VA
Smart C (1894) The connection of the army medical department with the development of meteorology in the United States. Bulletin 11:207–216, Department of Agriculture, Weather Bureau, Washington, DC
Thomas CG (1979) Volunteers gather valuable weather information. Weatherwise 32:200–201
U.S. Congress House (1869) Disasters on the lakes, papers relative to the losses of vessels on the lakes. 41st Congress, 2nd Session, Misc. Doc. 10
U.S. Congress House (1870) Disasters on the lakes, letter addressed to the hon. Halbert E. Paine. 41st Congress, 2nd Session, Exec. Doc. 10, pt. 2
Weber GA (1922) The weather bureau: in history, activities, and organization. Appleton, New York
Willis EP, Hooke WH (2006) Cleveland Abbe and American meteorology, 1871–1901. Bull Amer Meteor Soc 87:315–326

Spatial Metadata for Weather Stations and the Interpretation of Climate Data

Stuart Foster and Rezaul Mahmood

Abstract Observations made at weather stations are often assumed to be representative of their surrounding region, but they can be significantly influenced by highly localized forcings associated with the environmental exposure of instruments. The documentation of spatial metadata via digital elevation models, digital orthophotographs, site photographs, and descriptive narratives integrated within a geographic information system can provide key insights to aid the interpretation of climate data. Comparative analyses of climate data from proximate stations with documented spatial metadata help to reveal sources of observational bias associated with instrument exposures and contribute to a better understanding of the historical climate record.

Keywords GeoProfile · Geographic information systems · Site exposure · Statistics · Station move

1 Introduction

Climate data acquired from near-surface observations provide insights into climate variability and change, and therein support decision-making and policy formulation. Because observing stations provide, at best, sparse sampling over a region, users must infer how representative those data are to other locations in the surrounding area. The performance specifications of instruments and reliability of station

---

S. Foster (✉)
Department of Geography and Geology, Western Kentucky University, Bowling Green, KY 42101, USA
e-mail: stuart.foster@wku.edu

R. Mahmood (✉)
Department of Geography and Geology, Western Kentucky University, Bowling Green, KY 42101, USA
e-mail: rezaul.mahmood@wku.edu

Special acknowledgement to Ashley Littell and John Walker, graduate students at Western Kentucky University who developed GeoProfiles used in this paper.

L.-A. Dupigny-Giroux, C.J. Mock (eds.), *Historical Climate Variability and Impacts in North America*, DOI 10.1007/978-90-481-2828-0_12,

observers are among the factors that bear upon the quality of observations. In addition, the environmental exposure of instruments at a station can introduce complex biases that bring into the question the value of the observed climate record. The general notion holds that the observing stations should capture the influence of synoptic-scale forcings as expressed within a region, and should not be unduly affected by local- or micro-scale forcings. Examination of historical records suggests that this ideal is often not achieved (Davey and Pielke, 2005; Pielke et al., 2007a; Pielke et al., 2007b).

This chapter demonstrates the value of enhanced spatial metadata known as GeoProfiles (Mahmood et al., 2006), to document the physical environments in which climate observations are collected and aid in the interpretation of historical climate data. Our objective is to evaluate the effects of station site exposure characteristics on climate observations. Analysis using GeoProfiles draws upon visualization capabilities provided by a geographic information system (GIS) and supplemented by statistical summaries to characterize local terrain, land cover, and the nature and extent of development in the vicinity of an observing site. Daily data measuring nocturnal cooling, a refinement of diurnal temperature range, are analyzed from seven stations in Kentucky, USA to highlight unintended local- and micro-scale influences. GeoProfiles provide insights into these influences. Two case analyses are presented using similar methods based on pairwise comparisons, but highlighting different purposes. The first illustrates the content of GeoProfiles and shows how they provide valuable information to explain differences in historical climate records at two topographically distinct observing stations. The second employs multiple paired comparisons in an exploratory analysis for a more challenging scenario where an observing site characterized by complex small-scale climate forcings is compared with several proximate sites in an effort to identify analogs.

This chapter is divided into five sections. Following the introduction, a discussion of spatial metadata and illustration of GeoProfiles is presented. The third section provides background on the measurement of nocturnal cooling and defines a measure based on quantiles of daily nocturnal cooling that is used in subsequent analyses. The two case analyses are presented in the fourth section, followed by a conclusion and assessment of the need for better spatial metadata to aid studies of historical climate data.

2 Spatial Metadata

Spatial metadata are requisite for interpreting the historical climate record of an observing station and inferring how representative it is of other sites within a regional context. Mahmood et al. (2006) previously developed a methodology producing GeoProfiles to obtain and analyze spatial metadata for station climatic record assessment. Digital geospatial data produced by the US Geological Survey, including digital elevation models (DEMs), digital orthophotographs (DOQQs), and digital land cover (DLCs) data integrated within a geographic information system (GIS) provided the foundation for GeoProfiles. For this Chapter, we acquired data from the

Kentucky Division of Geographic Information (http://technology.ky.gov/gis/). Here we provide a brief discussion of these data.

DEMs are raster data containing elevation values tied to horizontal post positions. The vertical error is reported as a RMSE based on test points that are well distributed and representative of the terrain. Level 2 DEMs available for Kentucky provide 10-m spatial resolution based on interpolation from 7.5-min contour lines. DOQQs are rasters derived from aerial photographs using techniques that correct for image distortion caused by terrain relief and camera tilts. Photographs provide perspective from a flying height of 20,000 ft above mean terrain and offer sufficient detail to visualize landscape features, including many features of the natural landscape along with urban and infrastructure development. DLCs are rasters derived from Landsat imagery using techniques to correct for geometric and radiometric error that is characteristic of satellite imagery. The second-generation National Land Cover Database 2001 uses normalized Landsat 5 and 7 imagery for three time periods to provide a seasonally averaged representation of land cover. A decision tree classification system derives 29 land cover classes.

GIS provides a means to explore local and micro scale forcings that may be reflected in the climatological record of an observing station using a variety of visualization and quantitative summary tools that derive information from the digital representation of the physical environment, both natural and cultural, in the vicinity of an observing station. This has been achieved by integrating DEMs, DQQQs, and DLCs data within the GIS environment leading to GeoProfiles. A variety of standard GIS functions are useful for developing both static and dynamic perspectives. In an operational setting, these functions can be implemented in an interactive mode. Zooming and buffering are among the most basic functions. Zooming capability enables the user to visually evaluate site characteristics at different scales. Generation of buffers around observing stations and their use in overlays to partition raster or vector layers into concentric zones enhances visualization and facilitates quantitative summaries, topography and land cover. In cases where wind data are available, directional buffers may also prove useful. Dynamic perspectives on an observing station can be produced by developing three-dimensional views that implement panning or fly-through capabilities along with zooming.

Site visits are essential to validate the absolute location of an observing station and to document characteristics of the instrument exposure that may influence the climate record. In cases where a station has been closed or moved to another site, it is often difficult to reconfigure the historical site exposure of instruments. Where possible, digital site photographs can be integrated into a GIS and provide valuable information into micro scale forcings that may bias the climatological record.

A small set of stations from the National Weather Service Cooperative Observer Program (COOP) was selected for the analyses presented in this paper. Figure 1 identifies the stations used in the two separate case analyses and highlights their proximity. Two stations, Frankfort Lock 4 (Frankfort) and Williamstown 3 NW (Williamstown), used in the first case analysis provide examples of the visual

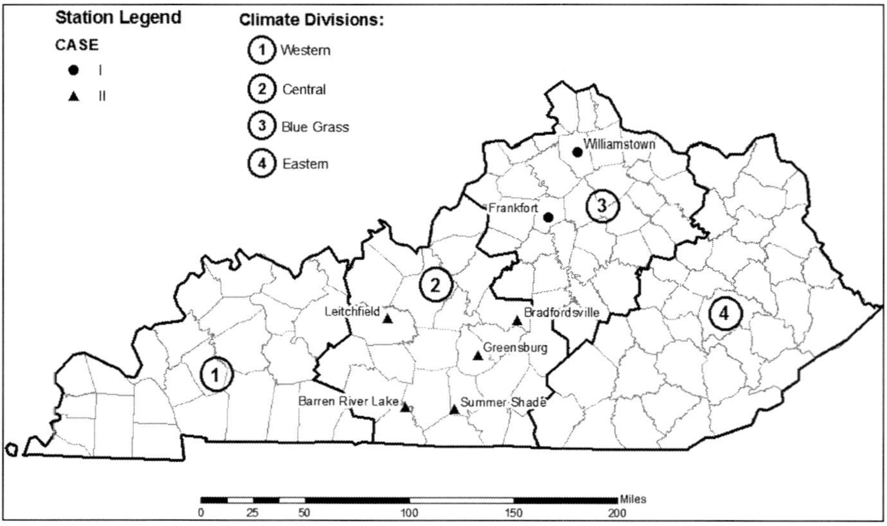

Fig. 1 Map of Kentucky's climate divisions and locations of stations used in two cases.

content of GeoProfiles (Figs. 2 and 3). Each displays layers of information that include aerial photography, land use, hill-shaded elevation, relative elevation with respect to the station location, slope, and aspect within a 1,500 m radius of the station. GeoProfiles can also be supplemented with tabular summaries derived from selected layers. Note that while this chapter incorporates static images for a preset radius, the operational GIS environment enables users to generate customized displays and summaries.

3 Data

GeoProfiles provide spatial metadata (Mahmood et al., 2006) that can help to identify characteristics of local- and micro-scale environments that may influence the climatological record of an observing site. Based on diurnal boundary layer characteristics, these forcings are likely to be more evident in the record of daily minimum temperatures than in that of daily maximum temperatures. Robeson and Doty (2005) used minimum temperatures as the basis for identifying rogue observing stations. Diurnal temperature range measured at observing stations can also capture the effects of land-atmosphere forcings (Durre and Wallace, 2001a, 2001b). More recently, Runnalls and Oke (2006) used a related measure based on nocturnal cooling to determine inhomogeneity in temperature time series. This study partly adopted their methods along with a pairwise comparisons approach shown in Mahmood et al. (2006).

Daily climate records were obtained from the Midwestern Regional Climate Center's MICIS database that reflects data archived in the "TD3200 Summary of

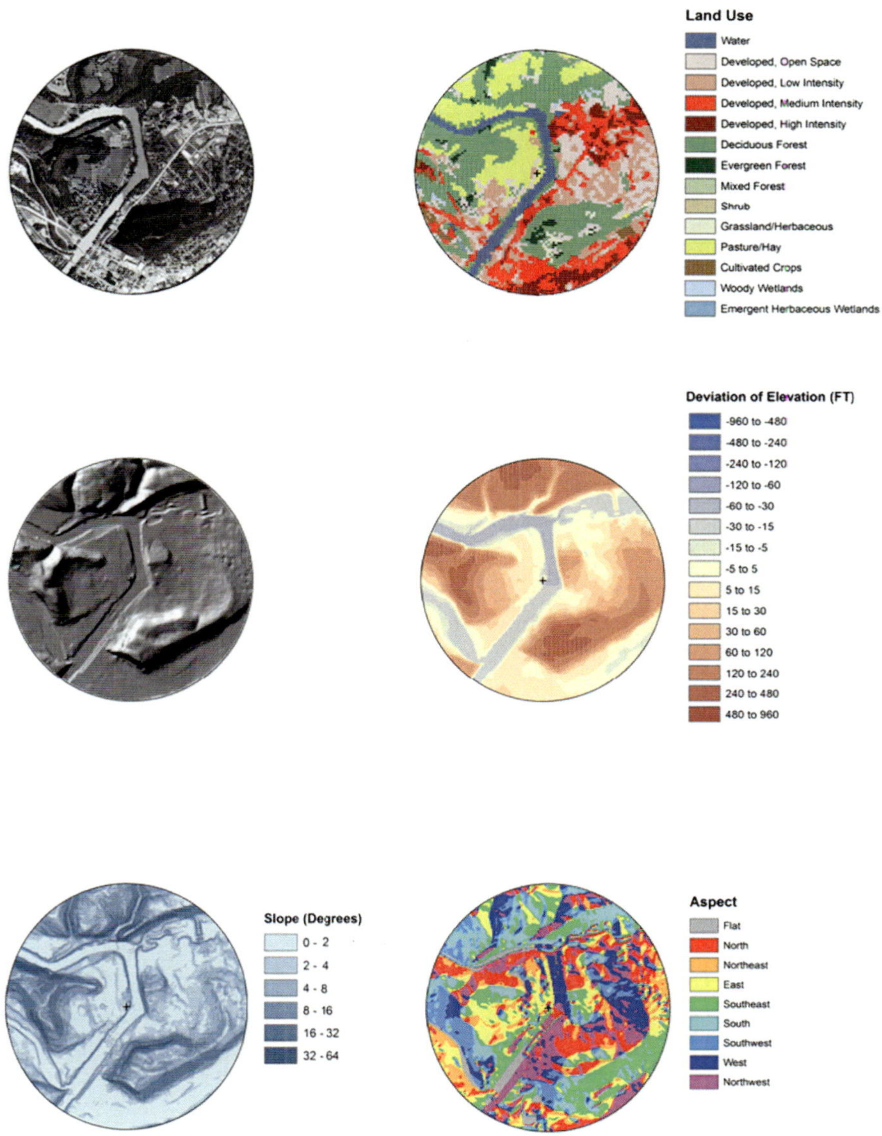

Fig. 2 GeoProfile of Frankfort cooperative observing station, including (left to right) aerial photography, land use, hill-shading of elevation, deviation of elevation from the site of the station, slope, and aspect. The radius of each image in 1500 meters

the Day" data set maintained by the National Climatic Data Center (NCDC). These records include daily maximum and minimum temperature, along with the time of observation. Following the approach of Runnalls and Oke (2006), we then calculated nocturnal cooling magnitude as

$$\Delta T_t = T_{\max(t-1)} - T_{\min(t)}. \tag{1}$$

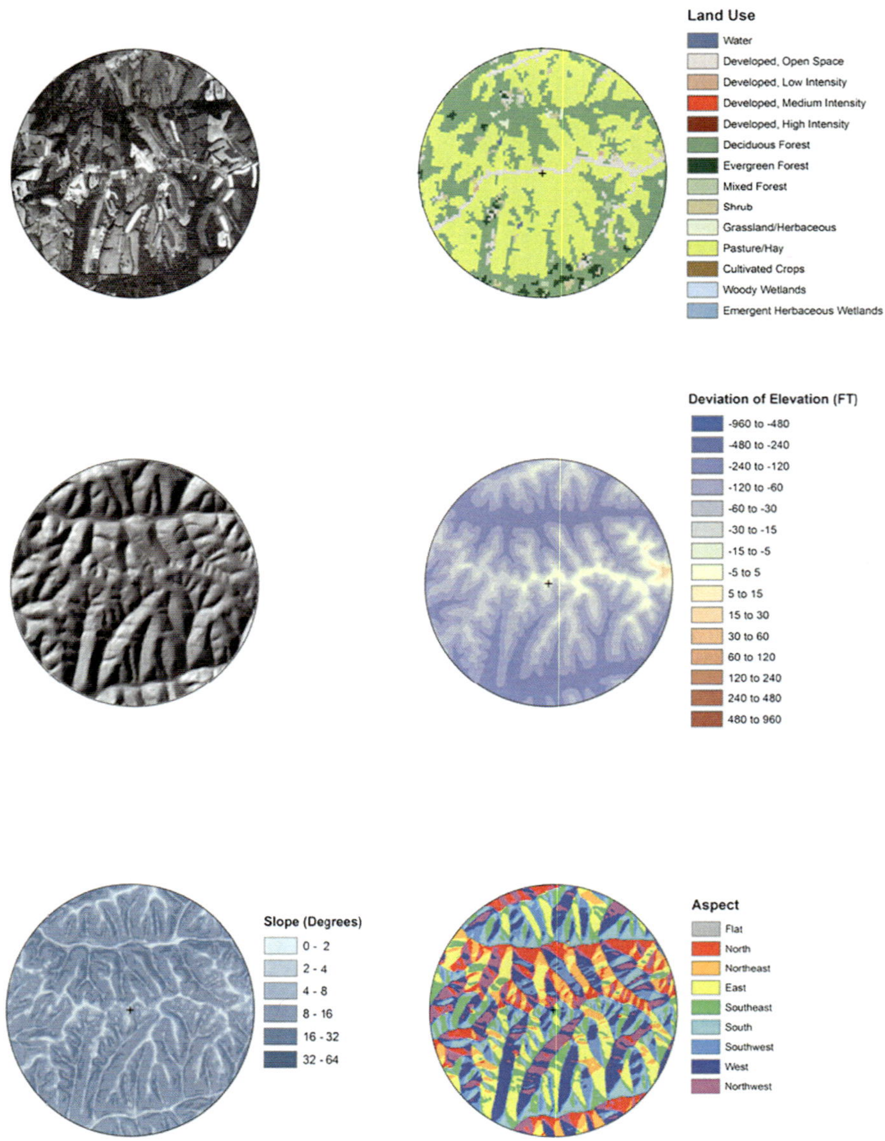

Fig. 3 GeoProfile for Williamstown cooperative observing station

Cooperative observer stations have a variety of observation times, including morning, evening, and midnight. In each case the observation reflects the 24-h period ending at the time of observation. Assuming a normal diurnal temperature cycle, the daily observation at a morning-reporting station includes the current day's low, $T_{\min(t)}$, and the previous day's high, $T_{\max(t\text{-}1)}$, necessary to calculate ΔT_t. To calculate ΔT_t for stations that report in the evening or at midnight, the current day's low is subtracted from the high reported on the previous day's observation.

Occasionally, due to synoptic weather patterns, the time of occurrence for the maximum and minimum temperatures will deviate from the normal diurnal pattern. While recognizing this, such occurrences are infrequent and do not introduce a systematic influence over a decadal time scale or in comparisons among proximate observing stations.

The magnitude of cooling that occurs at an observing station depends upon a combination of factors, including weather and site exposure characteristics. Effects of weather are particularly evident on cloudy and breezy nights. For stations where comprehensive meteorological measurements are taken, documented weather conditions can be used to subset observations that are associated with calm, clear nights. For most stations within the NWS cooperative observer network however, only daily temperature and precipitation measurements are taken, and information on weather conditions are not readily available.

We use percentiles of seasonal distributions of nocturnal cooling values as a surrogate approach to identify observations that reflect the effect of site exposure on station microclimate. Specifically, we partition nocturnal cooling values by season, where spring is March, April, and May, and the remaining seasons are defined in corresponding three-month intervals. Nocturnal cooling percentiles are then calculated for each season and year over the period of record. Since the largest ΔT_t values are likely to occur when skies are clear and winds are calm, we use the 90th percentile of the seasonal values during a given year as an indicator of the microclimate associated with the site exposure in the absence of weather effects. At the same time, the 90th percentile is expected to be robust with respect to outliers that might be associated with data errors or the occurrence of days, in which the normal diurnal temperature cycle is disrupted by a frontal passage.

4 Analysis

Is the magnitude of the difference in nocturnal cooling between observing stations consistent with random variability, or is it suggestive of inherent differences in the local and microclimates of stations? Assuming that instruments are accurate and observers are well trained, differences should be negligible and statistically insignificant for stations at proximate locations (at similar elevations). Meanwhile, large differences in nocturnal cooling between stations could be evidence of local or microclimate variability.

We use a matched pairs procedure to evaluate differences in the microclimates of two stations. The matched pairs procedures, also known as paired samples or paired differences, allows us to leverage the natural structure associated with spatial time series. It is based on a simple concept that is widely used in the design of statistical experiments: reducing background noise helps to clarify differences in a measured variable due to treatments applied to experimental units (Mendenhall, 1968). Our case involves the analysis of observational data, not conducting a controlled experiment. However the similar concepts still apply. Instead of experimental units, we have observational units, each defined as a particular season of a given year. In place of treatments, we compare microclimates associated with locations. Our measured variable is a chosen percentile, $p = 0.90$, of ΔT. The matched pairs procedure is implemented by first creating a homogeneous grouping of observational units. In this case, a comparison between two observing stations is made after selecting a matching time series and then calculating the observed difference, $D_t^{[p]}$, in $\Delta T^{[p]}$ for the paired observing stations A and B,

$$D_t^{[p]} = \Delta T(A)_t^{[p]} - \Delta T(B)_t^{[p]} \tag{2}$$

Station metadata available through NOAA's WSSRD (http://noaa.imcwv.com/) are examined to identify station moves. We truncate the historical record and use only the segment that represents the current (or most recent, for stations that are now closed) location that corresponds with our GeoProfile metadata. Comparisons are then based on the period of overlap between paired stations. If more than five percent of daily values are missing, a percentile is not calculated and the observation for that season is deleted from the analysis. Analysis of data from the 1800s and early 1900s is difficult, as metadata are often not available or else provide insufficient detail regarding station moves.

4.1 Case I: Highlighting Topographical Forcings Through Paired Comparisons

The first case demonstrates differences in small-scale topographical forcings for two observing stations, one in a valley and the other on a ridge. Data visualization tools highlight patterns and relationships involving nocturnal cooling data that are consistent with topographical information made available through GeoProfiles for the two stations.

We examined the period 1964–2001 where observations overlapped between the two stations. Unfortunately, several seasonal observations were omitted because more than five percent of daily observations were reported as missing at one or both stations.

The two COOP stations, Frankfort and Williamstown, KY are part of the US Historical Climate Network (USHCN) and are both located in the Central Climate Division of Kentucky (Fig. 1). Readings at Frankfort were made by the Army Corps of Engineers, which also managed a stream gauge on the Kentucky River at this site. While observations were taken at Frankfort as early as 1881, we utilize the segment of the historical record for the period from January 1948 through October 2001, during which no station relocations were reported. Observations at Williamstown date back as far as May 1902, and we use the segment of the historical record for the period from January 1964 through August 2004. The observer during this period was a farm operator. Neither station is currently active.

As indicated via its GeoProfile (Fig. 2), Frankfort is located in a valley and there is low-density urbanization throughout the valley. At this time, no photograph of the station was found to depict a precise observing location, although a sketch of the site from a National Weather Service metadata form indicates its proximity to a river and enables us to make a reasonable approximation of the station's location. The relative elevation of the land surface within the 1,500-m buffer ranged from 314 ft above the elevation of the station to 46 ft below it. Approximately 67% of the area within the buffer was at least 5 ft higher in elevation, while only 18% was more than 5 ft below the station in elevation.

The exposure of Williamstown was much different (Fig. 3). This station was located in a rural area characterized by gently rolling terrain. Valleys are generally wooded, while ridges stretching east to west have been cleared for farming. Hay and alfalfa are dominant in the area, and very little land is dedicated to cultivated crops. In contrast to Frankfort, Williamstown is located near a local elevation maximum. The relative elevation within the 1,500-m buffer ranged from positive 20 ft to negative 15 ft, and less than 1% of the area within the buffer was at least 5 ft higher in elevation, while 97% was more than 5 ft below the station in elevation.

Exploratory data analysis begins with visualization and is supplemented by statistical summaries. Figure 4 displays grouped box plots of the distribution of seasonal differences. Since these box plots show the distribution of seasonal values for the 90th percentile of nocturnal cooling, we assume that these values correspond to conditions observed on clear, calm nights. Under such conditions, a stable near-surface layer is conducive to the development of shallow, localized temperature gradients. The magnitude of nocturnal cooling that occurs is driven largely by the

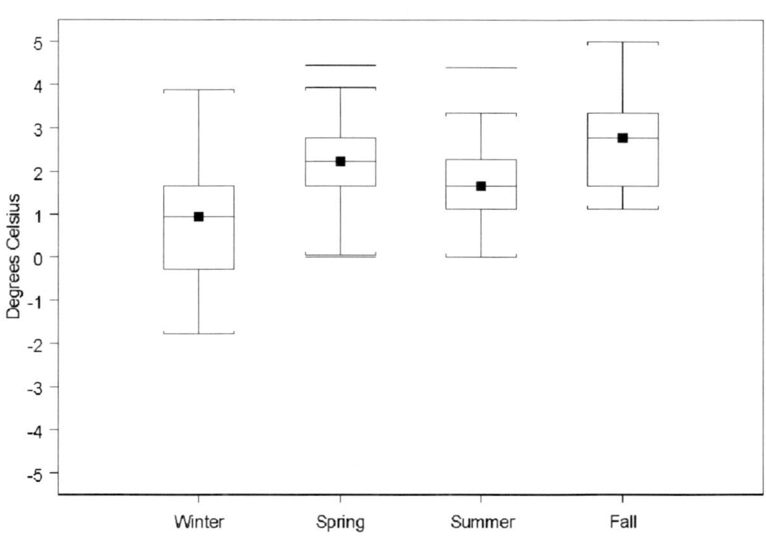

Fig. 4 Grouped box plots portraying the seasonal distributions of differences in nocturnal cooling between the Frankfort and Williamstown stations. Positive values indicate greater nocturnal cooling at Frankfort

thermal properties of soil, and these in turn are greatly influenced by soil moisture. It is well known that moist soils result in a lower surface-air temperature gradient. In addition, topography and land cover specific to an observing site are also expected to influence nocturnal cooling.

The magnitude of nocturnal cooling at a site is expected to show a distinct seasonal pattern. Precipitation exceeds evapotranspiration (ET) during the winter season. Soil moisture tends to be high throughout the season; hence the magnitude of nocturnal cooling is damped. The higher variability in nocturnal cooling differences during winter is consistent with variability in the frequency and extent of snow cover at Frankfort and Williamstown. Consistent with a lower magnitude of nocturnal cooling, pairwise differences between stations are expected to be smaller.

The spring, summer, and fall seasons each show a clear tendency toward greater nocturnal cooling at Frankfort than at Williamstown. The relative elevation of Frankfort contributes to the pooling of cool air during the overnight hours in contrast to Williamstown, and this is likely the dominant factor in each of these seasonal distributions. Other factors play a role too. Since summer is the peak of the growing season, the role of plants is to increase ET and hence the proportion of energy that is in the form of latent rather than sensible heat. In addition, vegetation canopy acts to reduce net surface heating during the day, while also limiting radiative cooling at night. The box plot for the fall season shows the greatest difference in nocturnal cooling. This is the dry season in Kentucky and soil moisture tends to be low, particular during the first half of the season. Dry soils increase the ratio of sensible to latent heat flux. Further, the end of the growing season means lower ET, again favoring sensible heat flux.

Evidence of the role of soil moisture in nocturnal cooling is observed in the relationship between seasonal precipitation and nocturnal cooling at Frankfort (Fig. 5) and Williamstown (Fig. 6). Here, precipitation refers to the climate division average seasonal precipitation while cooling magnitude is based on station data. The set of graphs and corresponding Pearson product-moment correlation coefficients show a tendency for increased nighttime cooling in dry years. In addition, the effect is greatest during the summer and fall when below normal precipitation would be an indicator of low soil moisture.

A paired t-test is reported for each of the seasonal differences (Table 1). The difference in the mean of the distributions of $D_t^{[90th]}$ ranges from 0.94°C for winter to 2.64°C for fall. The positive values indicate greater nocturnal cooling at Frankfort, and each difference is highly significant as indicated by the accompanying p-values and confidence intervals. Again, note that sample size differences reflect the effect of missing observations that are more prevalent in some seasons than others.

Following the spirit of exploratory data analysis, we evaluated the sensitivity of our results by using trimmed statistics and bootstrapping to generate confidence intervals. Those analyses produced qualitatively similar results, and hence they are not presented here.

The results from comparing Frankfort and Williamstown are not surprising. Obvious differences in the topographic exposure of each station, specifically in

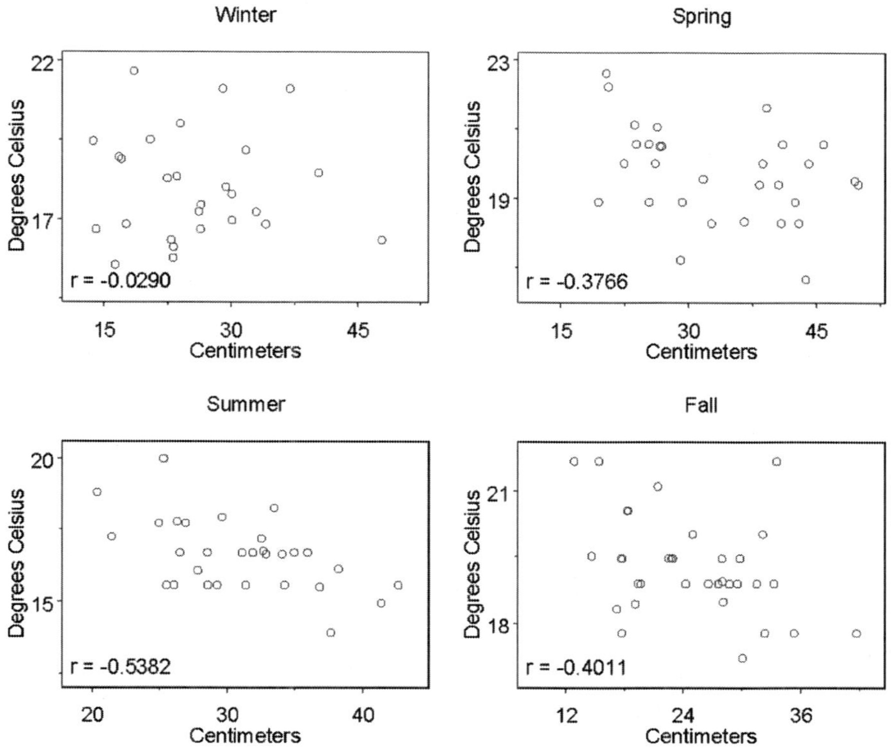

Fig. 5 Scatter plots and Pearson correlation coefficients showing the relationship between nocturnal cooling at Frankfort and precipitation in the Bluegrass climate division by season

terms of relative elevation, provide a quite plausible explanation for the large differences in nocturnal cooling. It is important to note, however, that without spatial metadata, the interpretation of these differences is not evident. The second case discussed below represents a more complex situation and further demonstrates how GeoProfiles can contribute to a greater understanding of historical climate records.

4.2 Case II: Searching for Analog Stations Using Multiple Comparisons

Stations used in the previous case were chosen because their exposures were well defined and represented an obvious contrast in terms of topography. Site visits that we have conducted to observing stations throughout Kentucky, however, have revealed numerous stations that are poorly situated. These stations have complex instrument exposures influenced by various competing local- and micro-scale forcings. Here, we use a similar methodology as in Section 4.1 but incorporate a statistical adjustment to facilitate multiple comparisons. Our purpose is also different.

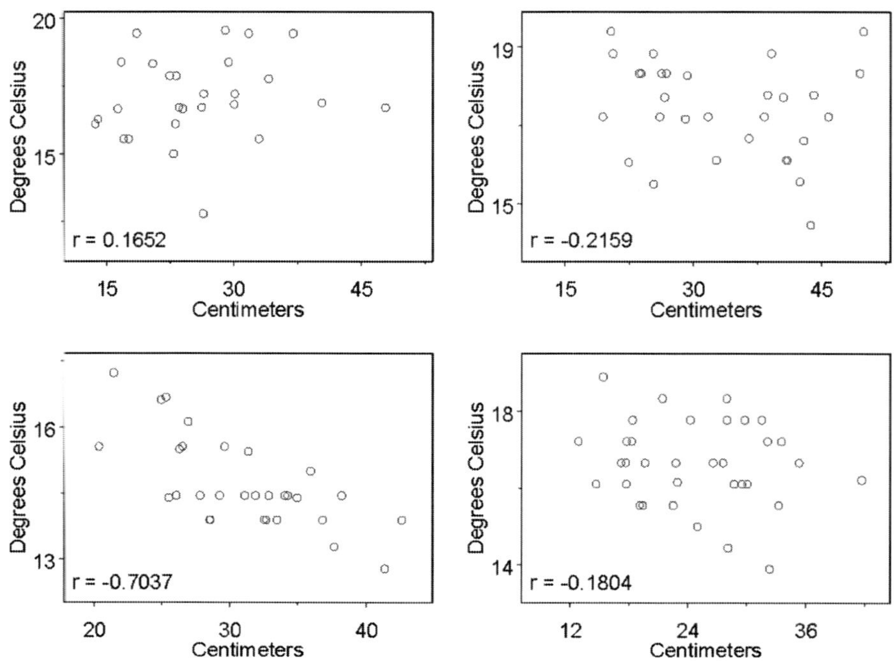

Fig. 6 Scatter plots and Pearson correlation coefficients showing the relationship between nocturnal cooling at Willliamstown and precipitation in the Bluegrass climate division by season

Table 1 Paired differences in seasonal means between Frankfort and Williamstown and their statistical significance

Season	Mean	Std. error	t-value	p-value	LCL (2.5%)	UCL (97.5%)	n
Spring	0.9424	0.2795	3.3718	0.0023	0.3679	1.5169	27
Summer	2.3000	0.1865	12.3345	0.0000	1.9186	2.6814	30
Fall	1.8907	0.1769	10.6894	0.0000	1.5290	2.2525	30
Winter	2.6373	0.1850	14.2550	0.0000	2.2609	3.0137	34

Instead of assessing the differential effects of two well-defined station exposures, we conduct multiple comparisons in search of analog stations. Pairwise comparison for a station of interest with multiple neighboring stations may help to identify analogs and contribute to a broader understanding of the historical climate record associated with the station of interest. Employing multiple comparisons, however, increases the complexity of the analysis and interpretation of results. Pairwise t-tests for $D_t^{[90th]}$ are performed as before, except here we compare a station of interest to each of the $m = 4$ neighboring stations. Fewer details of the GeoProfiles are presented, and the focus instead highlights the interpretation of results obtained when conducting multiple comparisons in exploratory data analysis.

The Greensburg observing station, part of the USHCN, is located in a small town (population near 2,500) on rolling terrain adjacent to the Green River. The

Fig. 7 Greensburg cooperative observer station. (source: Mahmood et al. 2006)

current location of the temperature sensor is on a small patch of grass surrounded by concrete and asphalt surfaces (Fig. 7). The station is located on a gentle slope facing south to southeast at an intermediate elevation relative to the buffered area. This complex exposure raises questions about how representative the site is with regard to the surrounding region. Hence, we compare Greensburg to four proximate stations: Barren River Lake, Bradfordsville, Leitchfield 2 N (Leitchfield), and Summer Shade (Fig. 1).

Unfortunately, none of these four comparison sites has a pristine instrument exposure. The temperature sensor at Barren River Lake is housed in a nonstandard shelter next to an asphalt parking lot on flat area next to a dam (Fig. 8a). The Bradfordsville site is a landscaped yard near a house in a small town located in a broad stream valley (Fig. 8b). Leitchfield is a USHCN station located in a rural area, but the temperature sensor is in a shaded location and immediately proximate to a brick building, concrete sidewalk, and asphalt parking lot (Fig. 8c). Summer Shade is located in a hamlet on a gentle south-facing slope. The temperature observations are taken near buildings and in a shaded location (Fig. 8d).

The interpretation of hypothesis tests for this case is more complicated because it involves multiple comparisons. When conducting simple hypothesis tests, α sets the threshold probability for making a Type I error, that is falsely rejecting the null hypothesis of no difference between paired stations. Since our analysis involves multiple comparisons, it is important to distinguish between α_{local} and α_{global}. We define α_{local} as the threshold for Type I error on the five individual, or *local*, matched pairs tests conducted independently of one another. Meanwhile, α_{global} is the overall, or *global*, threshold for falsely rejecting at least one of the local null hypotheses

Fig. 8 (**a**) Barren River Lake cooperative observer station. (**b**) Bradfordsville cooperative observer station. (**c**) Leitchfield cooperative observer station (source: Mahmood et al., 2006). (**d**) Summer Shade cooperative observer station (source: Mahmood et al., 2006)

among the four matched pairs tests under the assumption that all m of these are true. The relationship between the local and global values is defined by

$$a_{global} = 1 - (1 - a_{local})^m. \tag{3}$$

Hence, if we set $\alpha_{local} = 0.05$ as a conventional threshold, then $\alpha_{global} = 0.185$, which is arguably too large. Setting $\alpha_{local} = 0.013$ meanwhile, yields $\alpha_{global} = 0.05$, a value that is more appropriate in the context of exploratory analysis. Table 2 shows the relationship between local and global levels of significance for $m = 4$ comparisons. By setting α_{local} to a smaller value when conducting multiple comparisons, we counter the tendency to dredge data from neighboring stations in search of significant results.

Table 2 Relationship between local and global alpha (α) levels for $m = 4$ comparisons

α_{local}	α_{global}
0.100	0.344
0.050	0.185
0.026	0.100
0.013	0.050

Table 3 Paired differences in seasonal means between Frankfort and Williamstown and their statistical significance

Station	Season	Mean difference	Std. error	t-value	p-value	LCL (2.5%)	UCL (97.5%)	n
a. Winter								
Barren River Lake	Winter	0.26	0.16	1.63	0.1127	−0.07	0.59	34
Bradfordsville	Winter	−0.56	0.17	−3.36	0.0021	−0.91	−0.22	33
Leitchfield	Winter	0.69	0.21	3.28	0.0026	0.26	1.12	32
Summer Shade	Winter	0.13	0.42	0.32	0.7553	−0.73	1.00	22
b. Spring								
Barren River Lake	Spring	0.41	0.17	2.47	0.0187	0.07	0.75	35
Bradfordsville	Spring	−0.96	0.15	−6.61	0.0000	−1.26	−0.67	34
Leitchfield	Spring	0.37	0.21	1.81	0.0809	−0.05	0.79	30
Summer Shade	Spring	1.04	0.30	3.47	0.0024	0.41	1.66	21
c. Summer								
Barren River Lake	Summer	0.05	0.15	0.35	0.7299	−0.25	0.36	34
Bradfordsville	Summer	0.11	0.14	0.80	0.4321	−0.18	0.41	29
Leitchfield	Summer	0.11	0.14	0.80	0.4321	−0.18	0.41	29
Summer Shade	Summer	0.12	0.36	0.34	0.7390	−0.64	0.89	19
d. Fall								
Barren River Lake	Fall	0.61	0.15	4.05	0.0003	0.30	0.92	30
Bradfordsville	Fall	−1.17	0.11	−10.46	0.0000	−1.40	−0.94	29
Leitchfield	Fall	0.72	0.19	3.83	0.0007	0.34	1.11	27
Summer Shade	Fall	1.21	0.39	3.12	0.0062	0.39	2.02	18

Results from the paired t-tests are summarized by season for the comparison of Greensburg with each of the four proximate stations in Table 3. During the winter season, $D_t^{[90th]}$ of 0.69°C indicates significantly greater cooling at Greensburg compared to Leitchfield, but significantly less cooling (–0.56°C) than Bradfordsville. Contrasts involving Barren River Lake and Summer Shade are much smaller and do not reveal statistically significant differences. Unless otherwise indicated, significance levels are based on $\alpha_{\text{global}} = 0.05$ and the corresponding $\alpha_{local} =$ 0.013.

Spring season comparisons again show cooling at Greensburg that is significantly less than at Bradfordsville (–0.96°C). However, Greensburg documents significantly greater cooling than Summer Shade (1.04°C). Greensburg averages greater cooling in comparisons with Barren River Lake and Leitchfield, but neither difference is significant using the adjusted α for multiple comparisons.

Results from summer comparisons are striking. None of the pairwise comparisons produces a difference in nocturnal cooling that approaches any measure of significance. Synoptic conditions and land-atmosphere interactions during summer express themselves in a similar fashion among sites that have quite distinct exposures.

On the other hand, each of the fall contrasts is significantly different. Once again, Greensburg experiences less cooling than Bradfordsville (–1.17°C), but greater cooling than Barren River Lake (0.61°C), Leitchfield (0.72°C), and Summer Shade (1.21°C).

Results from this case reveal that station exposures can display distinct seasonal variability in comparisons with proximate stations. While the small-scale forcings affecting Greensburg have similar effects on nocturnal cooling compared to other proximate stations during summer, differing impacts are found in other seasons, particularly fall. Over the course of the year, these same stations are poor analogs for Greensburg in other seasons, particularly in spring and fall when small-scale forcings are expected to be most evident. None of the four proximate stations proves to be a suitable analog for Greensburg throughout the course of the year, and this raises concerns about the value of Greensburg as a representative site.

5 Conclusion

Climatological data acquired from in situ observing platforms have been widely used for both research and operational purposes. Ideally, the data recorded at a given observing site are broadly representative of a well-documented regional topography and landscape. Rarely, however, do those who use climatological data have access to adequate spatial metadata that adds critical contextual information to the data. In our work at the Kentucky Climate Center, we have found an alarming number of observing sites that are characterized by poor instrument exposures: stations located in narrow valleys, amongst groves of trees, near heat absorbing structures or surfaces, etc.. Such exposures contribute to local- and micro-scale forcings and create biases in observations that may no longer be representative of the surrounding region. In some landscapes that include highly variable terrain (i.e., elevation, slope, and aspect) and land cover, the concept of a regionally representative may simply not apply. Nonetheless, observations from sites with poor exposures can still provide useful data if the characteristics of those exposures are well documented. Indeed, one may argue that there is more to be learned from analyzing data from a network that includes a wide variety of instrument exposures than from one that includes only pristine observing sites.

This chapter has detailed how enhanced spatial metadata can aid in the interpretation of historical climate records. Two cases, one highlighting the content and application of GeoProfiles in comparing two topographically distinct observing sites, and the other demonstrating the use of multiple comparisons in the search for analog stations, have demonstrated exploratory approaches based on noise-reducing pairwise comparisons. The effects of local- and micro-scale forcings on the temperature record are most evident on calm, clear nights. Our approach based on percentiles is expected to yield more reliable results in areas where benign synoptic conditions are common in all seasons of the year. In areas where weather is dominated by frequent, strong thermal frontal passages, more detailed data regarding wind and sky conditions may be needed in order to isolate these forcings.

Efforts to homogenize station time series are an important prerequisite to a comparative analysis of local- and micro-scale forcings, and we used station histories to isolate time series segments for comparison. Unfortunately, the historical instability of observing networks, including station moves and closures makes it difficult to isolate long time series segments that can be analyzed for proximate stations. Nonetheless, analyses presented here have highlighted strong patterns of seasonality in the nature of small-scale forcings, and this provides a caution against the analysis of annual time series that can mask the strength and magnitude of differences in cooling magnitude between proximate stations.

Unfortunately, historical spatial metadata are poorly documented for many observing networks, including the National Weather Service Cooperative Observer Network. Given the scarcity of spatial metadata researchers often choose to conduct analyses of climate variability and change based on regional aggregates of observing stations, as illustrated by temperature and precipitation datasets produced for climate divisions. The rationale behind regional aggregates is that averaging individual stations together eliminates or at least minimizes what are assumed to be random biases. Without more detailed spatial metadata however, the validity of such a rationale is in doubt.

Archival research can provide valuable insights into possible sources of local- and micro-scale forcings that may compromise the value of an observing station as a site that is broadly representative of a large region. The Climate Database Modernization Program's 19th Century Forts and Voluntary Observers Database Build Project at the Midwestern Regional Climate Center (http://mrcc.sws.uiuc.edu/research/cdmp/cdmp.html) provides an example of the type of research to reconstitute historical spatial metadata that is necessary for developing a more comprehensive understanding of historical climates. If climatologists are to provide value-added services to their constituents, then an understanding of the context in which climatological observations are collected is essential.

References

Davey, C. A., Pielke Sr., R. A. (2005) Microclimate Exposures of Surface-Based Weather Stations: Implications for the Assessment of Long-Term Temperature Trends. Bull Amer Meteorol Soc, 86(4): 497–504.

Durre, I., Wallace, J. M. (2001a) The Warm Season Dip in Diurnal Temperature Range Over the Eastern United States. J. Clim, 14: 354–360.

Durre, I., Wallace, J. M. (2001b) Factors Influencing the Cold-Season Diurnal Temperature Range in the United States. J. Clim, 14: 3263–3278.

Mahmood, R., Foster, S. A., Logan, D. (2006) The GeoProfile Metadata, Exposure of Instruments, and Measurement Bias in Climatic Record Revisited. Int J Climatol, 26: 1091–1124.

Mendenhall, W. (1968) Introduction to Linear Models and the Design and Analysis of Experiments. Wadsworth, Belmont, CA.

Pielke Sr., R. A., Davey, C., Niyogi, D., Fall, S., Steinweg-Woods, J., Hubbard, K. G., Lin, X., Cai, M., Lim, Y.-K. , Li, H., Nielsen-Gammon, J., Gallo, K., Hale, R., Mahmood, R., Foster, S., McNider, R. T., Blanken, P. (2007a) Unresolved Issues with the Assessment of Multi-Decadal Global Land Surface Temperature Trends. Journal of Geophysical Research- Atmospheres, 112 D24S08, doi:10.1029/2006JD008229.

Pielke Sr., R. A., Nielsen-Gammon, J., Davey, C., Angel, J., Bliss, O., Doesken, N., Cai, M., Fall, S., Niyogi, D., Gallo, K., Hale, R., Hubbard, K. G., Lin, X., Li, H., Raman, S. (2007b) Documentation of Uncertainties and Biases Associated with Surface Temperature Measurement Sites for Climate Change Assessment. Bull Amer Meteorol Soc, 88(6): 913–928.
Robeson, S. M., Doty, J. A. (2005) Identifying Rogue Air Temperature Stations Using Cluster Analysis of Percentile Trends. J Clim, 8: 1275–1287.
Runnalls, K. E., Oke, T. R. (2006) A Technique to Detect Microclimatic Inhomogeneities in Historical Records of Screen-Level Air Temperature. J Clim, 19: 959–978.

Part IV
Methodologies and Other Analyses

A Seasonal Warm/Cold Index for the Southern Yukon Territory: 1842–1852

Heather Tompkins

Abstract Journals from three Hudson's Bay Company (HBC) posts from the Yukon Territory, Frances Lake, Pelly Banks and Fort Selkirk, were analyzed for weather information covering 1842–1852. Daily journal entries recorded both qualitative direct (e.g. temperature, cloud cover) and indirect (e.g. animal migration, ice activity) weather conditions. A hierarchical coding scheme was developed through content analysis that classified the entries into exclusive and unique categories. Monthly, seasonal and annual weighted averages were calculated for the post journals and culminated in a seasonal warm/cold index representing periods of normal and extreme weather conditions for the three post locations. Temperature readings taken by the HBC at Frances Lake from December 1842 to May 1844 were used to validate the index's reliability by comparison with climate normal data from a nearby Environment Canada weather station. Results show that 9 out of the 14 extreme seasons captured by the index were mild winters. The only prolonged period of extreme weather was a colder than normal six month period from the spring to the late fall of 1849.

Keywords Hudson's Bay Company · Forts data · Index · Yukon Territory · Content analysis

1 Introduction

Historical climatology research in Canada is relatively young and still under development. The first studies, published in the 1960 and 1970s (Mckay and Mackay, 1965; Moodie and Catchpole, 1975, 1976), focused on the use of the Hudson Bay Company's post journals from the Canadian north. During the 1980s, the number of Canadian studies published increased, furthering the idea that historical records

H. Tompkins (✉)
Geomatics Technician, Visual Heritage Division, Cartography, Architecture and Geomatics, Library and Archives Canada, Gatineau, Québec, Canada
e-mail: Heather.Tompkins@lac-bac.gc.ca

L.-A. Dupigny-Giroux, C.J. Mock (eds.), *Historical Climate Variability and Impacts in North America*, DOI 10.1007/978-90-481-2828-0_13,

from Canada could be used as primary sources for paleoclimatic reconstructions (Ball, 1983; Catchpole and Faurer, 1983; Wilson, 1982; Thomas, 1991). However, since then progress appears to have lagged in comparison to work from other countries. The limited work published since the 1990s seems split between reviews of previous work (e.g. Ball, 1995; Catchpole, 1995) and new forays into the field (e.g. Kay, 1995; Rannie, 1999; Slonosky, 2003). Paleoclimatic research in Canada has instead chosen to focus on the use of physical proxy records such as tree-rings, lake sediments and ice cores, in lieu of written records. Historical sources have been relegated to secondary sources used as supplemental evidence for physical proxy reconstructions (e.g. Gilbert and McKenna Neuman, 1988). However, by focusing on a previously unexplored region, the Yukon Territory, this chapter argues that there is much to still be accomplished within the field of paleoclimatology using historical documentary sources in Canada. It also introduces a new methodology for the creation of a seasonal warm/cold index that can be applied to historical documentary data.

2 Study Site

The Yukon Territory is located in northwestern Canada (Fig. 1) and was one of the last regions of the country to be explored. The first documented explorers in the region came in the early nineteenth century and the HBC'S Yukon posts represent some of the earliest permanent settlements in the region. The southern half of the Yukon Territory was selected as the study site for numerous reasons. Many of the earliest historical documents for the territory are from south of 65°N and also contain a prevalence of weather information. The majority of Canadian historical climatology studies have utilized the wealth of environmental and weather-related information stored in HBC post journals (Moodie and Catchpole, 1975, 1976; Rannie, 1983; Wilson, 1985, 1988). These studies have demonstrated that the detailed, daily journals can provide both reliable qualitative and quantitative climate data. Figure 1 provides an overview of the HBC posts' locations in the Yukon.

Additionally, paleoclimatic research using natural archives is relatively well established in the territory because it offers sites that are both climatically sensitive and free from human disturbance. The climate history of the Yukon, particularly the last 500 years, is reasonably well defined from tree rings, lake sediments and ice cores (e.g. Jacoby and D'Arrigo, 1989; Moore et al., 2001; Wake et al., 2002). However, there are still concerns with the records particularly in terms of temporal resolution and dating accuracy, which historical documents can help resolve.

Over 80 sources from the Library and Archives of Canada (LAC), the Yukon Archives (YA) and the Hudson's Bay Company Archives (HBCA) were initially consulted for the study. The texts included diaries, narratives, government records, letters and publications. The majority of the texts date from 1880 and later, when migration to the Yukon Territory increased, particularly due to the Gold Rush. Ultimately, journals from only three Hudson'S Bay Company (HBC) posts were used for the final analysis. The post journals from Frances Lake, Pelly Banks and

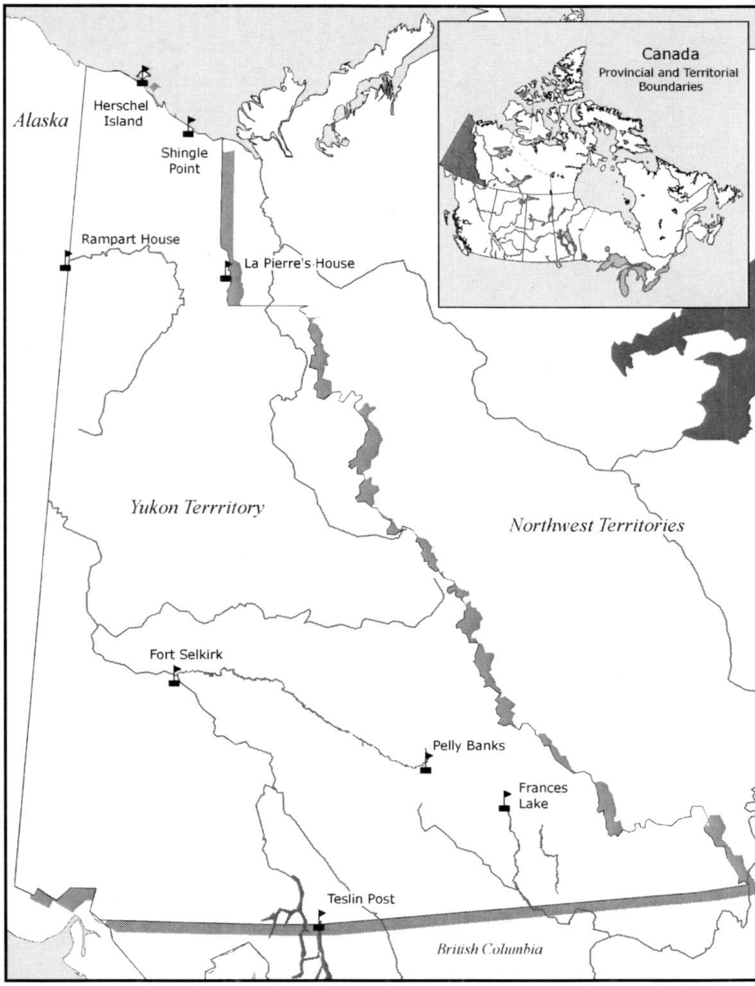

Fig. 1 Hudson's Bay Company posts in northwestern Canada. Post journals from Fort Selkirk, Frances Lake and Pelly Banks were used in this study

Fort Selkirk, represent the earliest, most reliable records and covered the 10 year period from 1842 to 1852. Temperature measurements were taken at Frances Lake from December 1842 to May 1844, providing an invaluable quantitative measure with which to compare the qualitative record.

2.1 Climate of the Yukon

The climate of the Yukon is one of the most extreme and varied in Canada due to its topography and geographic location. Kendrew and Kerr (1955) and Wahl et al.

(1987) both provide extensive analysis of the climate of the Yukon. In general, the climate is controlled by air masses from the west and the north, although there is a strong orographic influence in the western region due to the St. Elias mountain range. Therefore, the climate is considered to be continental sub-arctic characterized by long, dry, cold winters and short, dry, warm summers.

2.2 A Brief History of the Posts

Frances Lake was established in 1842 by Robert Campbell and was the first HBC post in the Yukon Territory (HBCA B.73/a/1). Frances Lake was located at the forks of the east and west arms of Frances Lake in the southeastern Yukon (Fig. 1). The post operated for close to 10 years, and the first four years of the post's operations are covered in near daily detail by the journals (HBCA B.73/a/1; B.73/a/2; B.73/a/3; B.73/a/4). The fort was successful during Robert Campbell's reign as its Postmaster. However, in the late 1840s, Frances Lake began to deteriorate under different leadership when Campbell was occupied with establishing Fort Selkirk and maintaining Pelly Banks. The post was abandoned in the winter of 1848/49, probably due to a lack of provisions and poor trading. By the fall of 1850, the fort had been rebuilt under the efforts of James Green Stewart who worked closely with Campbell at Fort Selkirk (Campbell, 1958). The Frances Lake journals used in this chapter cover the period July 1842–May 1846.

Temperatures readings, in Fahrenheit, were taken up to four times a day at Frances Lake for the period December 1842–May 1844 (Fig. 2) and were included

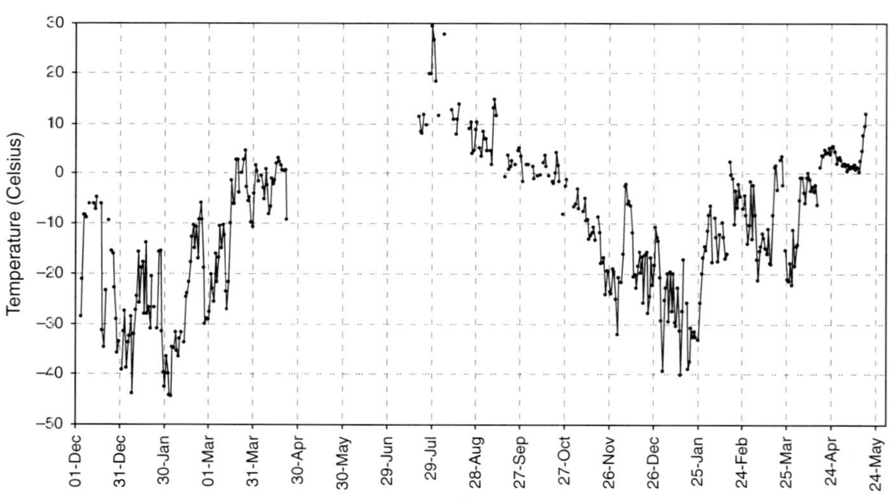

Fig. 2 Average daily temperatures for Frances Lake from December 1842 to May 1844. Temperature values may be an average of up to four daily readings. Multiple readings likely represented morning, noon and night temperatures

within the qualitative journal entries. From December 1842 to January 1843 additional notations in the journal entries indicated whether a reading was from the morning, noon or night. After this period, this information was not often given. After May 17, 1845, the number of temperature readings per month dropped dramatically and account for fewer than half a dozen days. Section 4 will discuss the temperature record in more detail.

Unfortunately, not much is known about the thermometers used at the Yukon posts. The journals make no reference to the manufacturer, the placement of the instruments, or whether more than one instrument was used for the readings, which could all influence the temperature readings in one manner or another. By the mid-1800s thermometer construction was relatively reliable, so it is unlikely that adjustments need to be made to the readings due to instrumentation alone (Ball, 1995).

Pelly Banks was established in 1845 on the east bank of Campbell Creek on the Pelly River. The post was less than 150 km from Frances Lake, allowing frequent travel between the two posts (Fig. 1). The Pelly Banks post's journals, entitled "Journal of Occurrences at Pelly Banks 1845–1847", begin in October 1845. The first volume of the journal (from October 1845 to April 1846) overlaps with the last part of the Frances Lake journal. However, it is sparse at best, and likely represents Campbell's visits to the site prior to its establishment as a trading post. The second volume, May 1846–April 1847, is considerably more detailed. Journals for the site are only available up to April 1847, although references to the post are made in the Frances Lake journals and in Robert Campbell's personal journals up to 1850, indicating it continued operation for a much longer duration. Pelly Banks suffered a notorious fate. On November 30, 1849, the post burned down; the men suffered starvation and some resorted to cannibalism (Campbell, 1958).

Fort Selkirk is the best documented HBC post in the southwestern Yukon (Fig. 1). The entire duration of the post's life from May 1848 to August 1852 is covered by the journals (LAC MG19-D13). The post was originally located on the east bank of the Yukon River, at the confluence of two rivers but in the spring of 1852, it was relocated three kilometers south to higher ground due to flooding problems. Life at the Fort Selkirk post was fraught with problems. The men regularly suffered from a lack of provisions because of the lengthy supply route (nearly 1,800 km) from the HBC post Fort Simpson in the Northwest Territories via Frances Lake and Pelly Banks. On August 22, 1852 Chilkat Indians attacked the post forcing its evacuation and its abandonment.

3 Creating the Warm/Cold Index

A four step process was employed in the creation of the seasonal warm/cold for the HBC posts (Tompkins, 2006). These steps build upon the methodologies developed by Moodie and Catchpole (1975) which were then later built upon by Pfister (1980), Baron (1982), and Bartholy et al. (2004) to name a few. Moodie and Catchpole

(1975) were the first to establish a set of content analysis procedures required to evaluate the reliability of documentary data in environmental research. Content analysis is used to determine the presence of words or concepts in textual, audiogenic or iconographic sources. There are two main categories of content analysis: conceptual analysis and relational analysis. This study focused on conceptual analysis which determines the presence and frequency of concepts and often generates "frequency counts". The greatest strength of content analysis is its ability to convert historical observation into numerical form, which facilitates comparison with contemporary data sets. Content analysis permits the application of mathematics and statistics to historical texts, thereby lending credibility to sources that have been often overlooked or discredited in the past (Moodie and Catchpole, 1975; Ingram, 1978).

The first step involved investigating the reliability of the texts by applying the rules of critical historiography. Second, the data were categorized into direct and indirect measurements of climate. Direct data include reference to actual climate parameters, such as temperature and precipitation. These references can be qualitative (e.g. "Weather warm and sultry." Frances Lake, June 28, 1844) or quantitative (e.g. "Weather remarkably warm. Thermometer rose to 83 [°F]." Frances Lake, June 14, 1844). Indirect documentary data refer to the impact of the weather either on the environment or humankind. Third, the categorized data were quantified using monthly, seasonal and annual frequency counts, averages and totals. Finally, the data were combined into a seasonal warm/cold index highlighting normal and extreme periods. The index was then examined in light of the current climate normal period (1971–2000) based on data collected from a nearby weather station (Environment Canada, 2004). It should be noted here that all quotations from the post journals presented in this chapter are left in the original format. No editing, other than the insertion of the occasional word or phrase for clarity or the truncation of a sentence, has occurred. These editorial changes are noted by the use of square [] brackets.

3.1 Step 1: Why, Who, When, Where and What?

In order to use the weather information stored within the post journals, the reliability of the texts had to be established, with regards to the "why", "who", "when", "where" and "what" of the journals.

Answering the question "Why were the post journals made?" was straightforward. The official HBC post journals were used to record daily events at each post, including work performed by employees, visits and trading by natives, and the weather. The HBC had a vested interest in the accuracy of the post journals. The information contained within, whether it was references to weather, trade or descriptions of the surrounding territory, gave the HBC the advantage in controlling the fur trade. Although no direct reference has been found in the three post journals under examination to indicate what specific instructions were given to the HBC authors regarding the information to be recorded, it can be assumed that the journals were treated as official documents and held to a certain standard. In the early days of the expansion of the HBC into Canada, the company specified what information should be recorded in both its post journals and ship's logbooks. It could be argued that by the time the posts in the Yukon Territory were established, the practice of maintaining journals was so engrained in the company that specific written instructions were not supplied to its Postmasters, such as Robert Campbell.

Ascertaining the post journal's authors proved more difficult. In the case of the Yukon posts, the author for every entry is not known except for the Fort Selkirk journals, which have undergone significant scrutiny (Johnson and Legros, 2000). Generally, the post's senior officer (i.e. the Postmaster) maintained the journals. If the senior officer was away, the next senior-level employee took over the duties. Therefore, it can be expected that the authors of the post journals were educated with significant fur-trading experience, reducing the likelihood that inexperience would make their entries biased.

Overall, the HBC journals satisfied the rules of contemporaneity and propinquity (i.e. the "when" and "where"). The journals were kept at the posts where the authors also resided. If the author left the post, the journal did not accompany him but remained there. Daily entries were recorded for the most part; entries may have been updated throughout the day, as in the case of Frances Lake where multiple temperature readings were recorded in the morning, noon and night. The only exception to this occurred in the period January 1852–September 9, 1852 for the Fort Selkirk journals. This period was covered by two different volumes of the Fort Selkirk journals: a co-written volume by Campbell and Stewart and an individual volume by Stewart only. There is some debate about whether the co-written volume was written at a later date (Johnson and Legros, 2000).

The final question addressed was "What type of information was recorded?" As indicated previously, both direct and indirect references to weather were recorded in the post journals, in addition to other information concerning the post's activities. The majority of the weather references were direct, qualitative references including information on temperature, precipitation, cloud cover and wind conditions. The post journals were not meteorological registers or weather journals specifically, as was the case with historical climatology studies based on other HBC records (e.g.Wilson, 1982; Ball, 1995).

3.2 Step 2: Coding Scheme

In order to quantify and analyze the textual weather information in the journals, a hierarchical coding scheme was developed using a software program called MaxQDA (Qualitative Data Analysis) (VERBI Software, 2005). Since digital versions of the three journals did not exist, the weather information was digitally transcribed into files of all the references to weather (both direct and indirect) for a given month of a given year. Due to time constraints, it was not feasible to transcribe the entire texts. Any information that could possibly indicate weather or environmental conditions was recorded. The coding scheme was used to separate weather references into categories which are unique, exclusive and representative of the weather information stored within the journals. The codes represent the main weather conditions as well as more extreme conditions. In order to represent differences in the intensity and type of weather, the codes are divided into subcodes. For the purpose of this study, the smallest coding unit of time was a day.

Some of the categories and subcodes were predetermined from the outset. It was assumed that a category for temperature, precipitation and wind conditions would be needed. Additional categories and sub-categories were based on the unique information from each of the post journals and took into account differences in terminology. Coding and categorization schemes used by other researchers were also considered. Baron (1982) utilized both direct and indirect (i.e. phenological) weather references in his rigorous study of eighteenth century diaries from Massachusetts. Bartholy et al. (2004) used only three categories (temperature, precipitation, and wind) in their extensive study of 15,000 records from the Carpathian Basin. Other studies have focused on single weather events, such as floods, droughts or storms, developing the classification scheme around the level of magnitude and frequency (Brazdil et al., 1999). Ge et al. (2003) divided their classification scheme for China into "natural evidence" which provided information on temperature directly (e.g. phenological and cryosphere data), and "impact evidence" which referred to the impact of cold/warm events on man and society. Ball (1995) developed 14 categories for his treatment of HBC data for the Hudson Bay region: instrumental temperature, wind direction, wind strength and type, precipitation, cloud cover, thunder, non-instrumental temperature, general weather, melting, frost, drift, and remarks.

The coding scheme used in this study is the result of extensive examination of the weather references in the HBC post journals. Initially, a number of years of texts were reviewed in their entirety and coded. Any reference to weather was coded using keywords. The codes were constantly revised in order to reduce redundancy and maximize the effectiveness of the coding structure. The coding scheme used in this study represents both direct and indirect data. Direct references to actual weather conditions were split into six categories: *Temperature*, *Rainfall*, *Snowfall*, *Cloud cover*, *Wind strength*, and *Wind direction*. Each category was further sub-divided to highlight the main weather conditions. For example, Temperature was divided into nine subcodes: *Cool*, *Cold*, *Very cold*, *Mild*, *Warm*, *Hot*, *Frost*, *Thaw*, and *No frost*. Indirect codes referred to the impact of the weather on the environment and human activity. It is represented by four sub-categories: *Biological*, *Ice activity*, *Human impacts*, and *Miscellaneous* remarks. Figure 3 shows the breakdown of the direct and indirect coding schemes by category and subcodes. Tompkins (2006) provides comprehensive examples of the terms and phrases included in codes and subcodes.

Of importance, the daily journal entries do not necessarily translate into daily weather references. At Frances Lake, weather was recorded, either directly or indirectly 77% of the time. Fort Selkirk boasted an even higher record at 79% in which 63% of the codes are attributed to direct weather references, indicating on average, the weather was mentioned directly 210 days of the year. Nonetheless, near daily weather references do not necessarily indicate that the same weather parameters

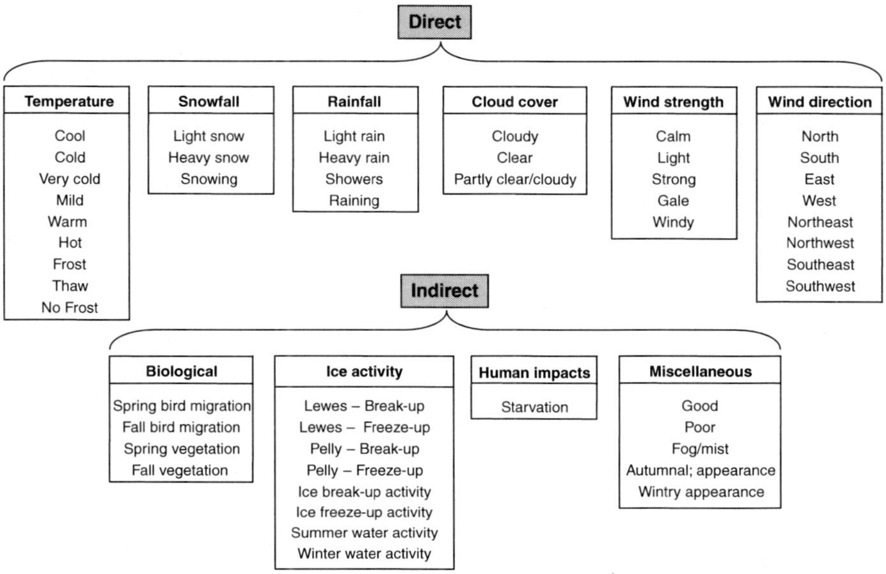

Fig. 3 Classification scheme used in study. Daily references are divided in direct or indirect categories and further sub-divided into exclusive codes

were always recorded. For example, temperature references were recorded on average only 179 days per year at Fort Selkirk.

3.3 Step 3: Quantification of the Qualitative Record

Once all the daily entries had been coded, monthly subcode totals were calculated for each month of each year by post. These totals were used for frequency counts (i.e. content analysis). In order to highlight trends, the daily subcode data were summed or averaged to a monthly and seasonal level. Viewing the data in seasonal formats was particularly useful for the identification of normal versus extreme conditions. The seasonal divisions used are based on seasonal divisions suggested by Wahl et al. (1987) and are as follows:

Spring = April–May
Summer = June–August
Fall = September–October
Winter = November–March

Totals and averages for the month, season and year were calculated for each post journal. Weighted averages were used to take into account missing months in the journals and to increase comparability between the records. The weighted averages implied that a lack of reference to a weather condition, such as temperature or cloud cover, did not necessarily indicate the lack of that actual weather condition.

3.4 Step 4: Creation of Seasonal Warm/Cold Index

The final step in reconstructing the climate from the post journals was to identify seasons with normal versus extreme weather through the use of a seasonal warm/cold index. Indices have become fairly widely used in historical climatology due to their ability to measure deviations from normal conditions and can be monthly, seasonal, annual or decadal in scale. Brooks (1926) was the first to apply indices to documentary data, resulting in winter wetness and severity annual indices for Europe for over 50 years of data. Lamb (1977) further refined the work and created a decadal winter severity index that was expressed by subtracting the number of unmistakably mild months from the number of unmistakably cold months per decade. Indices are particularly useful since they can incorporate frequency counts produced through content analysis, thus placing the information in a broader context. Indices themselves are relative in nature and can take a variety of factors into consideration. For example, Bartholy et al. (2004) determined codes of "cold", "cool", "frost", and "cold year" indicated normal conditions during the winter, while the codes "very cold", "severe winter" and "long winter" implied extreme conditions for that season. This type of coding is applicable to the temperature subcodes since words such as "cool" and "mild" can imply different conditions in different seasons. A ratio-based index is employed in this study since ratio indices are more robust and are not sensitive to missing data or changes in perception (Ingram et al., 1981). Ratio indices are used for shorter time series or when data are not available from multiple sources, as is the case with the HBC data. These indices present the total number of extreme references compared to the total number of references for a time period.

One of the drawbacks of an index is that a certain amount of subjectivity is involved in its construction and therefore researchers must strictly outline the criteria used since no standard methodology is available (Ingram et al., 1981). Regardless of the methodology chosen, it is advisable to incorporate both qualitative and quantitative data and present the findings in both ways. Extensive use of quotations can help illustrate the exact conditions experienced by the authors and serve as a reference point for the quantification of texts (Ogilvie and Jónssón, 2000; Nordli, 2001; Rannie, 2001). Indices alone lose meaning unless they are put into context.

The construction of the warm/cold index is a multi-step process. Two separate index calculations were required in order to determine whether a month was extremely cold or warm, since different subcodes were applicable to different extreme conditions depending on the season. Ratios of normal and extreme conditions were calculated for each month and a two-thirds "rule" was applied to the ratios when determining the index value (i.e. whether the month was extremely warm, extremely cold or normal). In other words, two-thirds of the total temperature references per month must fall into the extreme subcode category in order for a month to be assigned an extreme index value based solely on the temperature subcodes. The monthly index values were 1 (extremely warm), 0 (normal) and –1 (extremely cold). The warm/cold index construction is exclusive, meaning that a month cannot have both an extreme warm index value of +1 and an extreme cold index value of –1. The monthly index values were examined closely to establish reliability. Months with fewer than five qualitative temperature references were re-evaluated. At least two additional direct or indirect references were required in order to support the index value. If additional evidence was unavailable, the month was assigned a default index value of 0, since the month could not be satisfactorily confirmed as extreme.

Months with extreme index ratios that almost satisfied the 2/3 rule were also examined for further direct and indirect evidence. Indirect references that indicate extreme conditions were: *Rainfall* or *Winter water activity* in the winter season and *Snowfall, Frost, Ice activity* or *Bird migration* in the summer season. The monthly subcodes were averaged to create seasonal totals, resulting in an index that had a possible range from –1 to +1. The averages of the monthly values were used since the seasons were based on different numbers of months. It should be noted that the *Miscellaneous* references to *Good* and *Poor* were not included in the index calculation as indicators of normal and extreme conditions. References to "fine", "fair" and "good" weather were common, but it cannot be assumed that these are indicators of normal conditions. Also, the high number of miscellaneous subcodes did, in many cases, outweigh the number of qualitative thermal descriptors and could result in the false impression that a month was normal. For further information on the index construction see Tompkins (2006).

The thermal index is based mainly on the Temperature subcodes. Therefore, prior to constructing the index, it was important to examine the distribution of the temperature subcodes over the months before assigning a code as normal or extreme for a month. A consistently high relative frequency for a specific month over all years should imply that the code is likely representative of normal conditions. Conversely, the month with the lowest frequency over all years should indicate that the condition is extreme. However, this assumption is not absolute because if extreme conditions did occur during a season or longer over a number of years, the frequency values would be artificially higher. Therefore, examination of the references in context was required as well. Table 1 outlines the seasonal divisions used to categorize the temperature subcodes into normal versus extreme conditions. The warm/cold index does

Table 1 Seasonal division of normal vs. extreme temperature subcodes

Subcode	Normal season	Extreme cold season	Extreme warm season
Cool	Spring, Fall	Summer	Winter
Cold	Winter	Summer, Spring, Fall	
Very cold	Winter	Summer, Spring, Fall	
Mild	Spring, Fall	Summer	Winter
Warm	Spring, Summer, Fall		Winter
Hot	Summer		Spring, Fall, Winter
Thaw	Spring, Fall	Summer	Winter
Frost	Spring, Fall	Summer	Winter

Note: The subcode "No frost" was not included in this seasonal division since it indicates a lack of frost and should be represented inversely by the "Frost" code.

not capture extremely warm summers. References to "very hot" weather amounted to fewer than four days total for the period 1842–1852 and none of the references was from the same year.

The seasonal warm/cold index for the Yukon posts covers 117 months from 1842–1852 (with one year of missing data from May 1847–1848). The index is seasonally based to minimize the influence of individual months and highlight the impact of longer-term extreme conditions. The index is presented as representative for the entire southern Yukon Territory. It is assumed that the local extremes experienced at the posts are reflective of a more regional signal. Both Frances Lake and Pelly Banks do fall with the same mesoscale climate region, while Fort Selkirk falls within a different climate region; however, the main difference between the two climate regions is precipitation, not temperature related (Wahl et al., 1987). Eight months of overlap occur in the index between Frances Lake and Pelly Banks (October 1845–May 1846). The Pelly Banks record from October 1845 to April 1846 is very limited and therefore, the Frances Lake record takes precedence in assigning the index value. However, agreement between the two index records for the overlap period confirms the reliability of the index and the common signal between the two sites.

4 Results and Discussion

The seasonal warm/cold index is the culmination of this study, highlighting the seasonal normals and extremes for the historical period. The index is primarily based on qualitative thermal references although additional direct and indirect evidence of monthly conditions are incorporated. The index ranges from –1 (extreme cold),

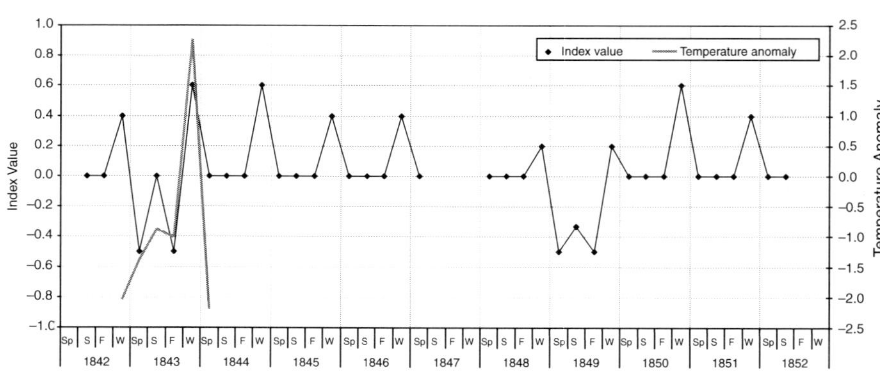

Fig. 4 Comparison of the seasonal warm/cold index with seasonal temperature anomalies (degrees Celsius) between the historical Frances Lake temperature data set and the Watson Lake modern climate data (1971–2000). Note the temperature anomaly for spring 1843 is based on April data only and summer 1843 is based on July and August data only (for both the historical and modern record)

0 (normal) and +1 (extreme warmth) (Fig. 4). There are a number of normal and extreme seasons for the period 1842–1852. The majority of the extreme seasons for the historical period were warm. However, two extreme cold periods were seen in the record: the spring and fall of 1843 at Frances Lake and an extended cold period at Fort Selkirk in 1849. Interestingly, the remaining extreme seasons were all warm winters. In fact, all winter seasons for the period were extremely warm according to the index. Index values for warm winters ranged from 0.20 (i.e. one month out of five) to 0.60 (i.e. three months). Not all winter seasons had extended periods of extremely warm temperatures. In some cases, months early and late in the winter season exhibited warmer tendencies (e.g. 1842/43), while other years had consecutive months with prolonged warm conditions (e.g. 1844/45, 1847/48, 1851/52). Some of these notable seasons will be examined in Sections 4.2.1, 4.2.2, 4.2.3, and 4.2.4, following a discussion on verifying the index reliability.

4.1 Verification of the Index

The indication that all winters during the historical period were extremely warm suggested that a bias could be present within the data, strengthening the need for the index verification. It is preferable to verify paleoclimatic data by using a period of overlap between the historical and modern data set. For example, the Kastellet et al. (1998) reconstruction of the weather in 1837/38 for southwestern Norway was facilitated by the availability of a temperature record from a nearby town for the same time period. However, this is impossible with the HBC's journals since the historical period ends in 1852 and the modern record does not commence until nearly 100 years later. As an alternative, the historical temperature record for the Frances Lake post (see Section 2.1 and Fig. 2) was used for verification. The historical temperature record for Frances Lake (December 1842–May 1844) was compared to the nearest modern climate normal station data for Waston Lake A (1971–2000), approximately 150 km away. Seasonal temperature anomalies (degree Celsius) were calculated using the same seasonal divisions as the index. A positive seasonal anomaly indicates warmer than normal conditions while a negative temperature anomaly indicates a colder than normal season. Figure 4 also compares the seasonal warm/cold index to the seasonal temperature anomalies over six seasons (from winter 1842/43 to spring 1844). Agreement between the index and temperature anomaly record is relatively good except in two instances. The index does not identify the winter of 1842 as extremely cold; rather, it is classified as extremely warm. Also, the summer of 1843 and spring of 1844 both have index values of zero (i.e. normal conditions) while the temperatures anomalies for the seasons indicate colder than normal conditions.

Comparison of the seasonal warm/cold index with the temperature anomaly data for Frances Lake shows that the index may have some limitations. Time of temperature readings and location of the thermometer could have biased the historical temperature record. Months and seasons with temperature anomalies less than 1°C may not be consistently identified as extreme years (e.g. October and November 1843, April 1844). There are some instances in the overlap period where the index was unable to identify months or seasons with much larger temperature departures (i.e. up to 7.5°C). However, these discrepancies between the index and the temperature anomaly record cannot be blamed on the index entirely. Missing temperature data from the historical Frances Lake record are probably responsible for some disagreement, particularly for August 1843, January 1844, and May 1845 (Tompkins, 2006).

The winter of 1842 is the only case where the index indicated extreme conditions that were opposite compared to the temperature anomaly record. In this case, the discrepancy is due to the authors at Frances Lake associating extremely cold temperatures (i.e. below the normal monthly mean) with the qualitative descriptor of cold rather than using terminology of very cold. This was not the case with days described as mild. It could be argued that a prolonged period of very cold weather in January 1844, which was –4.4°C colder than normal, could have biased the authors to under-represent the coldness of February (–7.5°C colder than normal) (Tompkins, 2006). Other factors, such as housing and nutrition, may have influenced the author's perception of the cold. As well, the HBC temperature readings at Frances Lake could include some unknown errors.

4.2 Notable Seasons from 1842–1852

Both the warm/cold seasonal index and the historical temperature record have shown that the same extreme conditions were experienced at the posts. This section will discuss some of the most notable seasons in more detail, providing excerpts from the journals.

4.2.1 Winter 1843/44

The winter of 1843/44 was notable for two reasons. First, the weather overall, was milder than normal (i.e. 1971–2000 average). December 1842, February and March 1843 were all warmer than normal according to the index. Daily temperature readings were often above –10°C (Fig. 2). Temperature records for the period indicate that December was 5.3°C warmer than normal and February was 9.4°C above normal – a monthly mean that has not even been recorded in the modern record. According to the historical temperature readings, March 1843 was actually 1.9°C colder than normal but the high number of references to mild conditions during the month actually resulted in an index value of +1. [See Nielsen-Gammon and McRoberts, this volume for an in-depth analysis of the weather of March 1843.]

The journal author for 1843/44 was assumed to be Robert Campbell for the majority of the entries. There is little indication that the journal author(s) recognized the post was experiencing a mild winter in December 1843, although references to mild weather were made, such as "clear and mild", "fine, clear and mild", "very fine and mild" and "the weather fine and mild" (HBCA B.73/a/2). By February 1844, the authors noticed the mild weather, writing on the 2nd, "Weather uncommonly mild". On the 15th of the month, thaw conditions are noted and on the 17th rain fell. No further comments reference the unusually warm conditions during the month except for days referencing "mild weather".

Second, the winter of 1843/44 was notable because the post was suffering through a period of very poor food supplies. This winter recorded the highest number of references to starvation (20) for any year and February 1844 recorded the single highest number of starvation references for any month (10 days). February 1844 also recorded the second highest number of references to south winds (6 days). The lack of food supplies and the mild weather were related but not directly. The poor food supplies were a result of the limited provisions supplied by the HBC. The company's belief that the post could be self-sufficient was misplaced on the idea that the fisheries, hunting and agriculture could supply enough sustenance for a year at a location that very little was known about. Fortunately, the mild weather of February heralded a change of fortune for Frances Lake by creating favourable hunting conditions.

Concerns regarding the food supply first arose in January 1844, although indications of the fisheries failing in November 1843 were noted. References were made in January 1844 to the men becoming weaker and to starving hunters. The reason for the lack of food was blamed on a combination of poor hunting weather and failing fisheries:

> Late last night Hoole arrived starving. In the three Lakes which he tried he could get no fish. All the Indians who are in that quarter have left it and are gone off in quest of animals towards the Pelly... the weather, till lately, has been too cold for hunting. We are now very low, for the last week, we have each 1 fish and ½ Pint of Barley per diem. All our fish are done now and our Barley which was only a veg at first and nearly so. Weather very mild for the season. 2, 20 and 12 [°F]. (January 31, 1844)

By February 1844 the weather had warmed considerably, creating favourable hunting conditions. However, it wasn't until the last week in the month that any of the hunters met with success. Food supplies improved throughout March as meat caches became restocked. On March 28th, 1,000 lbs of meat were brought to the post, effectively ending the food crisis.

4.2.2 Winter 1844/45

Three consecutive months (January–March 1845) of warmer than normal conditions tied with 1843/44 and 1850/51 as the warmest winter in the historical period. There were no starvation references in the winter of 1844/45. By now, Frances Lake had been in operation for two years. Lessons learned from the two previous winters had likely influenced the way food supplies were managed at the post for the better. The milder conditions also favoured better hunting. A few temperature measurements are available for this season. On November 24, 1844, the author noted, "The weather clear and uncommonly cold for the season. –32, –32 and –28 [°F]" (HBCA B.73/a/3). The average temperature for that day was –34.8°C, which is considerably lower than the normal monthly mean of –15°C. However, there is no indication that the cold weather lasted more than the one day. References to the mild weather in January 1845 indicated that the author(s) recognized the conditions were abnormal for the season: "Weather overcast and very mild for the season" (January 6), "Delightful weather for the season, calm, clear and sunshine" (January 9), and "Fine weather for the season, so mild and calm" (January 19). There are some references to very cold weather early in February 1845. Two temperature readings on the 2nd (–41.7°C) and 11th (–33.3°C) support the qualitative thermal descriptors (i.e. Very cold). However, the number of references to mild weather far outweighed the very cold and the one remaining temperature reading for the month, –10°C (February 17), supports the positive index value. By March 12, 1845, thawing conditions were noted and on the 13th the author wrote, "Southerly wind which freshened into a gale at the close of the day accompanied by rain. The snow fast disappearing".

4.2.3 Winter 1850/51

The winter of 1850/51 was the warmest winter recorded at Fort Selkirk. December, February and March were all warmer than normal according to the index. Campbell writes on December 23rd, "The weather uncommonly mild. Rain dropping from the houses- with strong southerly wind" (LAC MG19-D13). Interestingly, the winter of 1850/51 was the cloudiest winter for the historical period (i.e. 31 days compared to an average of 14 days). Unlike the winters at Frances Lake, the number of cloudy and clear days was almost equal during winters at Fort Selkirk. Also, the prevailing wind for most winter seasons at Fort Selkirk was northwest, not southeast as was common at Frances Lake and Pelly Banks. Campbell and Stewart both recognized that the winter of 1850/51 was unusual. On December 28th and 29th, Stewart wrote, "Never saw such continuation of mild weather at this season. Snow soft all day & no frost at night ... River all covered with water & snow nearly all gone in the evening. The wind went to the North." A period of colder weather in late January interrupted the mild spell. But by February, mild weather had returned along with references to reduced food supplies and "starving Indians". Heavier snow cover in January and February may have impeded the hunters. January 1851 reported eight days of snowfall and February recorded seven. The average number of historical snowfall days recorded at Fort Selkirk in January and February was 4.5 and 3.8 days respectively. On February 23rd, Stewart remarked, "Peter returned this afternoon having been unable to proceed to LaPie's Lake owing to the road being entirely blocked up by drifts". By the end of March, hunters were able to kill five caribou and four moose.

4.2.4 Coldness of 1849

The only prolonged period of cold weather in the index was recorded at Fort Selkirk during the spring, summer and fall of 1849. Each season had one month that was colder than normal (i.e. May, August and October). This prolonged cold period was unique in the index because it represented extreme conditions that persisted throughout several seasons. The ratio of extreme cold to normal conditions in the construction of the index did not originally yield a negative index value for May 1849. However, additional examination of the journal entries for the month indicated that the month was anomalous. There were numerous comments by Stewart referencing conditions that were unusual for the season (LAC MG19-D13):

> Weather colder for the season. The Wind after shifting to every point of the Compass remained North. (May 6, 1849)
> . . .went off to hunt ducks but only got a few owing to the bad weather. . .This has been the coldest day I have ever seen at this season, blowing hard with snow wh: was drifting as dry as in the middle of winter. (May 12, 1849)
> Water rose a little but the river holds fast still. We must be farther North than we think or else this is a very late season. (May 18, 1849)
> Extraordinary Weather for the season. We had a regular snowstorm this morning. . .. This is the latest season I have known in the North. (May 27, 1849)
> Barley in the upper field about an inch above ground. Here ends the month of May. Dreary and cold in the extreme, little or no vegetation & the banks of the river covered with Ice still. . . (May 31, 1849)

To further support a colder month, May 1849 had five snow days which is almost five times the number recorded at the nearby climate normal station, Pelly Ranch. May 1849 also had the latest first and last break-up dates for the Pelly and Lewes River (Tompkins, 2006). The month also had the latest first reference to spring migration at Fort Selkirk (i.e. May 9th). All the other references to spring migration from the posts indicated that migration commenced from April 16th to 29th.

August 1849 was the only cold summer month for the historical record, highlighted by indirect cryospheric and phenological evidence. Campbell made references to snowfall on the nearby mountains, stating on the 30th, "Pouring down rain all night but snowing in the Mts. They were coated with white to within a short distance of the water edge this morning, indicating an early winter. Flocks of geese passing" (LAC MG19-D13). This reference also indicated the earliest date of fall migration recorded at Fort Selkirk. Four days of frost were noted in August, double the number recorded in previous years. August 1849 was wetter than normal with 17 days of rain, well above the normal monthly mean of 12.2 days. Despite these additional indicators, Stewart and Campbell never referred directly to colder than normal temperatures.

The last month of colder than normal weather during this prolonged period was October 1849. Campbell noted 10 days of *Cold* weather. On the 29th he wrote "The ice set fast on the Pelly & drifting full channel from the Lewis. The day very cold for the season". October 1849 also recorded the earliest freeze-up dates for the Pelly and Lewes rivers. Even in September, which was assigned a normal monthly index value, Campbell commented on the poor weather conditions from spring through summer. On September, 27th he wrote: "Took up the potatoes in the upper field which but for the early frost & late spring would have yielded a tolerable return or had the season been as favourable as the last. As it is they are very small and but little more than the seed sown". The following day he stated, "Took up Potatoes in next field, say our entire crop. They are of larger size than what was digged yesterday & there is in all about a keg and a fourth or half, say 10 or 12 gallons, but for the very untowards season we had there would have thrice as much.... But few cranes passing today. Weather cold & raw". It appears that Campbell's comment in August predicting an early winter was confirmed by the end of October. By this point, it is unlikely that the authors were biased to record colder than normal conditions due to inexperience with the territory. Both Campbell and Stewart had already spent one entire year at Fort Selkirk and up to seven years in the Yukon Territory.

5 Conclusion

The warm/cold seasonal index revealed the presence of persistent mild winter seasons throughout the historical record. This winter warming may be linked to the end of the Little Ice Age (LIA) in the southern Yukon. Most paleoclimatic studies based in the Yukon Territory estimate the end of the LIA around the 1850s (e.g. Jacoby and Cook, 1981; Szeicz and MacDonald, 1995). The index suggests that the Yukon may have been exiting the LIA period as early as the 1840s due to the continuation of milder than normal winters. This is in agreement with some regional and hemispheric studies which suggested the warming trend began in 1840s (Overpeck et al., 1997) or earlier (Jones et al., 1998; Crowley and Lowery, 2000; Briffa et al., 2001). The index also suggests that much of the warming trend may have occurred during the winter instead of the summer, at least in the Yukon. The reality that some physical proxies (e.g. tree rings, ice cores) are poorly correlated with winter temperatures shows that historical documents offer the potential to fill in the "paleoclimatic gaps" by offering weather data from all seasons.

This study has demonstrated that historical documents from the southern Yukon Territory can be used for paleoclimatic reconstructions and supports the argument that there are untapped sources yet to be explored within the historical climatology field in Canada. One of the major challenges faced in the field is the amount of time required to fully investigate the sources and customize a methodology for analysis (Jones et al., 2001). This study offers a methodology that can be applied to short-term records (i.e. fewer than 50 years) that should be transferable to other regions with short-term, qualitative, historical climate records.

References

Ball TF (1983) The migration of geese as an indicator of climate change in the southern Hudson Bay regions between 1715 and 1851. Climatic Change 5: 85–93

Ball TF (1995) Historical and instrumental evidence of climate: Western Hudson Bay, Canada, 1714–1850. In: Bradley RS, Jones PD (eds.) Climate since A.D. 1500. Routledge, London, pp. 18–40

Baron WR (1982) The reconstruction of eighteenth century temperature records through the use of content analysis. Climatic Change 4: 385–398

Bartholy J, Pongracz R, Molnar, Z (2004) Classification and analysis of past climate information based on historical documentary sources for the Carpathian Basin. International Journal of Climatology 24: 1758–1776

Brazdil R, Glaser R, Pfister C, Dobrovolny P, Antoine JM, Barriendos M, Camuffo D, Deutsch M, Enzi S, Guidoboni E, Kotyza O and Rodrigo FS (1999) Floods events of selected European rivers in the sixteenth century. Climatic Change 43: 239–285

Briffa KR, Osborn TJ, Schweingruber FH, Harris IC, Jones PD, Shiyatov SG, Vaganov EA (2001) Low-frequency temperature variations from a northern tree ring density network. Journal of Geophysical Research 106: 2929–2941

Brooks CEP (1926) Climate through the ages. Benn, London

Campbell R (1958) Two journals of Robert Campbell, Chief Factor, Hudson's Bay Company, 1808–1853. Shorey Books, Seattle

Catchpole AJW (1995) Hudson's Bay Company ships′ log-books as sources of sea ice data. In: Bradley RS, Jones PD (eds.) Climate since A.D. 1500. Routledge, London, pp. 17–39

Catchpole AJW, Faurer MA (1983) Summer sea ice severity in Hudson Strait, 1751–1870. Climatic Change 5: 115–139

Crowley TJ, Lowery TL (2000) How warm was the Medieval Warm Period? Ambio 29: 51–54

Environment Canada (2004) Canadian climate normals or averages 1971–2000. http://www.climate.weatheroffice.ec.gc.ca/climate_normals/index_e.html. Accessed 1 February 2006.

Ge Q, Zheng J, Fang X, Man Z, Zhang X, Zhang P, Wang WC (2003) Winter half-year temperature reconstruction for the middle and lower reaches of the Yellow River and Yangtze River, China, during the past 2000 years. The Holocene 13: 933–940

Gilbert R, McKenna Neuman C (1988) Occurrence and potential significance of warm weather during winter in the eastern Canadian Arctic. Arctic and Alpine Research 20: 395–403

HBCA B.73/a/1 (1842) Journal of daily occurrences on the way to and at Frances Lake. Mackenzies River District Outfit 1842–1843. Hudson's Bay Company Archives, Winnipeg

HBCA B.73/a/2 (1843) Journal of daily occurrences on the way to and at Frances Lake. Mackenzies River District Outfit 1843–1844. Hudson's Bay Company Archives, Winnipeg

HBCA B.73/a/3 (1844) Journal of daily occurrences on the way to and at Frances Lake. Mackenzies River District Outfit 1844–1845. Hudson's Bay Company Archives, Winnipeg

HBCA B.73/a/4 (1845) Journal of daily occurrences on the way to and at Frances Lake. Mackenzies River District Outfit 1845–1846. Hudson's Bay Company Archives, Winnipeg

Ingram MJ (1978) Historical climatology. Nature 276: 329–334

Ingram MJ, Underhill DJ, Farmer G (1981) The use of documentary sources for the study of past climates. In: Wigley TML, Ingram MJ, Farmer G (eds.) Climate and history: Studies in past climates and their impact on man. Cambridge University Press, Cambridge, pp. 180–213

Jacoby GC, Cook ER (1981) Past temperature variations inferred from a 400-year tree-ring chronology from Yukon Territory, Canada. Arctic and Alpine Research 13: 409–418

Jacoby GC, D'Arrigo R (1989) Reconstructed northern hemisphere annual temperature since 1671 based on high-latitude tree-ring data from North America. Climatic Change 14: 39–59

Johnson LR, Legros D (2000) Journal of occurrences at the forks of the Lewes and Pelly rivers May 1848 to September 1852. Occasional Papers in Yukon History. Whitehorse, Heritage Branch, Government of Yukon, Yukon

Jones PD, Briffa KR, Barnett TP, Tett SFB (1998) High-resolution paleoclimatic records for the last millennium: Interpretation, integration and comparison with general circulation model control-run temperatures. The Holocene 8: 455–471

Jones PD, Ogilvie AEJ, Davies TD, Briffa KR (2001) Unlocking the doors to the past: Recent developments in climate and climate-impact research. In: Jones PD, Ogilvie AEJ, Davies TD, Briffa KR (eds.) History and climate: Memories of the future? Kluwer Academic, New York, pp. 1–8

Kastellet E, Nesje A, Pedersen ES (1998) Reconstructing the paleoclimate of Jaeren, southwestern Norway, for the period 182101850, from historical documentary records. Geografiska Annaler, Series A, Physical Geography 80: 51–65

Kay P (1995) An early nineteenth century meteorological register from the eastern Canadian Arctic. Polar Record 31: 335–342

Kendrew WG, Kerr D (1955) The climate of British Columbia and the Yukon Territory. Department of Transport, Meteorological Branch, Ottawa, p. 222

LAC MG19-A25 (1845–1847) Journal of occurrences at Pelly Banks 1845–1847. Library and Archives of Canada, Ottawa

LAC MG19-D13 (1848–1852) Journal of occurrences at the forks of the Lewes and Pelly rivers May 1848 to September 1852. Library and Archives of Canada, Ottawa

Lamb HH (1977) Climate: Past, present and future. Methuen, London

Mackay JR, Mackay DK (1965) Historical records of freeze-up and break-up on Churchill and Hayes Rivers. Geographical Bulletin 7: 7–16

Moodie DW, Catchpole AJW (1975) Environmental data from content analysis: Freeze-up and break-up of estuaries on Hudson Bay 1714–1871. Department of Geography, University of Manitoba, Winnipeg

Moodie DW, Catchpole AJW (1976) Valid climatological data from historical sources by content analysis. Science 193: 51–53

Moore JJ, Hughen KA, Miller GH, Overpeck JT (2001) Little Ice Age recorded in summer temperature reconstruction from varved sediments of Donard Lake, Baffin Islands, Canada. Journal of Paleolimnology 25: 503–517

Nordli PO (2001) Reconstruction of nineteenth century summer temperatures in Norway by proxy data from farmer's diaries. Climatic Change 48: 201–218

Ogilvie AEJ, Jónssón T (2000) Sea ice, climate and Icelandic fisheries in the eighteenth and nineteenth centuries. Arctic 53: 383–394

Overpeck JT, Hughen K, Hardy D, Bradley R, Case R, Douglas M, Finney B, Gajewski K, Jacoby G, Jennings A, Lamoureux S, Lasca A, MacDonald G, Moore J, Retelle M, Smith S, Wolfe A, Zielinski G (1997) Arctic environmental change over the last four centuries. Science 278: 251–1256

Pfister C (1980) The Little Ice Age: Thermal and wetness indices for central Europe. Journal of Interdisciplinary History 10: 665–696

Rannie WF (1983) Breakup and freezeup of the Red River at Winnipeg, Manitoba, Canada, in the 19th Century and some climatic implications. Climatic Change 5: 283–296

Rannie WF (1999) A survey of hydroclimate, flooding, and runoff in the Red River basin prior to 1870. Open-File Report 3705, Geological Survey of Canada, Ottawa

Rannie WF (2001) Awful splendor: Historical accounts of prairie fire in southern Manitoba prior to 1870. Prairie Forum 26: 17–46

Slonosky VC (2003) The meteorological observations of Jean-François Gaultier, Quebec, Canada: 1742–1756. Journal of Climate 16: 2232–2247

Szeicz JM, MacDonald GM (1995) Dendroclimatic reconstruction of summer temperatures in northwestern Canada since A.D. 1638 based on age-dependent modeling. Quaternary Research 44: 257–265

Thomas M (1991) The beginnings of Canadian Meteorology. ECW Press, Toronto, Ontario.

Tompkins H (2006) Historical Climatology of the southern Yukon Territory: Paleoclimatic reconstruction using documentary sources from 1842–1852. Masters Thesis, Queen's University, Department of Geography, Kingston

Verbi Software (2005) MaxQDA (Qualitative Data Analysis). http://www.maxqda.com/maxqda-eng/index.htm.

Wahl HE, Fraser DB, Harvey RC, Maxwell JB (1987) Climate of Yukon. Climatological Studies, No. 40. Environment Canada, Atmospheric Environment Service, Ottawa

Wake CP, Yalcin K, Gundestrup NS (2002) The climate signal recorded in the oxygen-isotope, accumulation and major-ion time series from the Eclipse ice core, Yukon Territory, Canada. Annals of Glaciology 35: 416–422

Wilson CV (1982) The summer season along the east coast of Hudson Bay during the nineteenth century: A sample study of small-scale historical climatology based on the Hudson's

Bay Company and Royal Society Archives. Part 2: the Little Ice Age on eastern Hudson Bay: Summers at Great Whale, Fort George, Eastmain, 1814–1821, Canadian Climate Centre. Report; 83-9. Atmospheric Environment Service, Climatological Services Division, Downsview, Ontario

Wilson CV (1985) The summer season along the east coast of Hudson Bay during the nineteenth century. Part 3: Summer thermal and wetness indices. A methodology. Canadian Climate Centre. Report; 85-3. Atmospheric Environment Service, Climatological Services Division, Downsview, Ontario

Wilson CV (1988) The summer season along the east coast of Hudson Bay during the nineteenth century. Part 3: Summer thermal and wetness indices, B. The indices 1800–1900. Canadian Climate Centre. Report 88-3, Canadian Climate Centre, Atmospheric Environment Service, Downsview, Ontario

Backward Seasons, Droughts and Other Bioclimatic Indicators of Variability

Lesley-Ann L. Dupigny-Giroux

Abstract Phenoclimatic fluctuations, biorhythms and agricultural patterns are intricately linked to the local meteorological conditions, and over time to the climate of a region. Many of these patterns were captured in the daily journal entries of diarists in the New England states of Vermont and New Hampshire in the pre-digital era. This chapter focuses on the data available from these states in the 1680–1900 time period. It presents an analysis of backward season characteristics and the concomitant influences on frost occurrences, sugar maple production and the onset of drought. The results demonstrate a unique application of historical data to reveal long-term spatial and topographic patterns. An indicator-based drought index also reveals spatio-temporal comparisons across the region, including some persistent severe droughts in the 1700s not apparent in the last century.

Keywords New England · Backward season · Drought · Phenology · Sugar maple · Frost

1 Introduction

New England's climate is often described as variable and changeable, a description that is as apt today as it was in three centuries ago. Physiographic and atmospheric factors from the local to synoptic scale are intricately linked to this variability, resulting in the precipitation and temperature fluctuations that characterise the region. The New England states of Vermont and New Hampshire are separated by the Connecticut River, with similar land cover and land use practices being found on either bank. Orographically, Vermont is dominated by the north–south trending Green Mountains, while the White Mountains run through central and northern New Hampshire. Lake Champlain exerts a moderating influence on western Vermont, while southeastern New Hampshire falls under the influence of the northern Atlantic

L.-A.L. Dupigny-Giroux (✉)
Department of Geography, University of Vermont, Burlington, VT 05405-0114, USA
e-mail: ldupigny@uvm.edu

L.-A. Dupigny-Giroux, C.J. Mock (eds.), *Historical Climate Variability and Impacts in North America*, DOI 10.1007/978-90-481-2828-0_14,

Ocean. Historically, Europeans first settled the lowlands, river valleys and other easily accessible regions, with later movement into the more mountainous interior. Journal accounts of weather and climate reflect this pattern of settlement.

Written accounts by farmers, schoolchildren and state officials in the 18th and 19th century are replete with entries about backward seasons, droughts, freshets and the influence of weather and climate on agroecosystems. As far back as 1860, Vermont officials recognized the need for an ongoing, standardized system of meteorological measurement as a way of discovering the relationship between these variables and agriculture, diseases and other phenomena such as sunspots. Even at that time, it was recognized that without ongoing records, human memory is often short and that a season that may have been perceived as cold, could actually be warm in terms of the instrumental record.

A backward season refers to one that is late and/or with weather that is inappropriate for that time of year. Perhaps the most famous of these occurred in 1816, when much of New England and Québec experienced the "Year without a summer" [See the chapter by M. Chenoweth in this volume]. Apart from 1816, ships' logs document the coincidence of backward weather conditions on either side of the North Atlantic Ocean at other times in the early 1800s, while United States Fish Commission bulletins chronicle the comparative effects of backward conditions on fisheries in Gloucester, Massachusetts in 1887 vs. 1886 (Wilcox, 1887). Planting delays, changes in bloom dates and other phenoclimatic fluctuations have also been noted in backward years. Yet despite these accounts, no formal definition of a backward season exists, nor are its meteorological characteristics well quantified.

Phenoclimatic studies use flora and/or fauna with well-established responses to atmospheric forcings. Many of these studies have focused on plant responses to spring weather when the forcing is most pronounced (Schwartz, 1998). One of the goals of this chapter was to select a series of phenological characteristics that were reproducible, spatially extensive and with a direct response to climate variations. Sugar maple tapping was selected as one such phenoclimatic indicator for a number of reasons. "Maple trees were quite abundant, and every family was enabled to supply itself with plenty of maple sugar" (Wilkins, 1871). In a similar vein, the Rutland Daily Herald of 2 April, 1873 reported that

> Those who have large orchards and the apparatus for evaporating and preparing the maple sap for syrup, or making it into sugar, often reap as much income from this operation, as from any single branch of their farm work.

Finally, the close coupling between daily temperature ranges and snow depth in late winter-early spring strongly influences the timing and length of the sugaring season as well as the resulting yields. Studies of recent climate fluctuations have shown seasonal shifts in the timing and duration of the sugaring season across New England.

Droughts are a cyclical hazard that affect not only agroecosystems, but various components of the hydroclimatic cycle as well. Early New England residents battled moisture deficits that often combined with disease and insect infestations to produce crop failure. The New England agricultural landscape of the eighteenth and nineteenth centuries was highly diversified and included such crops as English and Indian corn; spring, winter and Siberian wheat; potatoes; flax seed; cotton; oats; apples; turnips; thistles; rye; barley; peas; beans; pumpkins; grass and clover seed for hay; and grapes. Homesteaders also kept bees, raised sheep, hogs, beef and dairy cattle, and tapped sugar maple (*Acer saccharum*) trees for syrup. Although crop production and livestock statistics are available from the National Agricultural Statistics Service (NASS), declines in the late 1800s and early 1900s were related to intertwined socioeconomic and biogeophysical reasons that make the link between climate and crop production or animal yields tenuous at best. This chapter uses crop yield statistics and diary accounts as indicators of moisture extremes.

In focusing on the northern New England states of New Hampshire and Vermont (Fig. 1), the main objectives of this chapter were:

- to quantify the thermal, moisture and spatial characteristics of backward seasons from 1680–1900 and to determine the underlying synoptic and mesoscale patterns;
- to explore the changing occurrences of bloom dates during backward seasons;
- to create an indicator-based drought index using hydrometeorological observations and proxy data for quantifying the spatio-temporal extent of drought in the 1680–1900 period and;
- to use maple sap production as a bioindicator of climate fluctuations in the 1800s.

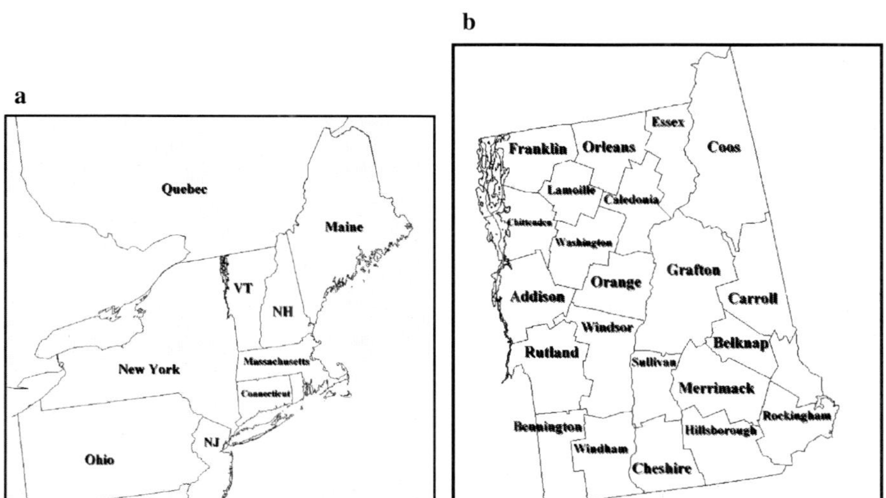

Fig. 1 Sketch map of northeastern North America showing the states of Vermont (VT) and New Hampshire (NH) (**a**) and the counties of Vermont and New Hampshire (**b**)

2 Data and Methods

Weather and phenology records were transcribed from diaries held at historical societies, the University of Vermont Special Collections and museums or acquired from the online transcriptions on the Historical Climatology of New England and New York, Pennsylvania, New Jersey (http://www.umaine.edu/oldweather). Journal entries also included bird sightings, agricultural production, killing frosts, crop damage and maple sap quality or sweetness. Special note was made of frost occurrences during the warm season. Drought was either identified as such by the diaries' authors or derived from observations about wells drying, low river or lake levels, stunted crops, soil moisture conditions and low atmospheric moisture ("drying winds"). Additional weather entries were extracted from the Index to Manuscript Vermont State Papers and New England County Gazetteers. Many of these data were recorded daily, but the records were often interrupted by travel, crop planting, etc..

Daily meteorological data were also acquired from the Climate Database Modernization Program (CDMP) 19th Century Forts and Voluntary Observers Database at the Midwestern Regional Climate Center (MRCC). To date, data from five stations in New Hampshire (Fort Constitution and Plymouth) and Vermont (Burlington, Lunenburgh and Strafford) have been digitized and manually keyed. The period of record at these sites varied in the number of variables recorded, times of observation and record length (1820–1853 Fort Constitution, NH; 1839–1868 Portsmouth, NH; 1832–1892 Burlington, VT; 1859–1892 Lunenburgh, VT; 1873–1892 Strafford, VT). The most frequently overlapping times (0700, 0900, 1200, 1300, 2100) of daily precipitation, temperature and wind directions were selected for this study. All data were converted to SI units and the station pressures corrected to sea level, using station histories and MRCC metadata (Doty and Dupigny-Giroux, 2005; Andsager et al., 2007). Data gaps were not filled due to the lack of surrounding stations from which to homogenize the records. Observed station pressures across New England were complemented by the daily mean sea level pressure series at Montréal, Québec, Reykjavik Iceland and Gibraltar, computed by the EMULATE (*E*uropean and North Atlantic daily to *MUL*idecadal clim*ATE* variability) program for the period 1850-present. Interrupted time series analysis was therefore applied. Figure 2 identifies all of the station locations used in this study.

Finally, present-day US Geological Survey topographic maps were used to extract the large scale topography of study sites. In some cases, exact site locations were found on Beers atlases of the 1800s.

3 Quantifying the Characteristics of Backward Seasons

For a cool season to be considered to be backward, it must have so chronicled in at least one diary, county gazetteer, Monthly Weather Review of the US Army Signal Service or other documents. The characteristics of these years are summarized on Table 1. The 1812–1820 time frame was noted for the number and spatial extensiveness of backward springs, summer droughts and summer killing

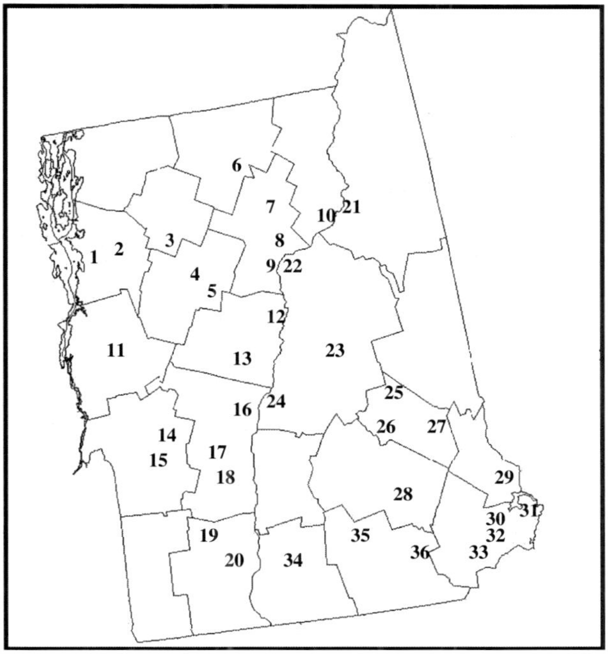

Fig. 2 Locations of the towns from which diaries and other documentary evidence were obtained

frosts. The extended coldness of 1816–1818 was related to the aftereffects of the eruption of Tambora in 1815 and highlights the fact that crop failures and other devastation lasted beyond the 1816 "Year without a summer" commonly found in the literature [see M. Chenoweth in this volume for a more complete description of the synoptic patterns in the northeast during 1816]. The coldness in 1812 may be related to a hemispheric anomaly that coincides with similar conditions across the Baltic Region in Europe, although not preceded by any known volcanic eruption(s) (Neuman, 1990). Abnormal and highly variable conditions continued in 1813 with frost observed on 25 June at Sanbornton, NH followed by 43.8°C (102°F) temperatures on 3 July (Joshua Lane diary).

Time series of the daily temperatures were plotted for the available years in the 1820–1892 time frame at Burlington, East Montpelier, Lunenburgh and Strafford, VT as well as Fort Constitution, Portsmouth, Hanover, Dartmouth, and Wentworth, NH. The most striking observation were the low January–June values observed across the century regardless of differences in observation times at the individual stations. Several spatial patterns also emerged. Landlocked stations such as East Montpelier and Lunenburgh, VT and Hanover/Dartmouth, NH were consistently colder (e.g. January sunrise temperatures of –30°C(–22°F)) than those moderated

Table 1 Location and seasonal characteristics of backward years in New Hampshire and Vermont: 1700–1899. Very wet seasons are shown in italics and very cold ones in bold. Killing frosts are underlined

Season	Location	Year(S)	Snow	Frost	Drought
Spring	Dover, NH	1703, 1706, *1708*, 1709			Summer, fall 17008, 1709
Spring	Hampton, NH	1737			
May	Durham, NH	1740			July
Spring	Sanbornton, NH	1748	April		Summer
Spring	Stratham, Exeter, Sanbornton, NH	1751			
Spring	Dover, Sanbornton NH	1758	March		
Spring	Bedford, Exeter NH	1759	April, June	June	Summer
Spring	Bedford, NH	1761	May		
Spring	Stratham, NH	**1762**, 1772			May, August Summer
Spring	Sanbornton, Bedford, Portsmouth, NH	1763	March, April, May		
Summer	Stratham, Bedford NH				
Spring	Portsmouth, NH	1769, 1777	May		
Spring	Concord, Stratham, NH	1780			May
Spring	Stratham, NH Pomfret, VT	*1781*	May		August
Spring	Stratham, Danville NH	1784			Summer
Spring	Stratham, NH	**1785** **1786** 1790 **1793** *1794* *1795*			Fall (1786), May–June 1793 summer, fall 1794
Spring	Danville, NH	1787			
Spring	Sanbornton, NH	1795 1798 1802 1803			July 1803
Spring	Dover, NH	1812 1813 1814 1815	April, May 1812, spring 1813	June	Summer 1815
Spring	Barnet, VT	1812, 1815	May		
Spring summer	Concord, Sanbornton, NH Stowe, Newbury, Middlebury, Townshend VT Essex County, VT	1816	June May, June, July, August	July June, July, September August	Summer
Spring	Barnet, Newbury, Middlebury, Wakefield, Essex County, VT	1817	May, April, May June	June June (all crops)	

Table 1 (continued)

Season	Location	Year(S)	Snow	Frost	Drought
May	Middlebury, VT	1818			
Spring	Lancaster, NH Middlebury, VT	1819	May		
Spring	Middlebury, Townshend, Wakefield VT	1820	May	May, June	Summer
Spring	Alton, NH	1832	April (ice storm)		
May	Burlington, VT		May		
June	Essex, VT	1833	June		
Spring	Essex, Newbury, Middlebury VT	1834	May		
Summer	Middlebury, VT	1836	September	September	Summer
April, spring, June	Monroe, NH	1837 1843 1844 1845 1850	June 1843 May 1850	Sept 1843 July 1844	July 1845 & 1850
May, August	Plymouth, VT	1838			Summer
April–May	Monroe, NH	1841	May,		June
May	Malborough, NH		May		
Spring	St. Johnsbury, VT				
Summer	Monroe, NH	1842	June	June, Sept	Summer
	Irasburg, Wheelock & St. Johnsbury		June		
	Bennington, VT		June		
Spring	Calais, Essex County VT,	1844	June April		
Spring	Alton, NH	1852 1863			Summer 1852
May–June	Monroe, NH				
Spring	Wentworth, Alton, Monroe NH	1854		June	August
Spring & summer	Wentworth, Monroe NH	1855			
Summer	Wheelock & northeast Vermont	1859	June	June, July	
Spring	Monroe, NH	1862 1866 1870			Summer 1862 1870
Spring	Newbury, VT	1874	May		
Spring	OCM	*1876*	May		Summer, fall
Spring summer	Lunenburg, VT	1884	May	August	
April–May	Windham, VT	1877	May		Summer
May–June		1878	May		
Spring		1888			
Summer	Craftsbury, VT	1879			

by the influence of a large body of water (e.g. Burlington, VT on the eastern shore of Lake Champlain or Fort Constitution/Portsmouth, NH near the Atlantic Ocean). Interestingly, the temperatures at Hanover/Dartmouth, NH were not moderated by proximity to the Connecticut River. Time series of the January–June temperatures were non-stationary, especially on a seasonal basis and for the aforementioned landlocked stations.

Temporal patterns were also evident. In the early part of the record (1820–1855), pronounced freeze/thaw cycles were observed in winter (January and February) with rapid temperature swings from –30°C to 10°C (–22°F to 50°F). The periodicity of these cycles was on the order of 11–14 days during the 1820s at Fort Constitution, NH and 1830s at both Burlington, VT and Fort Constitution. During the 1840s and 1850s the periodicity of the cycles declined to 5–7 days. Rapid swings during winter were also characteristic of the 1862–1888 period, but thaws were much less frequent. Instead, severe winter freezes (as low as –30°C (–22°F)) were very common and the 5–7 day periodicity was observed from January to June. During 1878–1880, backward conditions began around Julian date 150, shifting to around Julian date 110 in 1885–1886 before returning to around Julian date 150 in 1887.

Winter freeze/thaw cycles were an important predictor of backward conditions in the subsequent spring. When the cycles were superimposed on a lower frequency temperature oscillation and/or the freezes were severe (–10°C to –20°C (14°F to –4°F) or colder) and lasted for several days, a forward spring (i.e. gradual warming) usually occurred. However, when both the thaws and the freezes (10°C to –20°C (50°F to –4°F) respectively) were pronounced in January and February and followed by warming around Julian dates 135–140, backward conditions soon followed. As Ludlum (1976: 3) so rightly noted, “A summerish January, a winterish spring.” As the temperatures fell, often to around 0°C, precipitation would be in its frozen state – the snow and freezing rain in April, May and June documented on Table 1, including the snowstorms in April 1844 that left about 4 ft of snow on the ground in Essex County, VT (Child, 1887).

The timing of the majority of patterns (freeze/thaw cycles, warm anomalies, onset of backward conditions) was roughly coincident across the two states. Also consistent across stations and time frames was the decrease in the amplitude of the temperature fluctuations from winter into spring. Station temperatures were highly correlated (at least 0.89 with a standard error of 0.078) across observation times on a given day as well with other stations. These factors suggest that despite known data inhomogeneities (changes in observers and station height), mesoscale and/or synoptic spatial forcing functions produced similarities at the continental scale. For example, the backward spring of 1887 was also observed in southern New England, with the Bulletin of the United States Fish Commission noting “the almost continuous cold, foggy weather” during May in Gloucester, Massachusetts (Wilcox, 1887). Similarly, the spring of 1843 was backward across northern tier of US [see J. Neilsen-Gammon and B. McRoberts, this volume for the backward conditions of March 1843; H. Tompkins, this volume for cold anomalies in 1849]. With the exception of north/south flow at Burlington and Strafford, VT annual wind roses revealed a predominance of northwesterly flow at all other stations even those near the coast in New Hampshire, which today would display a strong easterly component. Such flow would be conducive to sustained cold advection into the region, especially under favourable upper level conditions. Predominantly northerly flow may have been an important factor in the development of drought conditions during or following backward seasons, by hampering the southerly or westerly flow which is conducive to moisture convection across New England.

It is important to note, however, that backward seasons were also localized in spatial extent. This was particularly true in the northeastern part of Vermont and adjacent northern New Hampshire where temperatures are least moderated by warming influences. Thus in some years, backward springs existed in southern Vermont (e.g. the snowy May of 1878 at Windham) while forward conditions existed in the northern Vermont county of Essex (Child, 1887).

3.1 Phenology Variations During Backward Seasons

A number of observations could be made about the bloom dates of apple and plum trees during backward seasons (Fig. 3). Full bloom dates occurred earlier in non-backward seasons. A station's location strongly influenced the julian date on which full blooms were observed such that they were consistently later at northern stations (e.g. Monroe, NH) than those further south in proximity to Lake Champlain

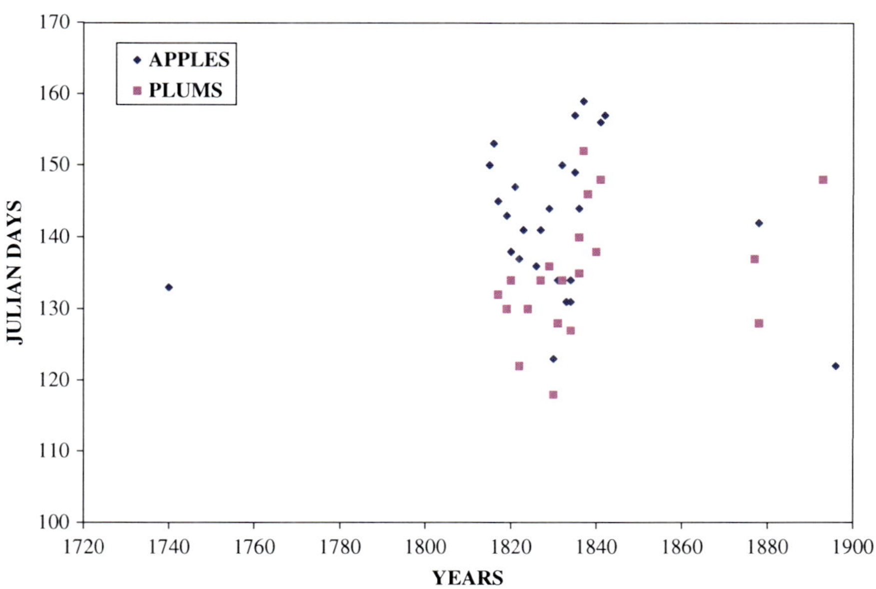

Fig. 3 Full bloom dates for apples and plums observed at Windham, Middlebury and Newbury, VT as well as Concord, Durham and Monroe, NH. Data extracted from the Emory H. Jones, Alexander Miller, Thomas Johnson, Timothy Walker, Nicholas Gilman and Albert Mason diaries respectively. Diary citations are given in the References

(e.g. Middlebury, VT). At Middlebury, VT there was a 5–7 day difference between the earlier blooming plums and the later blooming apples during normal years and 13–16 days during backward springs. Blooming delays in backward years were a function of spring temperatures, injury during cold winters (e.g. 1835), excessive snowfall in May and/or late snowmelt (e.g. 1841 at Monroe, NH). In some years (e.g. 1834 at Middlebury, VT), cold temperatures inhibited flowering even though budding had occurred.

3.2 Frost and Snow Occurrences During Backward Seasons

The average length of the growing season is usually defined as the interval between the last hard or killing frost in the spring and the first hard or killing frost in the fall. Hopp et al. (1964) have noted that the number of non-freezing days in turn affects the types of crops grown. Figure 4 summarizes 392 spring, summer and fall killing frost occurrences extracted from journal entries. Of special note was the timing of summer frosts relative to backward seasons and subsequent moisture extremes. The 1812–1820 period was marked by annual backward springs, followed by the largest number and spatial distribution of summer-long frosts, particularly across New Hampshire. The backward spring of 1834, observed across northern

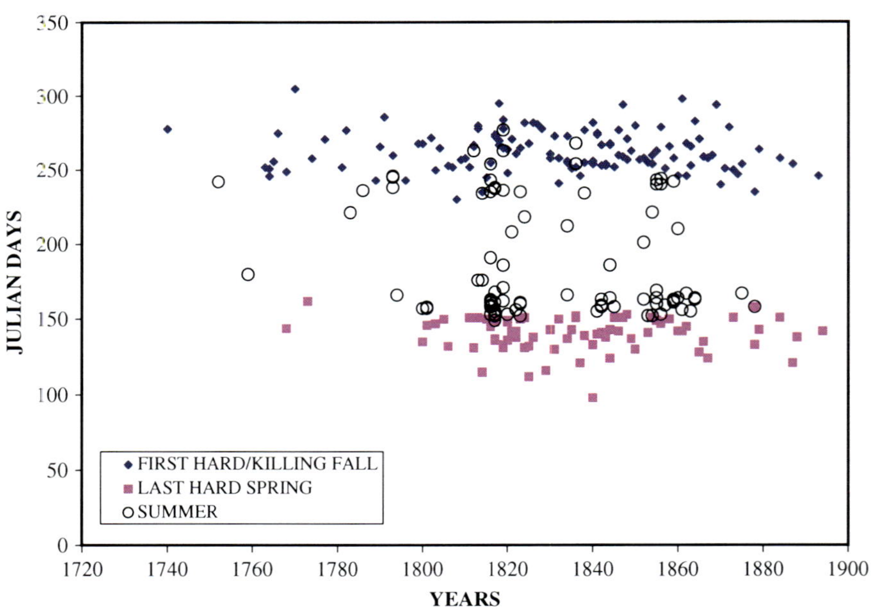

Fig. 4 Annual occurrences of the last hard/killing frost in the spring, first hard/killing frost in the fall, any hard/killing frosts in the summer at various locations in Vermont and New Hamsphire. Data extracted from the following diaries – Jonathan Carpenter, Sally and Pamela Brown, Hyde Leslie, Emory H. Jones, Allen, Sprague, Joshua Lane, Timothy Faulkner, Albert Mason, Dr. Peter Livingston Hoyt, Allen Varney, Alexander Miller, Matthew Patten, Clark, Abner Gover, Jeremy Belknap, Miriam Newton. Diary citations are given in the References

Vermont was followed by summer frosts at Middlebury, VT and Malborough, NH. Like the 1812–1820 period, 1841–1845 was marked by annual backward springs, with snow falling in May or June. Early summer frosts were less extensive spatially than during the 1812–1820 period, but were followed by summer droughts. Summer droughts were also observed following the backward springs of 1852–1855, a period characterized by a lack of snow in May/June, but frosts all summer in western New Hampshire. Finally, a large number of early summer frosts typified the 1860–1864 period, although few journal entries noted the existence of backward seasons.

Two types of frosts could be distinguished. Radiative frosts tend to localized and develop under clear, calm conditions such that freezing temperatures are confined to the ground, under a relatively shallow inversion. Advective frosts are associated more often with cold, polar air outbreaks and strong winds (Hopp et al., 1964). Radiative frosts were extracted from journal entries of air temperatures, the presence of radiation fog, the types of vegetation affected and the freezing of shallow water surfaces. Advective frosts were determined from the rapidity of onset of cool conditions, their duration and the height of the affected vegetation. Roughly 21 of the 392 frost occurrences were advective. They were more widespread and likely to affect plants at tree height, e.g. apple orchards in June 1817 across southern New Hampshire as well as plum and apple tree buds and blossoms at Middlebury, VT on 16 May, 1834.

The current definition of a killing frost is one that is severe enough to delay the start of the growing season or to bring it to an end. Of the roughly 50 killing frosts chronicled in journals, approximately 50% occurred during the summer (e.g. late 1700s at Stratham, NH; 1816 and 1817 across much of New Hampshire and western Vermont; 1821–1824 at Middlebury, VT; 1842 and 1859 at Monroe, NH). Middlebury, VT, Bedford, NH and Monroe, NH were particularly prone to killing frosts. Orographically, these locations/frost hollows had flat topography surrounded by steeply sloping land (at least to the east) that allowed for intense radiational cooling and katabatic cold air drainage. The June frosts of 1842 and 1859 at Monroe, NH which were accompanied by northerly winds, ice formation on standing water and the loss of corn, beans and leaves on trees, highlighted the role of the topography in enhancing advective frosts as well (Albert Mason diary). Other locations such as Wentworth, Sanbornton and Marlborough, NH accounted for most of the 392 frost entries, partly due to the length of the records at these stations. There, the topography was either that of a broad river valley (Wentworth and Marlborough, NH) or gently sloping uplands (Sanbornton, NH). Radiational cooling was less intense, but very frequent in these locations.

Killing frosts usually decimated potato vines, pumpkin vines, beans and melons. Cucumber stalks, corn and wheat plants were also affected. Although Landsberg (1958) and Thom and Shaw (1958) have discussed several limitations of using killing frosts for statistical analysis, some inference can be made about the existing ground temperatures from the types of crops affected. The critical or lethal temperature is the value that must be attained before plant damage occurs. For very tender crops such as cucumber, melon, pumpkins and beans, the critical temperature is 0° to –1°C (32°F to –30.2°F), but –1° to –2°C (30.2°F– to 28.4°F) for tender crops like potatoes, corn and apple or plum blossoms (Ontario Ministry of Agriculture, Food and Rural Affairs, 1985). The critical temperature depends upon the vegetation type, variety, phenology, as well as environmental factors such as soil conditions, freezing characteristics and duration, cloudiness and windiness (Cittadini et al., 2006; Ontario Ministry of Agriculture, Food and Rural Affairs, 1985).

4 Drought Characteristics, with Special Reference To Backward Seasons

In order to compare drought characteristics (including severity and duration) across time and space, a qualitative index was created using observations of precipitation deficits, soil moisture depletion, crop loss or yield reductions, atmospheric drought and the depletion of both surface water and groundwater. Atmospheric drought refers to an abnormal dryness of the atmosphere resulting from a high evaporative demand on vegetation and soil moisture amounts, where the intensity of the drying can be exacerbated by warm, moderate to strong winds (called *sukhovei* in the Soviet Union) (Dupigny-Giroux, 2001). In the historic record, *sukhovei* were pronounced under westerly flow.

The drought index created from the above indicators was qualitative in nature to account for the following sources of uncertainty:

- summer precipitation amounts often were not measured and snowfall amounts occasionally recorded, so that exact precipitation deficits, monthly totals and averages/normals could not be computed;
- daily maximum temperatures were not always recorded. The use of such terms as "warm", "very warm" or "exceedingly hot"made it difficult to identify heat wave durations. The descriptions were used to infer the persistence of high pressure conditions;
- some diary entries summarized conditions for the entire year on 31 December. No finer scale temporal information could be extracted on either daily to monthly time steps or from one year to the next;
- some locations lacked any direct or indirect reference to precipitation, necessitating the use of inferential information;
- some entries referred to "severe", "sharp" or "pinching" droughts, whereas others only noted "dry", "very dry", "earth is dry" conditions and never used the term drought;
- apart from drought, other stressors that led to crop losses were mildew, rust, frost, worm and grasshopper infestations, cool temperatures and moisture excess (sometimes consecutively in the same year);
- smokey conditions may have been due to land clearing practices in the spring and early fall (September–October), as well as local or regional forest fires;
- data gaps included entries for half the year only (e.g. some farmers only documented cool season or summer conditions), missing months or years, and changes in observers.

Table 2 An indicator-based drought index

Category	Characteristics	Indicator(S)
S1	a) 1–2 moth precipitation deficit b) precipitation deficit = precipitation in subsequent month c) no short-term or long-term effect on vegetation phenology or crop production d) local in scale	a) no rain/snow entries b & c) vegetation recovers completely
S2	a) 2–3 month precipitation deficit b) sukhovei may be present c) runs of high daily temperatures d) depletion of surface soil moisture e) some effect on vegetation phenology or crop production f) local and/or regional in scale	a) no rain/snow entries b) "drying wind" entries d) "earth is dry"entries, shallow rooted crops (e.g. grass for hay affected) f) number of coincident diary entries available
S3	a) 3–12 month precipitation deficit b) sukhovei may be present c) extended runs of high daily temperatures d) depletion of surface and subsurface soil moisture e) depletion of surface hydrology f) effect on vegetation phenology or crop production f) forest fires may occur * g) regional in scale	as for S2 d) "wells are very low" "river very low" entries e)"short crop" "middling crop" entries f) most fires observed in the woods
S4	as for S3 b) surface and subsurface waters exhausted c) large-scale forest fires including swamps and wetlands * d) regional or larger in scale	as for S3 b) "streams dry" entries c) "great fires in the country all round" entries
S5	as for S3 b) multi-year	few rain/snow entries in fall and winter, followed by summer drought with increasing number & severity of impacts

*summer and fall forest fires. Land cleared by burning in late spring.

Table 2 summarizes the five categories of the drought index which integrates both short-term moisture deficits and multi-year drought episodes. The shortest duration drought (S1 – one to two months) could be one of two types. The first were the droughts of the earliest period (1681–1730) when few details on the drought were given apart from its seasonality. The second category of S1 were "flash" droughts which exhibited a quick onset with no preceding atmospheric or land-surface moisture deficits, followed by an equally rapid termination due to precipitation in the following month. The timing of these flash droughts was critical in that spring occurrences had a temporary effect on crops, while late summer/early fall events affected surface waters and wells which tended to be at their lowest points of recharge at that time. S2 droughts tended to be summer occurrences.

One key difference between the S3 and S4 designations was precipitation effectiveness. In S3 conditions, soaking rains reversed soil moisture deficits resulting in less crop loss. S4 droughts combined both long-term moisture deficits with severity of impacts. Typical indicators included the lack of subsurface moisture and groundwater recharge (dry wells and water bodies; muck fires burning in swamps and wetlands) following moisture inputs that were sufficient for crop and vegetation requirements. S4 and S5 categories were also assigned to multi-year events observed over large spatial extents and were usually most pronounced in the late summer and early fall. It should be noted that long-term droughts (S3–S5) were not continuous periods of low to no precipitation, drying winds and high temperatures. Instead, at many locations soaking precipitation was received sporadically, although it was insufficient to either ameliorate crop losses or replenish soil moisture reserves. In extreme cases, these precipitation events included a "very great snow storm" on 7 June, 1759 at Exeter, NH and freezing rain on 30 July, 1847 at Malborough, NH.

The creation of the indicator-based drought index was not without uncertainty. There was a bias towards conditions observed in New Hampshire (Table 3). Due to the timing of settlement in both states, there was an abundance of diaries in New Hampshire during the early part of the record compared with those for Vermont. In addition, the latter part of the record exhibited a wetter climate regime such that drought and dry conditions were largely absent from the Vermont diaries of 1850–1900. This shift in climate regimes was reflected in Zadock Thompson's (1853:13) treatise on the Natural History of Vermont in which he noted that "very little damage is ever done by hurricanes or hail. The crops oftener suffer from an excess, than from a deficiency of moisture, though seldom from either" Finally, the varieties of crops grown may also have factored into the proneness to drought observed in the earlier vs. later years. Indian corn was often able to recover with sufficient effective precipitation from S1 to S2 droughts, while English crops and hay were often decimated. English crops were also more susceptible to mildew and rust which made them even more vulnerable to co-existing or subsequent droughts. It should be noted that not every drought could be classified without ambiguity. For the spring flash droughts (S1), the short term ranking often obscured the fact that excessive drying following snowmelt produced ideal fuel conditions, such that forest fires were often observed in April (e.g. in 1740 at Durham, NH). Similarly, some uncertainty exists for cases of long term dryness (S4) that caused mill ponds and streams to run dry (e.g. 1751 at Monroe/Lyman, NH; 1766 and 1773 at Bedford, NH; 1797 at Stratham, NH) in the absence of journal entries that would imply intermediate surface soil moisture deficits.

5 Maple Sap and Sugar Production

In the late 20th and early 21st century, maple syrup and sugar across northern New England are primarily made from sugar maple (*Acer saccharum*) and red maple (*Acer rubris*) trees that are tapped from late February to mid March. Sap flows best with a diurnal temperature range from about –4°C (25°F) to 4°C (40°F), and warm

Table 3 Indicator-based drought categories as extracted from New Hampshire & Vermont diaries. Split categories are shown in bold, italics (S1–S2, S2–S3, S3–S4)

Category	Location	Years
S1	Dover, NH	1683 1707 1815
	Exeter, NH	1738 1754
	Durham, NH	1740
	Concord, NH	1746 1780
	Bedford, NH	1760 1765 1769 1781
	Stratham, NH	1774 1775 ***1789*** 1791 1792
	Danville, NH	1791 1792 1797–1800 1803 1805 ***1826 1829***
	Sanbornton, NH	1817
	Middlebury, VT	1838 1850 1876 1885 1886
	Monroe/Lyman, NH	1854 1856 1876–1882
	East Montpelier, VT	1856
	Wentworth, NH	1863
	Alton, NH	
S2	Dover, NH	1709
	Stratham, NH	1746 1749 ***1757*** 1771 1772 1773 1774 1784 1793
	Exeter, NH	1748 1759
	Concord, NH	1764
	Bedford, NH	1771
	Danville, NH	1793 1796
	Sanbornton, NH	1793 1795 1796 1804 1806 1813 1818 ***1820 1823*** 1825
	Monroe/Lyman, NH	1836 1837 1842 1846 1847 1848 1851 1888
	Durham, NH	***1743*** 1840 1842
	Wentworth, NH	1852 1853
	East Montpelier, VT	1853 1859 1866
	Alton, NH	***1864*** 1865
	Weare, NH	1841
S3	Dover, NH	1708
	Stratham, NH	1748 1761 1770 1775 1780 1781 1792
	Exeter, NH	***1749***
	Lancaster, NH	1820
	Monroe/Lyman, NH	1841 1845 ***1852*** 1854 1864 1868 1870 1887
	Wentworth, NH	1854
	East Montpelier, VT	1860 1861 1862 1863 1864 1865 1868 1870–1874
S4	Portsmouth, NH	1762
	Stratham, NH	1762 1782 1786 1797
	Bedford, NH	1766 1773
	St. Johnsbury, VT,	1849
	Monroe/Lyman & parts of	1849
	northern NH	1849
	Monroe/Lyman, NH	1860
	Newbury, VT	1854
	Plymouth, VT	1887
S4–S5	Stratham, NH	1794 1798
S5	Sanborton, NH	1794

winds. In the 1800s, maple growers were also very attuned to the delicate balance between daily weather conditions and sap production. Good sap flow conditions included cloudy skies, easterly winds in the northern Connecticut Valley, as well as abundant snow late in early April in southern Vermont. Sap flow was inhibited by warm, rainy conditions in April and/or very sunny, mild and rainy conditions during the winter in the northern Connecticut valley (Albert Mason diary). In southern Vermont, little to no sap was produced under cold northwesterly winds, sleet or sub-freezing conditions even in the presence of abundant snow (Hyde Leslie diary and Emory H. Jones diary). One sugar maker in particular, Luther O. Weeks of Proctorsville, VT combined weather knowledge with his observation of sap flow from maples under varying conditions, and thus was able to produce "7,834 lbs. of good tub sugar" from only 520 white rock maples (a term used by New Hampshire residents for sugar maple trees (Cogbill, 2007, personal communication)) between 1868 and 1872. In an editorial in February 1873, he wrote

> Farmers in this county where the snow is apt to be deep in March and April, delay tapping for the snow to settle, and thus lose by far the best part of the sugar; as no frost is in the ground when the snow is deep; and as soon as the mercury stands at from 46 to 50 degrees above zero, in the shade, for a short time, sap commences to run freely. I have make 600 lbs. of sugar some springs ere a pound would be made in any adjoining sugar lot, the proprietors claiming that it was not sugar weather. (Rutland Daily Herald, February 1873, Vol. 12).

Table 4 summarizes diary accounts of maple sap and sugar production before 1916 when records of the US Department of Agriculture's National Agricultural Statistics Service (NASS) began. With the exception of Craftsbury and Monroe in north-central Vermont and northwestern New Hampshire respectively, all of the

Table 4 Maple sap and sugar production extracted from the following diaries – Jonathan Carpenter, Albert Mason, Allen Caldwell, Emory H. Jones, Henry Allen and Clara Jones. Diary citations are given in the References

Year	Location	Season duration	Trees tapped	Sap amount	Sugar (lbs)
1781	Pomfret, VT	13 March–16 April		82 pails	80
1874	Monroe, NH	21 March–1 May[a]			
1876	Windham, VT	10–25 April	1400[b]		1 ton[c]
1877	Windham, VT	3 February–5 April	3		100
1877	Wheelock, VT	31 March–21 April			
1878	Windham, VT	8 March–9 April		400 buckets	at least 500
1878	Craftsbury, VT	11 March–11 April		760 pails	1600
1878	Wheelock, VT	16 March–18 April		∃ 31 tubs	
1888	Windham, VT	23–26 April?			
1888	Tinmouth, VT	27 March–16 April		18 pails	
1893	Windham, VT	10–29 April			355
1894	Windham, VT	9 March–17 April	50		400
1895	Windham, VT	3–8 April	∃ 100		at least 125
1896	Windham, VT	26 March–13 April	50		300
1897	Windham, VT	last week March–24 April	50		450
1898	Windham, VT	20–26 March		500 lbs	

[a] only season duration records were available at Monroe, NH with an average length of 29 March–16 April during 1864–1876.
[b] records for the diary author's neighbour Mr. Dimicks.
[c] original entry given without conversion to pounds. If the short ton unit was used, the equivalent would be 2,000 lbs.

towns listed in Table 4 are in southern Vermont. Of note are the varying season start dates and durations. Maple sap production was, and still is highly variable spatially and from one season to the next. Delayed tapping was a function of deep snow or severe snowstorms such as the Blizzard of March 1888, when the season extended into late April. Similarly, the particularly stormy, snowy spring conditions of 1873 in southern Vermont retarded the start of the season until early April (Rutland Daily Herald, 5 April 1873 Vol. 12, No. 288). The extension of the sap flow season into April in the1800s corresponds with the March–April duration observed during the 1940s and 1950s (Taylor, 1956), but is decidedly later than the February–March season length of the late 20th century. This late excursion of sap flow into April is probably related to the extended freeze/thaw cycles discussed in Section 3 as well as the snowfall that often accompanied backward seasons. As Charles Nelson Morse wrote in letter to his brother Dane on 13 April 1861:

> There has been one snow storm since I wrote that sheet the 2nd day of April. It fell about 10 inches. Regular sugar snow and I believe it too for it has frozen every night and been warm every day, wind (what little there has been) in the North. Grand time for Still. I think he has made a lot of syrup, but have not heard. (Dwinell, 1996).

During some backward springs, e.g. 1815, maple sap "ran with great freedom and never before or since was so large a product gathered per tree" in the northern Vermont counties of Caledonia and Essex (Child, 1887: 25). In other years, the relationship was not straightforward. The spring of 1878 was a forward one in northern Vermont's Essex County that led to early May bloom dates for red plum, strawberry and apples (Child, 1887), while backward conditions prevailed in the southern town of Windham, VT. Between 21–25 March, journal entries in both northern (Craftsbury) and southern (Windham) Vermont recorded temperatures so low the sap froze, although the season extended into mid April.

The presence of ideal weather conditions was not a guarantor of good or abundant sap. As Emory H. Jones observed at Windham, VT on 26 March 1898, "A week of sap weather but not very good, about 500 lbs sap poor, watery." Such variability was also observed in the 1884 and 1885 U.S. government's chemical studies of maple sap (Wiley, 1885). Fluctuations in the timing, duration and quality of sap production at Windham, VT continued into the late 1890s with 1895 being a late and poor sugar season, followed by unseasonable warmth in mid-April of 1896 that curtailed the season (Emory H. Jones diary). Finally, journal entries indicate a relationship between the sweetness of the sap and the length of the season. Emory H. Jones at Windham, VT recorded "A poor season but sap very sweet wh[ich] is sign of short season" on 25 April 1876.

6 Conclusion

Backward springs and occasionally backward summers, were frequent occurrences in the late 1700s–1800s in New Hampshire and Vermont. They were often accompanied by snow in May or June, with frost and drought recurrences in the summer.

Backward seasons tended to be regional events that occurred under strong northwesterly flow into the region. Localized backward seasons were also observed in the northernmost interior parts of the study area which were removed from the moderating influences of a body of open water. This pattern of spatially extensive and frequent backward springs, summer droughts and summer killing frosts was particularly pronounced in the 1812–1820 time frame, with linkages to hemispheric anomalies in 1812 and the eruption of Tambora in 1815. Winter freeze/thaw cycles were an important of indicator of a backward spring with marked fluctuations in the amplitude and periodicity of these cycles across the 1800s.

The incidence of backward seasons also affected the phenology of common plants as well the viability of crops in the summer. Delayed full bloom dates were observed for apples, plums and other species as a function of low spring temperatures, preceding winter injury, excessive snowfall in May and/or late snowmelt. Summer frosts were particularly marked in the 1812–1820 and 1860–1864 periods. Of the 392 frost occurrences extracted from journal entries, about 21 were advective in nature and the remainder were radiative. Advective frosts tended to be widespread and affect tree species such as apple orchards and plum blossoms. Radiative frosts were more prevalent, especially in orographically conducive locales, and often resulted in the complete loss of crops such as corn, wheat, potato vines and melons.

The creation of an indicator-based drought index highlights how drought-prone the region was especially in the late 1700s. This may reflect the shift to a wetter climate regime that occurred around 1850, as well as the selection of crops grown in the later part of the study period. Multi-year, severe droughts (categories S4 and S5) were often interrupted by sporadic precipitation including a summer snowstorm, and were particularly evident in 1762, 1794, 1849 and 1864.

While maple sap production and collection in the 21st century are less reliant on the vagaries of the weather than was the case in the 1800s, key differences in climate regimes can be noted. The sap flow season from February/March into April in the 1800s is much longer and later than that observed in recent decades, although strikingly reminiscent of the 1940s and 1950s seasons. The extension of the season was probably a function of the extended freeze/thaw cycles in the spring as well as the snowfall often observed during backward seasons. It should be noted that the relationship between weather and sap production was not a straightforward one and the discovery of additional data sources and diaries may assist in better quantifying the spatio-temporal characteristics of the patterns observed.

Acknowledgments The author would like to thank the following persons for these data contributions: Sylvia Bugbee and the research librarians of the University of Vermont's Baily/Howe Library Special Collections; Gregory Sanford, Vermont State Archivist; Jason Shafer for transcriptions of New Hampshire and Vermont diaries; Rachel Cree Sherman for locating and transcribing Vermont diaries; Charles McGill for his assistance with base maps; Daniel Brown for data analysis; Gregory Zielinski for the Historical Climatology of New England and New York, Pennsylvania, New Jersey site; Timothy Perkins for his advice on sugar maple production; members of the Rutland, VT Historical Society; Eden Furtak-Cole, Allison Roy, for digital imaging and data transcription.

References

Abner Gover diary, 1796–1856, Weare, NH

Albert Mason diary, 1834–1886, Monroe and Lyman, NH

Alexander Miller Diary, 17 April 1816–27 December 1836, Middlebury, VT.

Allen Caldwell diary, February 1878–March 1879, Wheelock, VT.

Andsager K, Kruk MC, and Spinar ML (2007) DSI-3297. CDMP COOP Summary of the Day Forts, National Climatic Data Center, Asheville North Carolina.

Charles Mooney Diary, November 1865–May 25 1908, Alton, NH.

Child H (1887) Part First. Gazetteer of Caledonia and Essex Counties, VT. 1764–1887, the Syracuse Journal Company, Printers and Binders, Syracuse, New York.

Cittadini ED, de Ridder N, Peri PL and van Keulen H (2006) A method for assessing frost damage risk in sweet cherry orchards in South Patagonia. Agricultural and Forest Meteorology, 141: 235–243.

Clara Doty diary, 1888, Tinmouth VT.

Dartmouth College Meteorological register, November 1827–December 31 1890, Hanover, NH.

Diary of Henry Allen, January 1–December 10, 1878 Craftsbury, VT.

Doty SR and Dupigny-Giroux L-A (2005) Station History Report to the Climate Database Modernization Program of NOAA's National Climatic Data Center, Midwestern Regional Climate Center, 61pp. [Available from Midwestern Regional Climate Center, 2204 S. Griffith Drive, Champaign, IL 61820.]

Dr. Peter Livingston Hoyt, personal meteorology register, 1852–1856, Wentworth, NH.

Dupigny-Giroux L-A (2001) Towards characterizing and planning for drought in Vermont. Part I. A Climatological Perspective. Journal of the American Water Resources Association, 37(3): 505–525.

Dwinell JM (1996) Letters to Dane: November 1860–July 1862. Written by Charles Nelson Morse and Ellen Morse.

Emory H. Jones diary, 1875–1879, 1888–1906, Windam, VT

Fairbanks ET (1913) The Town of St. Johnsbury, VT. The Cowles Press, St. Johnsbury, VT, pp. 513–520.

Hopp RJ, Varney KE and Lautzenheiser RE (1964) Late Spring and Early Fall Low Temperatures in Vermont. Agricultural Experiment Station, University of Vermont – Burlington, Vermont, June, Bulletin 639.

Jeremy Belknap diary, 1758, Dover, NH.

Jonathan Carpenter diary 1781–1783 Pomfret Vermont. Transcribed and edited by Miriam and Wes Herwig, Greenhills Books, Randolph Center, VT, 1994.

Joseph & Charles Mooney Diaries, 1831–November 25 1865, Alton, NH.

Joshua Lane diary, 1788–1829, Sanbornton, NH

Landsberg H (1958) Physical Climatology, Gray Printing Co. Inc., DuBois, PA.

Ludlum DM and the editors of Blair & Ketchum's Country Journal (1976) The Country Journal New England Weather Book, Houghton Mifflin, Boston.

Matthew Patten diary, 1754–1798, Bedford, NH

Miriam Newton diary, 1820–1852 Malborough, NH.

Neuman J (1990) The 1810s in the Baltic region, 1816 in particular: Air temperatures, grain supply and mortality. Climatic Change, 17(1): 97–120.

Ontario Ministry of Agriculture, Food and Rural Affairs (1985) Freeze Protection Methods for Crops, Factsheet 85–116.

Rutland Daily Herald, 5 April 1873, Vol. 12, No. 288.

Schwartz MD (1998) Green-wave phenology. Nature, 394: 839–840 (27 August).

Seneca Ladd, meteorological register,1867–1891, Meredith, NH.

Sprague diary, 4 May–30 December 1884, Craftsbury, VT.

Taylor FH (1956) Variation in Sugar Content of Maple Sap. Vermont Experiment Station Bulletin 587.

The Diaries of Sally and Pamela Brown 1832–1838 [Jan.–Feb.1839] and Hyde Leslie 1887, Plymouth Notch, Vermont. Blanche Brown Bryant and Gertrude Elaine Baker (eds.), The William Bryant Foundation, Springfield, Vermont, 2nd edition, 1979.
Thom HCS and Shaw RH (1958) Climatological analysis of freeze data for Iowa. Monthly Weather Review, 86: 251–257.
Thomas Chandler diary in Eleanor Hutchinson (1961) History of the Town of Wheelock, VT. Emerson Publishing Co., Rochester, Vermont.
Thomas Johnson diary, 1761–1895, Newbury, VT.
Thompson Z (1853) Natural History of Vermont with Numerous Engravings and an Appendix. Stacy and Jameson.
Timothy Faulkner Diary, 1818–1821, Lancaster NH
Varney A, 1861–1865, Alton, NH
Weather Journal by Benjamin Wheeler, 1851–1883, East Montpelier, VT.
Wilcox WA (1887) The Fisheries of Gloucester Mass, in May 1887, with notes on those of other localities. Bulletin of the United States Fish Commission 1887.
Wiley HW (1885) The sugar industry of the United States. U.S. Bur. of Chem. Bull. 5.
Wilkins MN (1871) History of Stowe to 1869. As originally published in The Vermont Historical Gazetteer, a Magazine Embracing a History of Each Town Civil, Ecclesiastical, Biographical and Military. Vol. II (Burlington, Vermont, 1871) Edited and Published by Miss A.M. Hemenway.

The Challenge of Snow Measurements

Nolan J. Doesken and David A. Robinson

Abstract At times beautiful, at times annoying, the importance of snow cannot be diminished by fan or foe. Much has and continues to be written about this crystalline feature, particularly related to disruptive storms, its hydrologic significance and role in influencing and identifying climate variability and change. Many think that measuring snow is simple – nothing more than inserting a ruler and recording the depth. However, whether it is falling, accumulating or changing once on the ground, there are challenges in measuring snow at every stage of its existence. In this chapter we examine historical and current methods of recording the depth and water equivalent of snowfall and snow on the ground. Although the focus is on manual observations, a brief overview of some recent remote sensing methods is also included. Accurate manual snow observations require careful attention to guidelines and exercising careful judgment as one measures this ever-changing medium. Some observers have kept excellent records for extended periods of time and these valuable records are archived at various centers around the world. However, in too many instances both in the past and at present, snow observations are at best granted second class attention by those involved with weather observation, the training of observers and data archiving. This chapter will address the suite of challenges that continue to plague the accurate measurement of snowfall.

Keywords Snow · Snow measurement · Site exposure · Remote sensing

---

N.J. Doesken (✉)
Colorado Climate Center, Department of Atmospheric Science, Colorado State University, Fort Collins, CO 80523, USA
e-mail: nolan@atmos.colostate.edu

D.A. Robinson (✉)
Global Snow Lab, Department of Geography, Rutgers University, Piscataway, NJ 08854, USA
e-mail: drobins@rci.rutgers.edu

L.-A. Dupigny-Giroux, C.J. Mock (eds.), *Historical Climate Variability and Impacts in North America*, DOI 10.1007/978-90-481-2828-0_15,

1 Introduction – The Characteristics and Importance of Snow

Snow remains one of the truly incredible wonders of nature. Its delicate beauty, pure whiteness, endless variety and changeability delight children and adults. Its attractiveness is offset for some by the reality of the cold obstacle it presents to life's daily activities. The older we get, the greater an obstacle it seemingly becomes. How tiny and fragile crystals of ice totally transform a landscape (Fig. 1), even to the point of bringing temporary silence where the clamor of urban life usually prevails, is indeed remarkable. In January 2007, there were even reports of some sense of tranquility in the streets of Bagdad when rare snow fell over the city (BBC News, 2008). The process of snow formation has gradually been explained by generations of ardent scientists (Bergeron, 1935; Nakaya, 1954; Mason, 1957; Takahashi et al., 1991). Still, the reality of trillions of ice crystals forming and efficiently harvesting atmospheric water vapor and tiny cloud droplets as they fall through cloud layers on their path toward bringing moisture to the earth still seems miraculous to those who think and ponder.

The societal importance of snow cannot be understated. While it is loved by some and hated by others, it greatly affects economic activities in the mid and high latitude nations of the world. Each year millions travel long distances to ski, snowboard, snow mobile, or participate in other winter recreation. Millions of others spend their hard-earned income escaping the cold and snow. Hundreds of millions of dollars are now spent annually in the US alone clearing and treating sidewalks, streets, highways, parking lots and airport runways so that commerce and transportation are slowed as little as possible (Minsk, 1998; Kocin and

Fig. 1 A snow storm in progress

Uccellini, 1990). Despite these efforts, hundreds of lives are lost each winter to snow and ice-related accidents, while thousands more are injured. The economic cost of closed highways, blocked businesses, cancelled flights and lost time is enormous.

Snow is much more than an impediment to commerce. Just think of all the forts, snowballs and snowmen made each year. Snow is also a structural material providing practical temporary shelter and protection from extreme cold. When smoothed and compacted, it can make an effective temporary aircraft runway in cold, remote areas. Left uncompacted, snow is an excellent insulation material protecting what lies below it from the extreme cold that may exist in the air immediately above (Jones et al., 2001). The weight of snow is a necessary consideration in the design and construction of buildings. Almost every year some buildings are damaged or destroyed following extreme storms or periods of great or prolonged snow accumulation (DeGaetano et al., 1997).

The high albedo (ability to reflect light) of fresh snow has profound impacts on the surface energy budget of the globe, which, in turn, dramatically affects climate both locally and over larger areas. Snow-covered areas are significantly colder than adjacent land areas under most weather conditions. Interest in global snow accumulation patterns has risen greatly during the past two decades due to its great significance in the global climate system (Bamzai and Shukla, 1999; Gong et al., 2004; Groisman et al., 1994; Walsh et al., 1982) and the extent to which snow affects and is affected by large scale global climate change. This has in part resulted from the availability of satellite-derived mapping and the assembly of ground-based databases (Robinson et al., 1993; Hall et al., 2002; Dyer and Mote, 2006). Perhaps the primary driver is the role snow cover (or the lack thereof) seemingly plays in the recognition and regional amplification of climate change (Leathers et al., 1995; Mote, 2008).

Most important of all, snow is water. The hydrologic consequences of snow are so great that it is imperative to carefully track snow accumulation and its water content. Accumulating snowfall becomes a frozen reservoir that later releases water into the soil, into aquifers, into streams and rivers, and into reservoirs. This water source is critical to water availability and hydro-electric power generation throughout the year, especially in mountainous regions and where snow contributes a significant fraction of the year's precipitation. If snow melts too quickly or if heavy rains fall on melting snow or downstream of melting snow, flooding also becomes a possible consequence.

In some mountainous areas, the largest threats posed by snow are avalanches. Subtle changes in crystal structure within the snowpack are always occurring. Certain weather patterns, temperature changes and snowfall sequences lead to layering within the snowpack, leaving some layers "weak" and unable to adhere well to adjacent layers (Mock and Birkeland, 2000). With the addition of new snow, these unstable snowpacks can be very prone to avalanches that claim dozens of lives in North American, Europe and elsewhere each year. The study of avalanches is a field of its own, involving hundreds of scientists around the world and requiring some unique measurements of internal snowpack characteristics.

In recent decades, great improvements have been made in weather prediction on the scale of a few hours to a few days. Improvements are also evident in long-range forecasts. One of the biggest challenges in weather prediction today, however, remains the quantitative prediction of precipitation. Predictions of snow storms and the location of rain/snow/ice boundaries remain difficult, even just hours in advance. Almost every year there are examples of large snowstorms that catch us by surprise, or forecasted storms that never materialize. If forecasts are to continue to improve, adequate observations must be taken on the scale needed to track and model precipitation processes. Having adequate information regarding the location and quantity of snow on the ground preceding an event is an important piece of this forecasting puzzle.

The remainder of this chapter will discuss the various means of observing snowfall and snow cover, and will address how well these variables have been measured in the past. Mention will be made of remotely sensed methods of snow observation. However the focus will be on in situ observations, which cover a longer period and are critical to assessing the accuracy of remote methods. In keeping with the theme of this volume, United States surface observations will be highlighted.

2 Measurements of Snow

Because of the importance of snow within our natural and socioeconomic environments, measurements are essential for studying, learning, explaining and teaching others about the properties and impacts of snow. We also measure so that we can describe and document what has occurred and note changes that occur over time. This allows us to compare, prepare and predict so that our society can adapt as well as possible to the challenges and benefits derived from snow.

Much of what we know about snow, its spatial distributions and its contribution to the hydrologic cycle, comes from very simple observations taken at a large number of locations over a long period of time. Commonly measured snow properties are given below.

- Snowfall: The accumulation of new snowfall or other forms of frozen precipitation that has fallen and accumulated in the past day or other specified time period. Glaze from rain that freezes on contact is not included in this category. The measurement of snowfall is most often taken manually by trained observers using a simple measurement stick and their own good judgment.
- Precipitation amount: This refers to the water content of snowfall plus any other liquid, freezing or frozen precipitation falling during the same period such as rain, freezing rain or ice pellets (sleet). Measurements of precipitation amount are most often taken with a recording or non-recording precipitation gauge.
- Snow depth: This is simply the total depth of snow and ice, including both freshly fallen and older layers. For shallow snows, a ruler or longer measurement stick is all the equipment needed. For deeper snows, fixed snow stakes or special calibrated probing bars are used. Electronic methods for measuring snow depth have been developed in recent years and are used at some sites.

- Snow Water Equivalent: This term, commonly abbreviated as SWE, is the total water content expressed as an equivalent depth of the existing snowpack at the date and time of observation. Since snow does not accumulate or melt uniformly, the SWE is often the average of a set of representative measurements taken in the vicinity of the point of interest. The measurement of SWE is often taken by weighing a full core sample (snow surface down to ground surface) or averaging the weights from several samples.
- Density: The mass per unit volume is an important variable for describing the nature and potential impacts associated with snow (Judson and Doesken, 2000). A common but often incorrect assumption used in the US is that ten inches (25.4 cm) of new snow has a water content of one inch (2.54 cm), thus it has a density of 0.10. This could also be expressed as a percentage – 10%. Under ideal conditions (in particular little wind), the density can be obtained by dividing the measured precipitation amount (melted gauge catch) for the time period of interest, by the measured depth of new snowfall for that same period. However, a direct measurement of water content per carefully determined volume is often more accurate since gauge measurements of snowfall often do not catch all the precipitation that actually falls, especially under windy conditions.
- Precipitation type and intensity: For purposes of weather forecasting and verification as well as airport operations and other aspects of transportation, continuous monitoring of the type of precipitation (rain, freezing rain, ice pellets, snow pellets, snow, hail, etc.), its intensity (light, moderate, or heavy) and rate of accumulation, and how much the horizontal visibility is restricted are very important. For many years, airport weather stations in the US have used a simple definition of snowfall intensity based on the degree to which the snowfall reduces visibility (U.S. Dept. of Commerce, 1996). For example, unless the horizontal visibility is restricted to less than ¾ mile (1.21 km), the snowfall intensity can only be reported as "light". Precipitation type, intensity and visibilities were all determined manually until the mid 1990s. Electronic sensors have been introduced at many airport weather stations in recent years.
- Snow-cover extent. This is an assessment of how much of a specified land area is covered by sufficient snow to whiten the surface at any specified time. Before the implementation of satellites, this was accomplished simply by mapping individual weather station snow depth observations and approximating the location of the edge and area of snow-covered regions.

Other types of snow measurements are taken for basic research, special applications such as water quality assessments and avalanche prediction, and military applications in cold climates. These include:

- Albedo
- Crystal types and evolution
- Insulation
- Acidity (snow and the first flushes of snowmelt have been found to be among the most acidic forms of precipitation in some areas)

- Electrical conductivity
- Trafficability/compactability
- Layer structure and stability
- Forest canopy snow accumulation and sublimation

This is by no means an exhaustive set of measurements, and the reader is referred to the Handbook of Snow (Gray and Male, 1981) for additional information about snow properties and measurements. Measurements of snow water equivalent (SWE) in the mountains of the western US began systematically in the 1930s motivated by drought and the need to better anticipate water supplies provided by mountain snowpack (Helms, 1992). Measurements of snowfall, snow depth and precipitation date back to the 1800s and were a part of the original daily weather observation regimen of the US Signal Service and later the US Weather Bureau.

3 Obstacles to Snow Measurements

Amid the challenges and limitations of making any environmental measurements, snow's dynamic changing features provide challenges for observations. Snow melts, sublimates, settles and drifts. Its crystal structure changes from storm to storm and from time to time within a storm. Once on the ground, the crystals change again in the presence of surrounding crystals, temperature gradients, and vapor density gradients. Snow is not deposited uniformly on the ground and it melts even more unevenly depending on factors such as shading, slope, aspect, wind exposure, vegetation height, color and amount. For example, snow covered with a thin layer of dark dust will melt quicker than clean snow in the presence of bright sunshine. Traditional precipitation gauges that will function when measuring rain are often grossly inadequate for capturing and measuring the water content of snow. This is due to the feather-light crystals being easily deflected around the precipitation collector even by light to moderate winds, keeping some of the snow from falling into the gauge. Furthermore, snow may cling to the side of gauges, effectively changing the collection diameter of the instrument. Additionally, the compressibility of snow makes it difficult to gather the appropriate core samples. When all is said and done, measuring snow is easy. Measuring it accurately and consistently is the problem. It is a problem today just as it was in the 19th and 20th centuries.

4 Procedures for Measuring Snowfall, Snow Depth and Water Content

As with all other measurements of our environment, it is critical to find and preserve a consistent and representative location for measurement, and maintain strict standards for instrumentation and observing practices (Colorado State University, 2004). For comparing data from many locations, consistent procedures and representative measurement locations are essential (Doesken and Judson, 1996).

4.1 Precipitation Amount

The measurement of precipitation amount is arguably the most basic and useful of all meteorological variables. Here we will concentrate on the measurement of the water content of snowfall. In practice, the most accurate means of observing water content is by taking a core sample off a snowboard or some other surface that has captured a representative amount of the new snow and melting it down or weighing it. Unfortunately this is rarely a standard practice, either due to its lack of emphasis or introduction during observer training or as a result of the absence of a human observer.

As a result, the most common method of measuring the snowfall water content is a straight-sided cylinder of a sufficient diameter and depth to effectively catch rain, snow and other forms of precipitation. The National Weather Service's standard precipitation gauge has a diameter of eight inches (20.32 cm) and is approximately 2 ft (61 cm) tall. For capturing snow, the funnel and inner tube used to obtain accurate measurements of liquid precipitation are removed. Following a snow event, the standard observing procedure is to bring the gauge inside at the scheduled time of observation, to melt the snow either by setting the gauge in a container of warm water until the snow and ice in the gauge are melted, or by adding a known amount of warm water directly to the contents of the gauge to hasten its melt. Observers then pour the contents of the gauge through the funnel into the inner cylinder for measurement, carefully subtracting any volume of water that was added to hasten the melt. In very snowy locations, some observers may be equipped with specially calibrated scales for determining precipitation by weighing. This simplifies the observation process considerably, especially in locations where warm water is not readily available.

A variety of other precipitation gauges are also used to assess water content. Weighing-type recording rain gauges have been used for many years by the NWS for documenting the timing of precipitation. For winter operation, an antifreeze solution is required. An oil film on the surface of the fluid reservoir is also recommended to suppress evaporation losses. The use of oil and antifreeze may be an environmental hazard requiring care in the selection and use of these materials.

Storage precipitation gauges have been used for measuring total accumulated precipitation at remote locations. These large gauges can hold several feet of snow water content and require oil and antifreeze. The volume of additives must be accurately measured since their density differs from water. Tipping bucket precipitation gauges (popular due to their low cost, relative simplicity, and ease of use for automated applications), are not very effective for measuring the precipitation from snow (McKee et al., 1994). Heat must be applied to the surface of the funnel of these gauges in order to melt the snow. Since most snow falls at rates of only 1 or 2 mm per hour or less, even small amounts of added heat can lead to the sublimation or evaporation of much of the moisture before it reaches the tipping buckets. Furthermore, the addition of heat can create small convective updrafts above the surface of the gauge further reducing gauge catch. The alternative to using tipping bucket gauges for measuring the water content of snow is to wait until the temperatures rise and the snow melts on its own – assuming that fresh snow does not overflow the collection funnel. However, this reflects the water content when the snow melts and not when it actually fell.

As simple as it may seem, the measurement of precipitation amounts has yet to be perfected. Even if a perfect gauge were available, there is still a major problem limiting the accuracy and introducing uncertainty into gauge measurements of the water content of snow. The National Weather Service continues to search for a satisfactory and affordable all-weather precipitation gauge. Since 2000, the NWS has tested and deployed a new "All Weather Precipitation Gauge" at many of the larger airports across the US in order to overcome the known deficiencies of the heated tipping bucket gauge. Just as previous gauges have had advantages and disadvantages, the same is true with this next gauge. The problem is wind. Gauges which protrude into the air present an effective obstacle to the wind, resulting in deflection of lightweight snow crystals. The result is gauge undercatch of precipitation, the degree of which depends upon wind speed, snow crystal type, gauge shape and exposure. In the measurement of rain, gauge undercatch is not significant unless the winds are strong. However for snow, even a 10 km per hour breeze can result in significant gauge undercatch.

One approach to improving gauge catch efficiency is the installation of a wind shield surrounding the gauge to reduce the effects of wind-caused undercatches. Although the Alter shield is most commonly used in the US to improve gauge catch efficiency, it still has issues of concern. The Nipher shield has been a favorite in Canada, where snow most often has a low water content. Unfortunately, this shield does not perform well in the heavy, wet snows common in many regions of the world, nor does it adapt to other types and sizes of precipitation gauges, thus making it impractical for use with most precipitation gauges in use in the US. Currently, only a fraction of the US precipitation gauges are equipped with wind shields (Fig. 2).

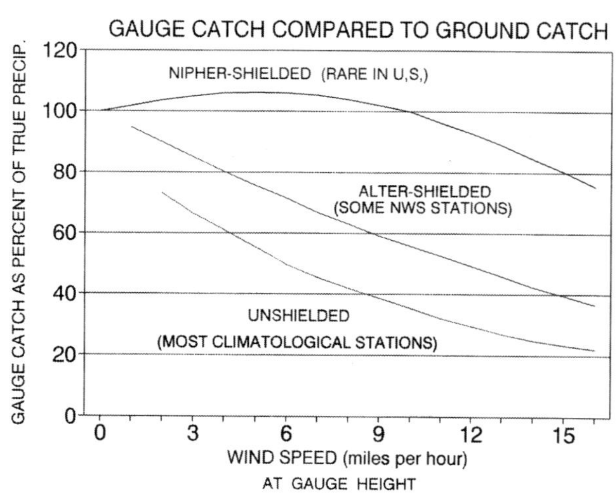

Fig. 2 Ground measurement of snowfall water equivalent from cores compared to that observed in various gauges under differing wind speeds (after Goodison, 1978; Doesken and Judson, 1996)

The World Meteorological Organization has been diligently investigating the challenge of measuring solid precipitation. An extensive international study completed during the 1990s thoroughly investigated the performance characteristics of a variety of gauges and wind shields used in snowy regions of the world (Goodison and Metcalf, 1992). Many consider the Double Fence Intercomparison Reference (DFIR) to produce the most representative gauge catch under a full range of snowfall and wind conditions (Yang et al., 1993). Unfortunately, the size of this windshield (12 m diameter) makes it too bulky and expensive for most common weather stations. No simple solution exists, and few countries show a willingness to change their long-standing measurement practices. The extensive documentation on the need to address gauge undercatch should prompt users to take appropriate action.

Adequate exposure and the potential for gauge undercatch should be carefully considered in the initial deployment of weather stations. The ideal exposure for a precipitation gauge is a delicate compromise between an open and unobstructed location and a protected site where the winds in the vicinity of the gauge are as low as possible during precipitation events. The center of a small clearing in a forest and an open backyard in a suburban neighborhood are examples of good sites. The closer to the ground the gauge is, the lower the winds will be due to surface friction, thus improving gauge catch. However, the gauge must also be high enough to always be above the deepest snow. Rooftop exposures are not recommended because of the enhanced wind problems and the potential for building-induced updrafts that will further reduce gauge catch.

4.2 Snowfall

The traditional measurement of snowfall requires only a measurement stick (ruler) and observer experience. While this may be the simplest meteorological measurement, in practice, it may also be the most inconsistent. Although public interest in snowfall measurements is always great, their qualitative nature is not often perceived. The inconsistency is a direct result of the dynamic nature of fresh snow, which often falls, accumulates and melts unevenly, moving with the wind, and settling over time. Snow measurement are also affected by the location where they are taken, the time of day, the time interval between measurements, the length of time since the snowfall ended, the temperature, the cloudiness and even the humidity.

For climatological and business applications, spatial mapping, and station-to-station comparisons, the best functional definition of snowfall is "the greatest observed accumulation of fresh snow since the previous day prior to any significant amounts of melting, settling, sublimation or redistribution". The typical observer may only go out to measure snowfall once per day at a designated time which depending upon ambient weather conditions and timing, may or may not coincide with the time of greatest accumulation. For example, suppose there were four inches (10.16 cm) of fresh snow on the ground early in the day but only one inch still remains at the scheduled observation, did it snow four inches (10.16 cm) or one inch (2.54 cm)? Four inches (10.16 cm) is obviously the better answer, but if the observer was not there to see it, he/she wouldn't know for sure. Ideally, the observer would be available to continuously watch the snow accumulate, note the greatest accumulation, and then note the settling and melting that occurs later. In reality, however, this may not be the case, since much of the historic snowfall data in the US has come from the National Weather Service Cooperative Observer program where most observations are made by volunteers who can only be expected to make one observation each day (National Research Council, 1998).

Although perfectly consistent measurements may not be possible due to the nature of fresh snow, the following criteria can produce a high degree of consistency.

- The location for taking measurements is critical. An unobstructed yet relatively protected location (such as a forest clearing or open back yard away from buildings, trees and fences) is best to ensure uniform and undrifted snow accumulations that are representative of the average.
- The use of a snowboard (a square or rectangular flat, white surface positioned on the ground and repositioned daily on the top of the existing snow surface) for measuring the accumulation of new snowfall provides a smooth, solid surface on which to measure and from which core samples can be taken (Fig. 3).

Fig. 3 Snow measurement being made using a snow board

- When snow is falling, observers should periodically check the accumulation of new snow on the snowboard and note the greatest amount before settling or melting begins to reduce the depth of fresh snow. The greatest accumulation will typically occur just before the snowfall diminishes or changes to rain. Observers should only clear the new snow from their measurement surface at the scheduled observation time and then reposition the snowboard on the top of the new snow surface.
- To account for blowing and drifting snow and the resulting uneven accumulation patterns, observers should assess the representativeness of the snowboard measurement by taking an average of several measurements in the surrounding environment making sure to include only that snow which has fallen since the previous observation. If the snowboard measurement is found to be unrepresentative (due to snow being partially or totally blown clear; drifts forming in its immediate vicinity; or snow melting on the snowboard but not on most ground surfaces), the reported snowfall should be the average of as many readings as are needed to obtain an appropriate average over an area including both moderate drifts and moderate clearings (avoiding the largest and least representative drifts).

With the creation and expansion of airport weather stations in the US from the 1930s through the 1950s, came a new emphasis on weather forecasting for air transportation and safety. Hourly weather observations were initiated that included the manual recording of precipitation type and intensity, visibility and many other weather elements. These became the foundation of the surface airways weather observations and provided further details about snowfall characteristics than had been previously available. Every hour a complete report of weather conditions was gathered in a consistent manner from a large number of stations across the US. Conditions were also monitored between hourly reports, and any significant changes were reported in the form of "Special" observations. During snowfalls, depths were measured every hour and special remarks were appended to observations whenever snow depth increased by an inch or more. Although these "SNOINCR" remarks always caught the attention of meteorologists since they signaled a significant storm in progress. However, they also introduced a new complexity into the observation of snow. Instead of observing snowfall once daily, some weather stations reported more frequently. The instructions to airways observers stated that snowfall was to be measured and reported every 6 h. The daily snowfall was then the sum of four 6-h totals. Some weather stations then used the seemingly appropriate procedure to measure and clear their snow boards every hour and add these hourly increments into 6-h and daily totals.

For some applications, short interval measurements are extremely useful. However, for climatological applications, snowfall totals derived from short intervals are not the same as measurements taken once daily. For rainfall, a daily total can be obtained by summing short interval measurements. However, for snowfall, the sum of accumulations for short increments often exceeds the observable accumulation for that period. To demonstrate this, volunteer snow observers were recruited from several parts of the country and measured snowfall for several winters at several locations in the US in the late 1990s and early 2000s. Each observer deployed several snowboards, and measured snow accumulations on each board. One board was cleared every hour, one every three hours, one every 6 h, one every twelve hours and finally one was only measured and cleared every 24 h. While results varied from storm to storm, it was very clear that snowfall totals are consistently and significantly higher based on short-interval measurements (Doesken and McKee, 2000). Based on 28 events where measurements taken every 6 h throughout the storms were summed and compared to measurements taken once every 24 h, the 6-h readings summed to 164.4 in. (417.6 cm), 19% greater than the 138.4 in. (351.5 cm) total from the once-daily observations. When hourly readings were summed and compared to once-daily measurements, the total was 30% greater. This implies that two adjacent stations measuring the same snow event at different time intervals may report greatly different values.

From the 1800s to the late 1900s, manual snowfall observing proceeded with little evidence of significant change or improvement in instrumentation or methodology. The use of "snow measurement boards" was recommended, but there is no evidence that stations were issued standardized measurement boards. Inconsistencies noted included frequent reports of snow to water ratios of exactly ten to one. At airport weather stations, some stations summed hourly snowfall measurements to form daily totals, while others measured every 6 h and others still, only once daily. In an effort to standardize procedures among different types of weather stations, the National Weather Service issued revised snow measurement guidelines in 1996 (U.S. Dept. of Commerce, 1996). These guidelines stated that observers should measure snowfall at least once per day but could measure and clear their snowboards as often as, but no more frequently than, once every 6 h (consistent with long-standing airways instructions). These guidelines were promptly put to the test when an extreme lake-effect snowstorm over the Tug Hill Plateau east of Lake Ontario in January 1997 produced a reported 77 inches (195.6 cm) of snowfall in 24 h. Subsequent investigations by the US Climate Extremes Committee found that this total represented the sum of six observations, and thus were for intervals of less than 6 h. While the individual measurements were taken carefully, the summation did not conform to the national guidelines and hence could not be recognized as a new record 24-h snowfall for the US (Leffler et al., 1997).

Another incorrect method of observing snowfall that still appears to occur occasionally is the use of the change in snow depth between consecutive days as the value of new snowfall. This likely leads to a reduced fall amount due to settling of the older, underlying covering of snow. Also, there are instances where the measured water equivalent of falling snow (whether it melts on contact within the gauge or if a heating element within the gauge melts the snow) is used to estimate snowfall. A multiplicative factor that is a function of temperature or simply a factor of 10 may be used for the estimate.

The main conclusion here is that where station-to-station comparability and long-term data continuity are the goals, stations must measure in a consistent manner and that has not always been the case. There may be justification for different observation times and increments, but data from incompatible observation methods should not be interchanged or compared. Many of the data sets in common use today are not fully consistent, so end users may have the responsibility of determining the compatibility and comparability of various station data.

4.3 Snow Depth

The measurement of snow depth is not subject to as many qualitative and definitional issues that plague the measurement of snowfall. Basic equipment is used. For areas with shallow or intermittent snowpacks, a sturdy measurement stick is normally carried by an observer to the point(s) of observation and inserted vertically through the entire layer of new and/or old snow, down to ground level. In areas of deep and more continuous snow cover, a fixed snow stake that is either round or square, is often used. It is clearly marked in whole inches or in centimeters, permanently installed in a representative location and read "remotely" by an observer standing at a convenient vantage point such that the snow near the stake is not disturbed by foot traffic.

The key to useful comparable measurements of snow depth lies in identifying and maintaining representative locations for taking measurements. Blowing and drifting snow are inevitable challenges. Uneven melting adds further complications. For uneven snow accumulation, measurements should be taken from several locations representing both the deeper and shallower areas. Snow cover persists considerably longer in the shade or on north-facing slopes in the Northern Hemisphere, particularly compared to south-facing ones. Thus it is best to take measurements in a level area that is exposed to the sun for most of the day. The underlying surface should be natural and short grass if possible. The observer must factor out from their measurements any air space within the grass. Also, when the observing site and surrounding open areas are less than 50% snow covered the depth should be recorded as a trace until such time that natural snow accumulations are absent. Since most traditional weather stations are widely spaced, it is imperative that each measurement represent the predominant conditions in the vicinity of each station and be comparable with observations made at other sites.

4.4 Snow Water Equivalent

Snow Water Equivalent (SWE) is an extremely important measurement for hydrologic applications. Both flood and overall water supply forecasting rely on its accurate measurement from as many locations as possible to represent the spatial patterns of snow water content that will contribute to subsequent runoff.

The process of measuring SWE is much like the fresh snowfall core measuring procedure described in Section 4.2. The difference here is that the both the new snowfall or existing snow (i.e. the entire snowpack) is captured for measurement. Only a fraction of NWS stations measure SWE and it is not a requirement at NWS Cooperative Observing Stations. It is usually accomplished by taking core samples using the 8-in. diameter precipitation gauge. Some stations are equipped with special scales that make it relatively easy to take a core sample and immediately estimate the SWE from the weight of the sample. Under deep snow conditions,

Fig. 4 A Natural Resources Conservation Service SNOTEL site in the western United States; includes a snow pillow (foreground), a shielded standpipe storage precipitation gauge and radio telemetry equipment. Photo from NRCS

the melting of snow cores requires considerable amounts of warm water and is tedious and time consuming for observers. When snow depths exceed 2 ft, the NWS overflow can is inadequate for effectively coring the snowpack.

The US Department of Agriculture Natural Resources Conservation Service (NRCS) SNOTEL (Snow Telemetry) network over western mountains is the world's most extensive one (Fig. 4). The NRCS and other water resources organizations have a long history of measuring SWE in high snow accumulation areas. Records date back to the 1930s throughout the western mountains in the US with even longer records from a few sites. A wealth of existing literature (much of it informal and non peer-reviewed) outlines years of experimentation and field testing of devices and techniques to measure snow water in the deep snow regions of North America. Among the more formal outlets are the Proceedings of the Western Snow Conference and the Eastern Snow Conference.

Over time, two devices have emerged as the standard SWE measurement tools in deep snow accumulation regions. The first is the Federal Snow Sampler, a portable set of tubes, handle, and a cutter to cleanly penetrate deep snow and ice layers. Core samples of the snow pack are extracted and weighed in situ with a specially calibrated scale to determine the snow water equivalent of the core. To account for the non-uniform accumulation of snow, several cores are taken at each site. Each measurement site is called a "snow course". Measurements taken across the snow course are averaged to produce the final SWE value for the site and core measurements are taken at the same approximate set of individual points. Snow courses have been traditionally read once or twice a month beginning in mid winter and continuing into the spring until all the annual snow has melted.

The snow pillow represents the second instrument in widespread use since the late 1970s. It is a scale built at ground level to measure the weight of the snowpack as it accumulates and melts. Snow pillows were developed to provide remote measurements of SWE without requiring the investment of time and effort involved in sending teams of scientists and hydrologic technicians into the back country every month.

As with snowfall and snow depth, the utility and comparability of SWE observations are only as good as the representativeness of the measurement location and the averaging process. Snow pillow measurements have their own set of challenges including, "bridging" and other non-uniformities in load-bearing characteristics within the snowpack.

5 Snow Data Continuity

One of the most important issues in conducting analyses of climate variability and change is the longevity and consistency of the data. Over 100 years of snowfall, snow depth and water content measurements are available at a number of locations in the US. These observations provide a remarkable resource for meteorological, hydrological, environmental, engineering and societal applications. The data, however, are far from perfect. All of the aforementioned observational challenges have been addressed with varying success from the beginning of the records. Examination of US station snow records reveals variations in observational methods both amongst observers at adjacent stations at a given time, as well as amongst observers at a given station through time. Some of these differences appear to be related to changes in observational directives in manuals or by regional observation program managers of the National Weather Service and its forerunners. Other variations result from observing options provided in the directives. Finally, personal idiosyncrasies of observers themselves could also play a role. Despite this, however, legitimate snow time series are attainable if one scrutinizes historical records, although the majority of long-term station records contain too many inconsistencies to be of use in studies of climate variability and change.

Observational consistency has been difficult to maintain partly as a result of the volunteer nature of the observations in the US by volunteers who have received only modest training and who often can take only one observation per day. Even at primary weather stations in the US, observational consistency has been affected by the changes that occurred over time. The large natural variability in snowfall sometimes hides the impact of observing changes. Yet, in historical perspective, seemingly small observational changes such as station exposure, time and frequency of observation, and observing procedures do have profound impacts on historical time series. A comparison of three studies below reveals the importance of accounting for such fluctuations in the instrumental record.

Robinson (1989) evaluated the quantity and quality of snowfall and snow-on-ground observations at 7637 NWS Cooperative Observer Network (COOP) stations. A step-wise series of three tests was applied to monthly snow data for every station in operation in the December–March period of 1985–1986, 1986–1987 and 1987–1988. The first test required snowfall observations to be acceptable in at least nine of the twelve study months. Of the remaining stations, the same evaluation criteria were applied to monthly maximum snow depths. Few observers make satisfactory snow depth observations while making inferior snowfall measurements. Finally, the stations that passed the first two tests needed at least nine of twelve acceptable months of data for the number of days with 1 in. (2.54 cm) or more of snow on the ground. This served to identify those stations that tend to record snow depth only on the day of a snowfall event and ignore snow cover on subsequent days. Thirty-seven states were evaluated. Of the 4,960 stations considered, 43% failed one of the three tests (28% for the snowfall test; 13% with poor maximum depth data but acceptable snowfall observations; and 2% with inadequate days-on-ground data). Considerably more stations would have failed had the 9 of 12 months criterion been tightened. Snow measurements were better in places where it was quite common. However differences in the quality of snowfall data were often substantial in neighboring states. For instance, Missouri stations often failed to record satisfactory snowfall and snow cover data, in contrast with others like Vermont and Alaska. Iowa was especially noted for having stations with credible snowfall data, but with far fewer providing accurate snow cover observations. Wisconsin stations displayed relatively good snowfall and snow depth data, but 10 stations there failed to collect accurate days-on-ground information. Differences among states may have been a function of varying emphasis about snow observations during volunteer observer training or to inconsistencies in quality control as data are processed and archived.

In a more recent study, Kunkel et al. (2009) examined station observations assembled from a long-term snow data set of the NWS COOP network. As with the previous studies, a number of consecutive criteria were applied to the dataset. Of the over 10,000 initial stations, a small percentage had fewer than 10% of the total number of days missing during the October–May snow seasons from 1930 to 2004. Winter-centered annual snowfall totals were then calculated for these stations for snow seasons from 1900–1901 to 2006–2007. Extreme values relative to station normals and observations were either omitted or corrected with data on the original observation forms. Each station time series needed at least five non-missing winter-centered years in the last decade of the time series to capture any recent changes in snowfall extremes. Finally, only stations with the 1971–2000 mean annual snowfall over 12.5 cm, were selected to focus on areas where snow often fell. Only 1,124 stations met the above criteria and were manually examined by the study team for homogeneity. Figure 5 is an example of the graphs prepared for each long-term station. It highlights (a) a time series of annual snowfall for a given station and its 14 nearest neighbors with at least 30 years of data and (b) the difference between the annual snowfall anomaly for the targeted station minus the annual snowfall anomaly of the neighboring stations. The timing of station moves

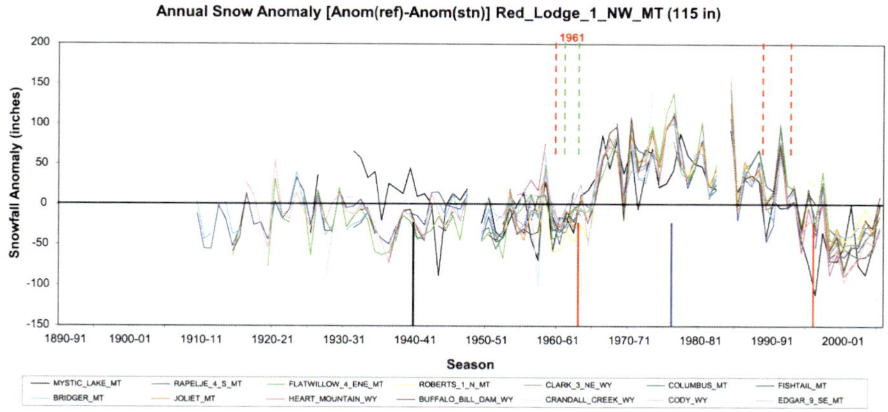

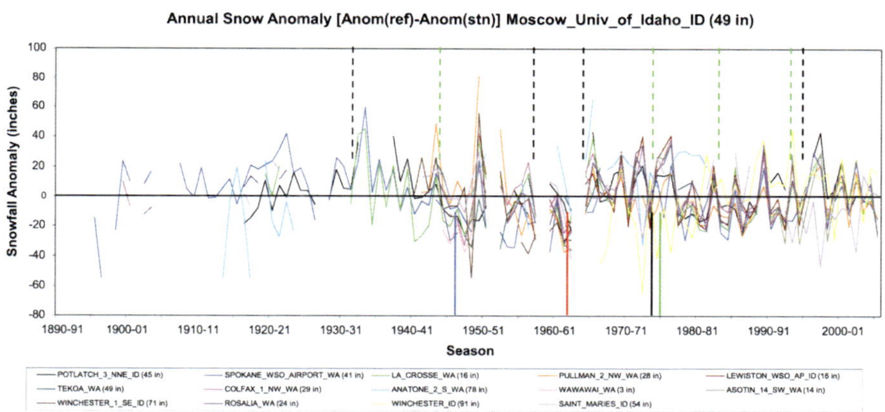

Fig. 5 Annual anomalies (inches) of the differences between the target reference station snowfall anomalies at (*top*) Red Lodge, MT, and (*bottom*) Moscow, ID, and the snowfall anomalies of the closest 14 stations with at least 30 years of records. Dashed vertical lines extending downward from the upper margin of the graph indicate known station or observer changes at target or comparison stations, whereas both *solid vertical lines* extending upward from the lower margin and years printed above the upper margin indicate discontinuities objectively identified using separate methods. The change point indicators are color coded by station, with *red* assigned to the target station (from Kunkel et al., 2009)

or observer changes recorded in station histories were noted on the graphs. Two statistical change point detection tests were also applied. A central assumption in this assessment is that multi-year fluctuations in snowfall will be spatially coherent and detectable by multiple stations.

Kunkel et al. (2009) found that 440 stations were homogenous and suitable for trend analyses back to 1930. Of the selected stations, 314 contain five or more years of snowfall data in the 1920s, 260 have five or more years of data in the 1910s, and 194 have five or more years of data in the 1900s (Fig. 6). The 440 stations are heavily concentrated in the central US, with less dense coverage in the eastern and western

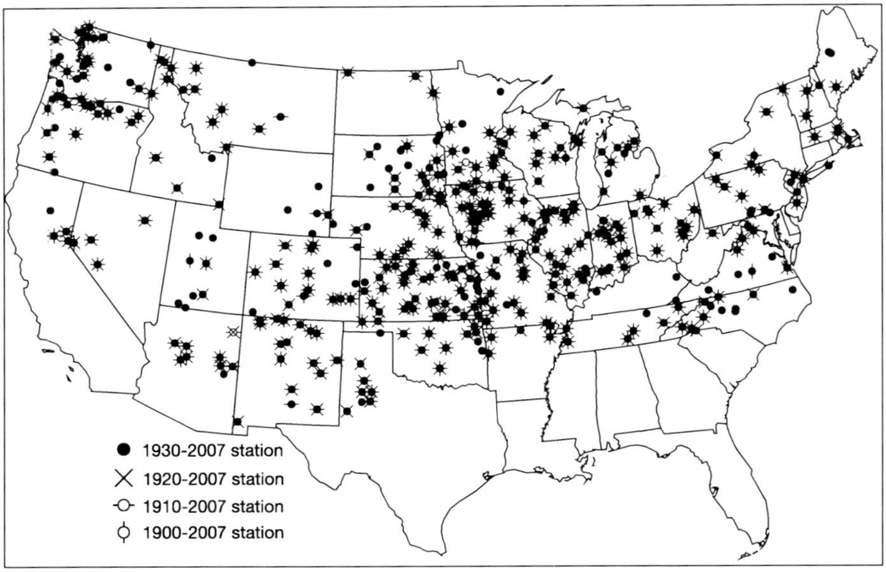

Fig. 6 Homogeneous snowfall stations selected by a plurality of evaluators, with fewer than 5 years missing in each of the beginning decades and in the ending decade for trend analyses. The symbols overlap at the locations of stations available for multiple trend periods (from Kunkel et al., 2009)

US, especially in the northern Great Plains. In mountainous areas, particularly in the western US, cooperative observer network stations are systematically located at lower elevations where most observers reside. Thus, the results presented here do not necessarily reflect the possibility of valley behavior different from that of adjacent snow-dominated higher elevations.

The final method of assessing snow data continuity involves examining the widespread change in surface observations that resulted from the deployment of the Automated Surface Observing System (ASOS) by the National Weather Service during the 1990s. This was part of a very extensive national modernization effort in the US that included changes in precipitation gauges and the automation of the measurement of visibility and precipitation type that resulted in discontinuous records at several hundred stations. In addition, the measurement of snowfall was discontinued completely at many stations because it was not a requirement of the Federal Aviation Administration at that time and did not lend itself to automation.

6 Remote Sensing Approaches to Snow Measurement

Many aspects of snow measurement continue to use only the simplest of instrumentation by trained and experienced observers. However, scientists and practitioners have increasingly turned to new technologies and measurement techniques to better understand and apply our knowledge of snow for improving forecasts. Remote sensing provides data sets showing greater areal coverage and finer spatial resolution about the distribution of snow. This leads to improved models of snowmelt processes and water supplies. At the same time, improved physical models have pointed out the deficiencies in surface data, motivating greater efforts to employ technology to gather more and better data about snow. The operational use of remotely sensed data in snow studies has been greatly augmented in recent years by low cost computing power and significant advances in instrumentation.

6.1 Satellite Remote Sensing of Extent, Depth and Water Equivalent

Visible and passive microwave sensors onboard geostationary and polar orbiting satellites record information used to monitor snow cover on regional to hemispheric scales. Beginning in the 1960s when satellites first orbited the earth, it was apparent that snow cover was rather easily detectable from space. Snow cover extent is best identified on visible imagery by recognizing characteristic textured surface features and brightness. Shortcomings of this method include the inability to detect snow cover when solar illumination is low or when skies are cloudy and the lack of all but the most general information on pack depth. Recent decades have seen improvements in sensor resolution and in the frequency of coverage. This permits finer spatial and temporal monitoring of snow extent, along with continuing decades-long monitoring on broad regional to continental scales (Robinson and Frei, 2000).

Energy in the form of microwaves is continuously emitted from the earth's surface. Snow crystals within a snowpack scatter and attenuate these microwaves. The recognition of snow using microwave techniques results from differences in the emissivity of snow covered and snow free surfaces across several different frequency ranges. Information on water equivalent can be obtained, although generally not with the accuracy necessary for climatological or hydrological studies. Clouds and low solar illumination are not problems when using microwave data to chart snow cover, however there are difficulties in identifying shallow or wet snow.

High resolution mapping of the elevation of the earth's surface is leading to the opportunity to do similar mapping of snow depth by computing the difference between the elevation of a current surface with the known elevation of the ground from previous satellite measurements. The accuracy of this method requires extremely high-resolution background data and nearly perfect navigation of the data. It is not yet in common use.

6.2 Meteorological Radar

Radar refers to the remote sensing technique of transmitting microwaves of a specified wavelength and receiving, processing and displaying that portion of the transmitted energy reflected back to the transceiver. Rain and mixed-phase precipitation reflect microwave energy relatively effectively. Snow crystals can also be detected but, depending on crystal structure and temperature, are not detected as well as "wetter" forms of precipitation. The NWS routinely uses radar to monitor the development, movement and intensity of snowfall. Improvements in radar realized by the NWS WSR-88D allow estimates of snowfall intensity (Holyrod, 1999) and potential accumulation rates. However, ground truth data remain essential for radar calibration. Detection efficiency varies greatly with distance from the radar, cloud height, and other factors. Still, this technology affords opportunities for studying snowfall processes in action.

6.3 Gamma Radiation Remote Sensing

The soil near the surface of the earth constantly emits radiation to space in the form of gamma waves. Snow cover attenuates this radiation in proportion to the water content of the snow on the ground. This form of radiation is best detected over relatively narrow bands by receivers mounted on aircraft. Levels of background gamma emissions must be measured in the fall prior to snow accumulation and then along the identical flight path at different times throughout the winter. This method of mapping snow water equivalent is used operationally in several parts of the US where large river flooding from snowmelt is a common problem. It is not practical to use this methodology over broad expanses.

6.4 Acoustic Snow Depth Sensing

Point measurements of snow depth can be taken continuously and remotely. Sound waves from an above-ground transmitter reflect off the snow surface. By measuring the time for the reflected wave to reach the receiver, the distance can be measured that corresponds to a snow depth. The depth of older snow is easiest to measure since the surface tends to become smoother and harder with time. But recent improvements in signal processing have led to accurate measurements of the depth of fresh snow as well. While heavy snow is falling or snow is drifting, measurements may be compromised as a portion of the sound wave is reflected by ice crystals in the air.

6.5 Portable Depth/Water Content Sensors

New sensors are being developed for use by ski areas and others concerned about detailed spatial patterns of snow depth and water content. These devices can be sled-mounted and pulled by a skier. Using Geographical Positioning Systems to automatically map the sensor's location, detailed maps of snow depth/water content can be made.

7 Conclusions

While some might imagine that measurements of snowfall or snow on the ground are among the simplest of meteorological observations to make, they are anything but straight-forward. And despite all the changes in technology over recent decades, the bulk of our current and historic point data about snow comes from manual measurements using very basic instrumentation. Snow is a dynamic substance, constantly changing as it falls, is deposited, and metamorphoses in place or is transported once on the ground. Manual observations of snow vary in kind and quality as functions of how often observations are made, where they are taken and how well an observer has been trained and subsequently follows acceptable practices. Even the best of observers are permitted to follow different practices when it comes to the frequency with which they may measure new snowfall. Many observers follow rules and use excellent judgment for extended periods. These valuable records are archived at various centers around the world. However, in too many instances, both past and present, snow observations are at best granted second class attention by those doing weather observing, observational training and data archiving.

It is fascinating to note that with all of our supposed progress and increase in knowledge, nearly all of the concerns raised here were clearly identified and well described more than 120 years ago. A treatise on Meteorological Apparatus and Methods by Cleveland Abbe (Abbe, 1888) concisely described the importance of consistent snow measurement, the variability of snow types, the problems with gauge undercatch, and the concern over assuming a ten to one ratio of snowfall to water content. Alas, the more things change the more they stay the same.

Still, with careful scrutiny, often requiring the judgment of experts, credible manual observations can be identified and employed in short term evaluations of snow depth, extent and water equivalent and in long term investigations of snowfall and snow cover variability. Assisted greatly by remotely sensed observations of snow at local to global scales, the challenge of snow monitoring is being met better than ever whether it is near real time assessments or better understanding past conditions.

References

Abbe, C., 1888. 1887 Annual Report of the Chief Signal Officer of the Army under the direction of Brigadier-General A. W. Greeley, Appendix 46. US Government Printing Office, Washington, DC, 385–386.

BBC News, 2008. Baghdad wakes up to rare snowfall http://news.bbc.co.uk/2/hi/middle_east/7183881.stm

Bamzai, A.S. and J. Shukla, 1999. Relation between Eurasian snow cover, snow depth, and the Indian summer monsoon: an observational study. *Journal of Climate*, 12, 3117–3132.

Bergeron, T., 1935: On the physics of cloud and precipitation. *Proceedings of the Fifth Assembly of the International Union of Geodesy and Geophysics*, Lisbon, Portugal, IUGG, 156–178.

Colorado State University, 2004. Measuring Snow. Instructional video prepared for the National Weather Service, Office of Climate Water and Weather Services by the Department of Atmospheric Science at Colorado State University. 23 minutes.

DeGaetano, A.T., T.W. Schmidlin and D.S. Wilks, 1997. Evaluation of East Coast snow loads following January 1996 storms. *Journal of Performance of Construction Facilities*, 90–94.

Doesken, N.J. and A. Judson, 1996. The Snow Booklet: A Guide to the Science, Climatology and Measurement of Snow in the United States. Colorado State University Department of Atmospheric Science, 85pp.

Doesken, N.J. and T.B. McKee, 2000. Life after ASOS (Automated Surface Observing System) – Progress in National Weather Service snow measurement. *Proceedings, 68th Annual Western Snow Conference*, Port Angeles, Washington, 69–75.

Dyer, J. and T.L. Mote, 2006. Spatial variability and trends in observed snow depth over North America. *Geophysical Research Letters*, 22, L16503, doi: 10.1029/2006GL027258.

Gong, G., D. Entekhabi and J. Cohen, 2004. Orographic constraints on a modeled Siberian snow – Arctic Oscillation teleconnection pathway. *Journal of Climate,* 17, 1176–1189.

Goodison, B., 1978. Accuracy of Canadian snow gauge measurements. *Journal of Applied Meteorology*, 27, 1542–1548.

Goodison, B.E. and J. R. Metcalfe, 1992. The WMO solid precipitation intercomparison: Canadian assessment. *Proceedings WMO Technical Conference on Instruments and Method of Observation*, Vienna, Austria, World Meteorological Organization, 221–225.

Gray, D.M. and D.H. Male (eds.), 1981. Handbook of Snow. Pergamon Press, Toronto, Canada, 776pp.

Groisman, P.Ya., T.R. Karl and R.W. Knight, 1994. Observed impact of snow cover on the heat balance and the rise of continental spring temperatures. *Science*, 263, 298–200.

Hall, D.K., G.A. Riggs, V.V. Salomonson, N.E. DiGirolamo and K.J. Bayr, 2002. MODIS snow-cover products. *Remote Sensing of the Environment*, 83, 181–194.

Helms, D., 1992. Snow surveying comes of age in the West. *Proceedings of the Western Snow Conference*, Jackson Hole, WY, pp. 10–17.

Holyrod, E., 1999. Snow Accumulation Algorithm for the WSR88D Radar: Supplemental Report, R-99-11, U.S. Bureau of Reclamation Report, 39pp.

Jones, H.G., J.W. Pomeroy, D.A. Walker and R.W. Hoham (eds.), 2001. Snow Ecology: An Interdisciplinary Examination of Snow-Covered Ecosystems. Cambridge University Press, Cambridge, UK, 378pp.

Judson, A. and N.J. Doesken, 2000. Density of freshly fallen snow in the central Rocky Mountains. *Bulletin of the American Meteorological Society*, 81, 1577–1587.

Kocin, P.J. and L.W. Uccelini, 1990. Snowstorms Along the Northeastern Coast of the United States: 1955–1985. Meteorological Monograph No. 44, American Meteorological Society, 280pp.

Kunkel, K.E., M. Palecki, L. Ensor, K.G. Hubbard, D.A. Robinson, K. Redmond and D. Easterling, 2009. Trends in 20th century U.S. snowfall using a quality-controlled data set. *Journal of Atmospheric and Oceanic Technology*, 26, 33–44.

Leathers, D.J., A.W. Ellis and D.A. Robinson, 1995. Characteristics of temperature depressions associated with snow cover across the northeast United States. *Journal of Applied Meteorology*, 34, 381–390.

Leffler, R.J., R.M. Downs, G.W. Goodge, N.J. Doesken, K.L. Eggleston and D.A. Robinson, 1997. Evaluation of the Reported January 11–12, 1997, Montague, New York, 77-inch, 24-hour Lake-effect Snowfall. Special Report, National Weather Service, NOAA, DOC, 41pp.

Mason, B.J., 1957. The Physics of Clouds. Clarendon Press, Oxford, UK, 481pp.

McKee, T., N.J. Doesken and J. Kleist, 1994. Climate Data Continuity with ASOS – 1993 Annual Report (for period September 1992–August 1993). Climatology Report 94-1, Colorado Climate Center, Department of Atmospheric Science, Colorado State University, Fort Collins, CO, 95pp.

Minsk, L.D., 1998. Snow and Ice Control Manual for Transportation Facilities. McGraw-Hill, New York, 289pp.

Mock, C.J. and K.W. Birkeland, 2000. Snow avalanche climatology of the western United States mountain ranges. *Bulletin of the American Meteorological Society*, 81, 2367–2392.

Mote, T.L., 2008. On the role of snow cover in depressing air temperature. *Journal of Climate*, 47, 2008–2022.

Nakaya, U., 1954. Snow Crystals: Natural and Artificial. Harvard University Press, Cambridge, MA, 510pp.

National Research Council, 1998. Future of the National Weather Service Cooperative Observer Network. National Academy Press, Washington, DC, 65pp.

Robinson, D.A., 1989. Evaluation of the collection, archiving and publication of daily snow data in the United States. *Physical Geography*, 10, 120–130.

Robinson, D.A., K. F. Dewey and R. Heim, Jr., 1993. Global snow cover monitoring: an update. *Bulletin of the American Meteorological Society*, 74, 1689–1696.

Robinson, D.A. and A. Frei, 2000. Seasonal variability of northern hemisphere snow extent using visible satellite data. *Professional Geographer*, 51, 307–314.

Takahashi, T., T. Endoh, G. Wakahama and N. Fukuta, 1991. Vapor diffusional growth of free-falling snow crystals between −3°C and −23°C. *Journal of the Meteorological Society of Japan,* **69**, 15–30.

U.S. Dept of Commerce, 1996. NWS Snow Measurement Guidelines (revised 10/28/96). National Weather Service Handbook No. 7.

Walsh, J.E., D.R. Tucek and M.R. Peterson, 1982. Seasonal snow cover and short-term climatic fluctuations of the United States. *Monthly Weather Review*, 110, 1474–1485.

Yang, D., J. R. Metcalfe, B. E. Goodison and E. Mekis, 1993. An evaluation of the double fence intercomparison reference gauge. *Proceedings of the Eastern Snow Conference, 50th Meeting*, Quebec City, PQ, Canada, Eastern Snow Conference, 105–111.

Index

L.-A. Dupigny-Giroux, C.J. Mock (eds.), *Historical Climate Variability and Impacts in North America*, DOI 10.1007/978-90-481-2828-0_BM2,

M

N

O

P

R

S